Studies in Computational Intelligence 469

Editor-in-Chief

Prof. Janusz Kacprzyk
Systems Research Institute
Polish Academy of Sciences
ul. Newelska 6
01-447 Warsaw
Poland
E-mail: kacprzyk@ibspan.waw.pl

For further volumes:
http://www.springer.com/series/7092

Jan Rauch

Observational Calculi
and Association Rules

 Springer

Author
Prof. Jan Rauch
University of Economics, Prague
Prague
Czech Republic

ISSN 1860-949X e-ISSN 1860-9503
ISBN 978-3-642-44533-0 ISBN 978-3-642-11737-4 (eBook)
DOI 10.1007/978-3-642-11737-4
Springer Heidelberg New York Dordrecht London

Printed on acid-free paper

Springer is part of Springer Science+Business Media (www.springer.com)

To the memory of my parents

Foreword

In a broader sense, the present book deals with a long-standing problem in science and philosophy—the problem of scientific discovery. The question of the nature, the rules, and the "logic" of the process of scientific discovery occupied the minds of great thinkers for centuries. With the advent of computers and computerized data processing, the particular question of whether and to what extent it is possible to automate the process of discovery of knowledge became of greater interest and practical importance. The possibility to run computer programs and process large amounts of data on computers as well as the availability of relevant mathematical methods paved way to new advances in the development of methods for knowledge discovery.

In the early 1960s a group of researchers, which included most importantly Petr Hájek, an expert in mathematical logic, started to develop the GUHA method. The aim of GUHA is to compute from data describing objects and their attributes all interesting hypotheses. The hypotheses are in the form of certain logic formulas describing relationships between the attributes. The nowadays well-known association rules, which were rediscovered in the area of data mining some twenty-five years later, are but a particular case of GUHA hypotheses. To illustrate the point, let me recall that for two sets of attributes, A and B, the association rule $A \Rightarrow B$ is considered valid in data if (1) at least c (confidence) percent of objects in the data that have all the attributes in A have also all the attributes in B, and (2) at least s (support) percent of objects have both the attributes in A and B. In GUHA, the meaning of hypotheses, such as $A \approx B$, depends on which generalized quantifier, an important concept in GUHA, we choose to use. The quantifier $\approx$ determines the meaning of $A \approx B$. With the so-called supported implication quantifier, $A \approx B$ has essentially the meaning described above. Importantly, there are many other generalized quantifiers and they lead to further types of hypotheses that can be explored using the the GUHA method. This example is just one illustration of the broadly conceived conception of GUHA.

Several people later joined the research in GUHA, including Tomáš Havránek, an expert in mathematical statistics. The research in GUHA led to new, original problems in logic, statistics, and computational complexity. In 1978, Hájek and

Havránek wrote "Mechanizing Hypothesis Formation, Mathematical Foundations for a General Theory", a book published by Springer. This important book developed, in a general setting, the mathematical foundations for formation of hypotheses and the GUHA method. The book examines in detail the process of discovering hypotheses from data from a methodological point of view and, most importantly, develops fine logico-statistical foundations of this process. I believe the book, particularly the clean conception and powerful blend of logic and statistics, has not yet been fully appreciated. In addition to the development of theoretical foundations, further research in GUHA had been conducted, including the development of algorithms, software, and applications of GUHA to real-world problems.

Since the publication of Hájek and Havránek's book, many new results were obtained and many things happened in this area of research. Most significant is probably the surge in interest in association rules since the early 1990s. There is perhaps no person better qualified than Jan Rauch to report comprehensively on the research related to the GUHA method and its development till the present days. Since the early 1970s, when Jan Rauch joined Professor Hájek as his student, he has been an active member of Hájek's research group and later became the leading figure in the research related to GUHA, contributing by new methods and theoretical results, as well as his continuing effort to develop software for GUHA.

In the present book, Professor Rauch covers thoroughly the foundations of association rules and the GUHA method, including the recent developments, and puts these into the context of the current research in data mining. The emphasis of the book is on theoretical foundations—the observational calculi in the conception of Hájek and Havránek—which may be regarded as representing the logic of association rules. In addition, the book contains quite a useful part on LISp-Miner, the software implementing the methods described in the book, real-word applications of the methods which have been carried out, and selected current research projects related to these methods. Overall, the book presents a carefully written advanced material that will surely be regarded as a valuable source of information in the years to come by researchers as well as by teachers and students involved in knowledge discovery from data.

Binghamton, NY, October 2012   Radim Belohlavek

Preface

Associational rules are very popular in the area of data mining. Their development and research began in the early 1990's in relation to market basket analysis. Observational calculi are much less known, even if they are closely related to association rules and data mining. The basic idea of this book is to introduce observational calculi as a useful tool for dealing with association rules.

Observational calculi were introduced by Petr Hájek and Tomáš Havránek in the book *Mechanizing Hypothesis Formation - Mathematical Foundations for a General Theory* (Springer, 1978, see also http://www.cs.cas.cz/hajek/guhabook/). The original book starts with two questions: (1) *Can computers formulate and justify scientific hypotheses?* (2) *Can they comprehend empirical data and process them rationally, using the apparatus of modern mathematical logic and statistics to try to produce a rational image of the observed empirical world?*

A theory developed in the book to answer these questions is based on a schema of inductive inference: $\dfrac{theoretical\ assumptions,\ observational\ statement}{theoretical\ statement}$. This means that if we accept theoretical assumptions and verify a particular statement about the observed data, we accept a conclusion – a theoretical statement. Statistical approaches are used to bridge the gap between observational and theoretical statements. The theoretical statements refer to systems of "possible worlds" and the observational statements talk about observed data. It is crucial that an intelligent statement about the observed data leads to theoretical conclusions, not the data itself.

Formulas of observational calculi correspond to suitable statements about the observed data. The most studied observational calculi are observational predicate calculi. These are the results of modifications of predicate calculi – only finite models are allowed and generalized quantifiers are added. Finite models correspond to the data resulting from observation and generalized quantifiers make it possible to express suitable observational statements.

Generalized quantifiers also make it possible to express minimal thresholds for confidence and support and thus the association rules defined and studied since the 1990's can also be understood as formulas of observational calculi. The association rule – a formula of observational predicate calculus – consists of two derived predicates and of a generalized quantifier. There are generalized quantifiers corresponding both to simple statements about observational data and to various statistical hypothesis tests, e.g. to Chi-square test. In this way a general relation of two derived predicates in a finite $\{0, 1\}$-data matrix can be expressed as an association rule – a formula of an observational calculus.

Theoretically interesting and practically important results about formulas of observational predicate calculi which we can consider as association rules are achieved in the book by P. Hájek and T. Havránek. Several classes of association rules are defined and studied, e.g. classes of implicational or symmetrical rules. The results concern namely the deduction rules among association rules, dealing with missing information, and the definability of association rules in classic predicate calculus (i.e. only with the $\forall$ and $\exists$ quantifiers).

There is often a one-one correspondence between theoretical and observational statements. Thus there is also a one-one correspondence between tasks of finding interesting theoretical statements and tasks of finding interesting observational statements. The task of finding interesting observational statements is solved by the GUHA method. The method is realized by the GUHA procedures. The input of the GUHA procedure consists of analysed data and several parameters defining a large set of relevant observational statements. The output is a representation of a set of all relevant observational statements which are true in the input data.

The most used GUHA procedure is the ASSOC procedure mining for association rules, which are however understood as general relations of Boolean attributes derived from the columns of the analysed data matrix. ASSOC is also introduced in the book by P. Hájek and T. Havránek. It was implemented several times and applied many times. Additional information on early development and applications of the GUHA method can be obtained in two special volumes of the International Journal of Man-Machine Studies – *Volume 10, Issue 1, January 1978* and *Volume 15, Issue 3, October 1981*, devoted to the GUHA method.

The intention of applying the GUHA method to data in databases led to a deeper study of association rules and new results were achieved in the mid 1980's. These involve new and more general deduction rules among association rules and deeper criteria for the definability of association rules in classic predicate observational calculi. Many-sorted observational predicate calculi were also defined and studied.

The boom of association rules in the 1990's was the start of a new effort in the study of association rules as formulas of observational calculi. The syntax of used formulas of observational calculi has been significantly simplified, and only calculi with monadic predicates are further studied. Free and bound variables are omitted and basic Boolean attributes are used instead of predicates. The basic Boolean attribute says that a value of an attribute for a given object is part of a given subset of all the possible values of the attribute. New classes of association rules are defined and studied and results on their mutual relations were obtained. Additional results on deduction rules were also achieved. The new results can be understood as the logic of association rules.

Development of an academic software system called LISp-Miner for data mining research was launched in 1996. The 4ft-Miner procedure, which is an extended

ASSOC procedure, was implemented first. The long-term experience with implementation of the GUHA method was applied including a representation of analysed data by suitable strings. The well known a-priori algorithm was not used. This approach proved to be very efficient and six additional GUHA procedures were implemented in the LISp-Miner system. Two of them mine for interesting couples of association rules. The couples of association rules can be used, for example, to express the differences between two subsets of objects or to point to a suitable action.

The experience with the LISp-Miner system led to several research projects with the goal of studying possibilities of application of suitable formalized items of domain knowledge in the data mining process. There are possibilities to use such items of knowledge to formulate reasonable analytical questions or to filter out uninteresting association rules which can be considered as consequences of known domain knowledge. The possibilities of automatic formulation of analytical reports answering formulated analytical question are also studied as well as the possibility of automating the whole data mining process. Observational calculi play a very important role in this research. In addition, this research leads to new research challenges in the field of observational calculi.

The goal of this book is to introduce observational calculi as a useful tool for dealing with association rules. It includes the following partial goals:

- to introduce logical calculi of association rules as a special observational calculi suitable to study association rules – general relations of two Boolean attributes including relations defined by thresholds for values of various association measures
- to introduce theoretical results on logical calculi of association rules in the form *definition – theorem – proof*
- to draw attention to possible applications of presented theoretical results including emerging applications in research projects related to current challenges in data mining research
- to present several open problems and research challenges related to association rules and observational calculi.

The book is intended for researchers interested in data mining and knowledge discovery in databases, especially in association rules. The book could also be of interest to researchers working in the field of theoretical foundations of data mining as well as logicians interested in applications of mathematical logic. The book can be also used as a source of topics for seminar works and diploma theses for computer science students.

My sincere thanks are to Prof. RNDr. Petr Hájek, DrSc., who started the research of observational calculi and the GUHA method and who taught me these topics. My sincere thanks also go to Assoc. Prof. Ing. Milan Šimůnek, Ph.D. who is an excellent software engineer and author of the software of the LISp-Miner system. My thanks are also to all colleagues who took part in the activities of the Prague informal GUHA circle for discussions and co-operation which helped me to better understand observational calculi and the GUHA method. I would also like to stress my gratitude to colleagues and students taking part in the Knowledge Engineering

Group at the Faculty of Informatics and Statistics of the University of Economics, Prague for a lot of fruitful discussions and cooperation related to the LISp-Miner system and its applications.

The work on this book was supported by the grant 201/08/0802 of the Czech Science Foundation and by the project ME913 of The Ministry of Education, Youth and Sports, of the Czech Republic.

Prague, September 2012   Jan Rauch

Contents

Acronyms, Symbols, and the Like

GUHA   General Unary Hypotheses Automaton
LAR   Language of Association Rules
TPC   Truth Preservation Condition
OPC   Observational Predicate Calculus
MOPC   Monadic Observational Predicate Calculus
CMOPC   Classical Monadic Observational Predicate Calculus

$\langle \mathrm{Sent}, \mathrm{M}, V, Val \rangle$ semantic system

$\mathcal{M} = \langle M, f_1, \ldots, f_K \rangle$ data matrix
$\mathcal{T} = \langle t_1, \ldots, t_K \rangle$ type of a logical calculus of association rules
$\mathrm{M}_{\mathcal{T}}$ set of all data matrices of type $\mathcal{T}$

$A_1, \ldots, A_K$ basic attributes
$\diamond(1), \ldots, \diamond(u)$ categories
$A_i(\diamond(v_1), \ldots, \diamond(v_k))$ basic Boolean attribute
$\diamond(v_1), \ldots, \diamond(v_k)$ coefficient of basic Boolean attribute $A_i(\diamond(v_1), \ldots, \diamond(v_k))$
$\approx$ 4ft-quantifier
$\varphi \approx \psi$ association rule
$\mathscr{L}_{\mathcal{T}}$ language of association rules of type $\mathcal{T}$
$\mathscr{BS}(\mathscr{L}_{\mathcal{T}})$ basic symbols of language $\mathscr{L}_{\mathcal{T}}$
$\mathscr{BA}(\mathscr{L}_{\mathcal{T}})$ set of Boolean attributes of language $\mathscr{L}_{\mathcal{T}}$
$\mathscr{AR}(\mathscr{L}_{\mathcal{T}})$ set of association rules of language $\mathscr{L}_{\mathcal{T}}$

$\mathfrak{I}(A_i)$ interpretation of attribute A_i
$\mathfrak{I}(\diamond(v_1), \ldots, \diamond(v_k))$ interpretation of coefficient $\diamond(v_1), \ldots, \diamond(v_k)$
$f_{A_i(\alpha)}$ interpretation of basic Boolean attribute $A_i(\alpha)$
f_{φ} interpretation of Boolean attribute φ
$Fr(\omega, \mathcal{M})$ frequency of Boolean attribute ω in $\mathcal{M} = \langle M, f_1, \ldots, f_K \rangle$
$4ft(\varphi, \psi, \mathcal{M})$ 4ft-table of φ and ψ in $\mathcal{M}$
$F_{\approx}$ associated function of 4ft-quantifier $\approx$

$Val(\varphi \approx \psi, \mathcal{M})$ value of association rule $\varphi \approx \psi$ in data matrix $\mathcal{M}$

$\mathcal{LC}_{\mathcal{T}}$ logical calculus of association rules of type $\mathcal{T}$

$\Rightarrow_p$ 4ft-quantifier of p-implication (i.e. defined by confidence / precision)

$\Rightarrow^!_{p,\alpha}$ 4ft-quantifier of likely p-implication i.e. upper critical implication

$\Rightarrow^?_{p,\alpha}$ 4ft-quantifier of suspicious critical implication i.e. lower critical implication

$\Leftrightarrow_p$ 4ft-quantifier of double p-implication (i.e. defined by Jaccard)

$\Leftrightarrow^!_{p,\alpha}$ 4ft-quantifier of lower critical double implication

$\Leftrightarrow^?_{p,\alpha}$ 4ft-quantifier of upper critical double implication

$\equiv_p$ 4ft-quantifier of p-equivalence (i.e. defined by accuracy / success rate)

$\equiv^!_{p,\alpha}$ 4ft-quantifier of lower critical equivalence

$\equiv^?_{p,\alpha}$ 4ft-quantifier of upper critical equivalence

$\sim_q$ 4ft-quantifier of simple deviation (i.e. defined by odds ratio)

$\sim^1_\alpha$ Fisher's quantifier

$\sim^2_\alpha$ χ^2-quantifier

$\sim^+_q$ 4ft-quantifier of above average

$\sim^-_p$ 4ft-quantifier of below average

$\equiv^E_p$ E-quantifier

$\Rightarrow^\neg_{p,q}$ 4ft-quantifier of above negation

$\sim^{\circ,\bullet}_p$ pairing quantifier

$\oplus_{Base}$ 4ft-quantifier Base

$\odot_s$ 4ft-quantifier support i.e. support

$\overline{\oplus}_{Base}$ 4ft-quantifier ceil

$\overline{\odot}_s$ 4ft-quantifier ceil support

$\rightarrow^C_p$ 4ft-quantifier defined by coverage

$\rightarrow^P_p$ 4ft-quantifier defined by prevalence

$\Rightarrow^R_p$ 4ft-quantifier defined by recall

$\Rightarrow^S_p$ 4ft-quantifier defined by specificity

$\approx^{L+}_q$ 4ft-quantifier defined by lift / interest

$\approx^L_u$ 4ft-quantifier defined by leverage

$\approx^A_u$ 4ft-quantifier defined by added value / change of support

$\approx^R_s$ 4ft-quantifier defined by relative risk

$\approx^C_u$ 4ft-quantifier defined by certainty factor

$\approx^Q_u$ Yule's Q quantifier

$\approx^Y_u$ Yule's Y quantifier

$\approx^V_p$ 4ft-quantifier defined by conviction

$\Rightarrow^L_p$ 4ft-quantifier defined by Laplace correction

$\approx^{V1}_t$ 4ft-quantifier defined by one-way support

$\approx^{V2}_t$ 4ft-quantifier defined by two-way support

$\approx^P_u$ Piatetski-Shapiro quantifier

$\Leftrightarrow_p^C$ 4ft-quantifier defined by cosine

$\approx_u^G$ Loevinger quantifier

$\approx_t^I$ 4ft-quantifier defined by information gain

$\Rightarrow_q^S$ Sebag-Schoenauer quantifier

$\Rightarrow_u^C$ 4ft-quantifier defined by least contradiction

$\approx_s^O$ 4ft-quantifier defined by odd multiplier

$\Rightarrow_y^E$ 4ft-quantifier defined by example and counterexample rate

$\approx_t^Z$ Zhang quantifier

$\Rightarrow_{p,Base}$ founded p-implication

$\Rightarrow_{p,\alpha,Base}^!$ founded lower critical implication

$\Rightarrow_{p,\alpha,Base}^?$ founded upper critical implication

$\Rightarrow_{p,Base}^L$ founded Laplace correction

$\Rightarrow_{q,Base}^S$ founded Sebag-Schoenauer quantifier

$\Rightarrow_{u,Base}^C$ founded least contradiction

$\Rightarrow_{v,Base}^E$ founded example and counterexample rate

$\Leftrightarrow_{p,Base}$ founded double p-implication

$\Leftrightarrow_{p,\alpha,Base}^!$ founded lower critical double implication

$\Leftrightarrow_{p,\alpha,Base}^?$ founded upper critical double implication

$\equiv_{p,Base}$ founded p-equivalence

$\equiv_{p,\alpha,Base}^!$ founded lower critical equivalence

$\equiv_{p,\alpha,Base}^?$ founded upper critical equivalence

$\sim_{q,Base}$ founded simple deviation

$\sim_{\alpha,Base}^1$ founded Fisher's quantifier

$\sim_{\alpha,Base}^2$ founded χ^2-quantifier

$\rightarrow_{p,s}$ supported p-implication

$\rightarrow_{p,\alpha,s}^!$ supported lower critical implication

$\rightarrow_{p,\alpha,s}^?$ supported upper critical implication

$\rightarrow_{p,s}^L$ supported Laplace correction

$\rightarrow_{q,s}^S$ supported Sebag-Schoenauer quantifier

$\rightarrow_{u,s}^C$ supported least contradiction

$\rightarrow_{v,s}^E$ supported example and counterexample rate

$\leftrightarrow_{p,s}$ supported double p-implication

$\leftrightarrow_{p,\alpha,s}^!$ supported lower critical double implication

$\leftrightarrow_{p,\alpha,s}^?$ supported upper critical double implication

$\equiv_{p,s}^{\odot}$ supported p-equivalence

$\equiv_{p,\alpha,s}^{\odot!}$ supported lower critical equivalence

$\equiv_{p,\alpha,s}^{\odot?}$ supported upper critical equivalence

$\sim_{q,s}^{\odot}$ supported simple deviation

$\sim_{\alpha,s}^{\odot 1}$ supported Fisher's quantifier

$\sim_{\alpha,s}^{\odot 2}$ supported χ^2-quantifier

$C(a,b,c,d,a',b',c',d')$ truth preservation condition

$TPC_\Rightarrow$ truth preservation condition for implicational quantifiers

$TPC_\Leftrightarrow$ truth preservation condition for double implicational quantifiers

$TPC_\equiv$ truth preservation condition for equivalence quantifiers

TPC_{SYM} truth preservation condition for symmetrical quantifiers

TPC_{adSYM} truth preservation condition for a-d symmetrical quantifiers

TPC_F truth preservation condition for quantifiers with the F-property

$TPC_\Rightarrow^\mathcal{M}$ truth preservation condition for weakly implicational quantifiers

$TPC_\Leftrightarrow^\mathcal{M}$ truth preservation condition for weakly double implicational quantifiers

$TPC_{\Sigma,\Leftrightarrow}$ truth preservation condition for Σ-double implicational quantifiers

$TPC_{\Sigma,\Leftrightarrow}^\mathcal{M}$ truth preservation condition for weakly Σ-double implicational quantifiers

$TPC_\equiv^\mathcal{M}$ truth preservation condition for weakly equivalence quantifiers

$TPC_{\Sigma,\equiv}$ truth preservation condition for Σ-equivalence quantifiers

$TPC_{\Sigma,\equiv}^\mathcal{M}$ truth preservation condition for weakly Σ-equivalence quantifiers

$4ft[C]$ class of 4ft-quantifiers defined by truth preservation condition C

$4ft[TPC_\Rightarrow]$ class of imlicational quantifiers

$4ft[TPC_\Leftrightarrow]$ class of double imlicational quantifiers

$4ft[TPC_\equiv]$ class of equivalence quantifiers

$4ft[TPC_{SYM}]$ class of symmetrical quantifiers

$4ft[TPC_{adSYM}]$ class of a-d symmetrical quantifiers

$4ft[TPC_F]$ class of quantifiers with the F-property

$4ft[TPC_\Rightarrow^\mathcal{M}]$ class of weakly implicational quantifiers

$4ft[TPC_\Leftrightarrow^\mathcal{M}]$ class of weakly double implicational quantifiers

$4ft[TPC_{\Sigma,\Leftrightarrow}]$ class of Σ-double imlicational quantifiers

$4ft[TPC_{\Sigma,\Leftrightarrow}^\mathcal{M}]$ class of weakly Σ-double implicational quantifiers

$4ft[TPC_\equiv^\mathcal{M}]$ class of weakly equivalence quantifiers

$4ft[TPC_{\Sigma,\equiv}]$ class of Σ-equivalence quantifiers

$4ft[TPC_{\Sigma,\equiv}^\mathcal{M}]$ class of weakly Σ-equivalence quantifiers

$\mathcal{N}$ set of all non-negative integer numbers

$\mathcal{N}^+$ $\mathcal{N} \cup \{\infty\}$

$Tb_{\Rightarrow^*}$ table of maximal b for implicational quantifier $\Rightarrow^*$

$Tb_{c,\Leftrightarrow^*}$ table of maximal c for double implicational quantifier $\Leftrightarrow^*$

$Tb_{c,\Leftrightarrow^*,b}$ table of maximal c for double implicational quantifier $\Leftrightarrow^*$ and frequency b

$Tb_{b,\Leftrightarrow^*}$ table of maximal b for double implicational quantifier $\Leftrightarrow^*$

$Tb_{b,\Leftrightarrow^*,c}$ table of maximal b for double implicational quantifier $\Leftrightarrow^*$ and frequency c

$Tb_{\Sigma,\Leftrightarrow^*}$ table of maximal $b+c$ for Σ-double imlicational quantifier $\Leftrightarrow^*$

$Tb_{E,\equiv^*}$ table of maximal $b+c$ for Σ-equivalence quantifier $\equiv^*$

$Tb_{|b-c|,\approx}$ table of minimal $|b-c|$ for symmetrical 4ft-quantifier $\approx$ with the F-property

$\frac{\varphi \approx \psi}{\varphi' \approx \psi'}$ deduction rule concerning association rules $\varphi \approx \psi$ and $\varphi' \approx \psi'$

$\tau \vdash \omega$ Boolean attribute ω logically follows from Boolean attribute τ

$\overline{A_i(h)}$ propositional variable related to basic Boolean attribute $A_i(h)$

$\mathrm{Val}(\overline{A_i(h)})$ concrete value of the propositional variable $\overline{A_i(h)}$

$\mathscr{PV}(\mathscr{LC}_{\mathscr{T}})$ set of propositional variables related to logical calculus $\mathscr{LC}_{\mathscr{T}}$

$\kappa_{i,h}(A_i)$ propositional formula of uniqueness of category h of basic attribute A_i

$\lambda(A_i)$ propositional formula of consistency of basic attribute A_i

$\pi(A_i(\alpha))$ propositional disjunction associated with basic Boolean attribute $A_i(\alpha)$

$\pi(\varphi)$ propositional formula associated with Boolean attribute φ

$\mathscr{M}^X = \langle M, f_1', \ldots, f_K' \rangle$ data matrix with missing information

$\mathrm{M}^X_{\mathscr{T}}$ set of all data matrices of type $\mathscr{T}$ with missing information

$f'_{A_i(\alpha)}$ interpretation of basic Boolean attribute $A_i(\alpha)$ in a data matrix with missing information

f'_{φ} interpretation of Boolean attribute φ in a data matrix with missing information

$9ft(\varphi, \psi, \mathscr{M}^X)$ 9ft-table (i.e. nine-fold table) of φ and ψ in data matrix $\mathscr{M}^X$ with missing information

$F^X_{\approx}$ secured extension of associated function $F_{\approx}$ of 4ft-quantifier $\approx$

$\mathrm{Val}^X(\varphi \approx \psi, \mathscr{M}^X)$ value of association rule $\varphi \approx \psi$ in data matrix with missing information

$\mathscr{LC}^X_{\mathscr{T}}$ logical calculus of association rules of type $\mathscr{T}$ with missing information which is a secured extension of $\mathscr{LC}_{\mathscr{T}}$

$F^O_{\approx}$ optimistic extension of associated function $F_{\approx}$ of 4ft-quantifier $\approx$

$F^D_{\approx}$ deleting extension of associated function $F_{\approx}$ of 4ft-quantifier $\approx$

$t = \langle t_1, \ldots, t_n \rangle$ type of OPC

$\mathscr{M} = \langle M, f_1, \ldots, f_n \rangle$ V-structure of type t of OPC

$\underline{M}^V_t$ set of all V-structures $\mathscr{M}$ of type t of OPC

$P_1, \ldots, P_n$ predicates

$x_0, x_1, x_2, \ldots$ variables

q quantifier

$\langle 1^s \rangle$ quantifier type

$P_i(u_1, \ldots, u_{t_i})$ atomic formula

$(qu)(\varphi_1, \ldots, \varphi_s)$ formula with quantifier q of type $\langle 1^s \rangle$ and variable u created from formulas $\varphi_1, \ldots, \varphi_s$

$\mathscr{L}$ predicate language of a type of OPC

Asf_{q_i} associated function for quantifier q

$\mathscr{P}$ observational predicate calculus

$\underline{M}^{\{0,1\}}_t$ set of all $\{0,1\}$-structures $\mathscr{M}$ of type t of OPC

$FV(\varphi)$ set of free variables of formula φ

$\|\varphi\|_{\mathscr{M}}[\varepsilon]$ $\mathscr{M}$-value of φ for mapping ε of $FV(\varphi)$ into M where $\mathscr{M} = \langle M, f_1, \ldots, f_n \rangle \in \underline{M}^{\{0,1\}}_t$

$T_{\mathscr{M}} = \langle a_{\mathscr{M}}, b_{\mathscr{M}}, c_{\mathscr{M}}, d_{\mathscr{M}} \rangle$ 4ft-table of $\{0,1\}$-structure $\mathscr{M} = \langle M; f_1, f_2 \rangle$

$\mathscr{N}_{\langle 1,1 \rangle}$ $\langle 1,1 \rangle$-shortened structure $\mathscr{N} = \langle M; f_1, f_2, f_3 \rangle$ of the type $\langle 1,1,2 \rangle$ of OPC

$T_{\mathscr{N}}$ 4ft-table of $\mathscr{N} = \langle M; f_1, f_2, f_3 \rangle \in \underline{M}_{\langle 1,1,2 \rangle}^{\{0,1\}}$

Asf_q^{4ft} 4ft-associated function of quantifier q of the type $\langle 1,1 \rangle$

$(\approx x)(\varphi(x), \psi(x))$ association rule in monadic observational predicate calculus $\mathscr{P}$

$C_{\mathscr{M}}(o) = \langle p_1(o), \ldots, p_n(o) \rangle$ $\mathscr{M}$-card of o, $\mathscr{M} = \langle M, p_1, \ldots, p_n \rangle$

$Tbp_{\approx,c,d}$ partial table of maximal b for equivalence quantifier $\approx$ and for the couple $\langle c, d \rangle$

Chapter 1
Introduction

This book deals with observational calculi and association rules. The term association rules is well known, it was coined in the early 1990's in relation to market basket
analysis [1]. The goal of market basket analysis is to better understand the purchase
behavior of customers in supermarkets. Transaction data recorded by point-of-sale
systems in supermarkets is analysed. We assume there is a set $I = \{i_1, \ldots, i_n\}$ of
possible items of goods and set $D = \{b_1, \ldots, b_m\}$ of market baskets; it is $b_i \subset I$ for
$i = 1, \ldots, m$. An association rule is commonly understood as an expression of the
form $X \to Y$, where $X \subset I$, $Y \subset I$ and $X \cap Y = \emptyset$.

There are two important measures of interestingness of association rules: *confidence* and *support*. The confidence $conf(X \to Y)$ of the rule $X \to Y$ is defined as
$conf(X \to Y) = \dfrac{\text{number of baskets containing both } X \text{ and } Y}{\text{number of baskets containing } X}$. The support $supp(X \to$
$Y)$ is defined as $supp(X \to Y) = \dfrac{\text{number of baskets containing both } X \text{and } Y}{m}$.

A task of mining association rules is understood as a task of finding all association rules $X \to Y$ satisfying $conf(X \to Y) \geq minconf$ and $supp(X \to Y) \geq minsupp$
in a given set of market baskets D. Here $minconf$ and $minsupp$ are user-specified
minimum confidence and support. This task is usually solved by the apriori algorithm which was implemented and modified in various ways many times. The apriori algorithm for mining association rules is considered the fourth most influential
algorithm in data mining [95].

Observational calculi were introduced in the book [18] as a tool for logic of
discovery. The book is inspired by two questions:

Q_1: Can computers formulate and justify *scientific* hypotheses?
Q_2: Can they comprehend empirical data and process them rationally, using the
apparatus of modern mathematical logic and statistics to try to produce a rational
image of the observed empirical world?

It is assumed that scientific hypotheses describe the whole potentially infinite set of
objects and that they are formulated on the basis of empirical data resulting from
observation of a finite subset of all objects. There are several tools based on mathematical logic and statistics developed in [18] to enable computers to formulate and

J. Rauch: *Observational Calculi and Association Rules*, SCI 469, pp. 1–13.
DOI: 10.1007/978-3-642-11737-4_1 © Springer-Verlag Berlin Heidelberg 2013

justify scientific hypotheses. They involve formally described theoretical and observational languages and the GUHA method.

Formulas of theoretical languages correspond to hypotheses concerning the whole set of objects and formulas of observational languages concern the empirical data. The rational inductive inference rules that bridge the gap between observational and theoretical languages are based on statistical approaches, i.e. estimates of various parameters or statistical hypothesis tests are used. This leads to theoretical statements about state dependent systems. A theoretical statement is justified if a condition concerning an estimate of a used parameter or a statement given by a statistical hypothesis test is satisfied in the analysed data. Observational calculi are special logical calculi formulas which correspond to observational statements justifying useful theoretical statements. Important results on properties of observational calculi are achieved in [18].

There is usually a one to one correspondence between observational and theoretical statements. Thus the task of finding theoretical statements relevant to the given research problem can be converted to the task of finding a suitable set of observational statements. One method for finding such sets of observational statements is the GUHA method, which is realized using GUHA procedures. The input of each GUHA procedure consists of analysed data and of several parameters defining a large set of relevant patterns. The output is the set of all prime patterns. A pattern is prime if it is true in the analysed data and if it does not logically follow from another simpler true pattern which is already a part of the output.

The GUHA procedure used most is ASSOC, also introduced in [18]. It produces interesting pairs of general Boolean attributes derived from columns of an analysed data matrix. Criteria of interestingness based on statistical approaches (e.g. on the χ^2-test or Fisher's test) can be used. The ASSOC procedure was implemented several times and applied many times, see [25, 28, 56, 80].

Association rules of the form $X \to Y$ can be seen as a very simplified case of pairs of Boolean attributes produced by the ASSOC procedure. Implementation of the ASSOC procedure does not use the well known apriori algorithm; it is based on a representation of analysed data by suitable strings of bits [61, 80]. The pairs of Boolean attributes produced by the ASSOC procedure are also called association rules here.

There are special observational calculi defined and studied in [18]. The formulas of these calculi correspond to association rules – interesting pairs of general Boolean attributes derived from columns of a data matrix. Results concerning these calculi are applied in the ASSOC procedure. The boom of association rules related to market basket analysis in the 1990's was the start of a new effort in the study of association rules as formulas of suitable observational calculi. The new results can be understood as the logic of association rules [68]. There are many publications dealing with observational calculi and the GUHA method including [12, 13, 14, 15, 16, 23, 20, 21, 27]. Some of these deal also with relations of observational calculi and the GUHA method to association rules and data mining. However, the goal of this book is not to describe the history of these topics and all their mutual relations. This history is introduced in [17], some details are also in [18, 26] .

The goal of the book is to show observational calculi as a useful tool for dealing with association rules. To describe the goal of the book in more details we need to outline the main features of association rules – pairs of general Boolean attributes as well as of the ASSOC procedure and logic of association rules. This is done in Sects. 1.1 – 1.3. The goal of the book is described in details in Sect. 1.4.

1.1 Association Rules

We deal with association rules – pairs of general Boolean attributes derived from columns of a data matrix. The columns of the data matrix are usually called *attributes*. The rows of the data matrix correspond to observed objects. We assume there is a finite number of possible values for each attribute. The possible values are called *categories*. In the data matrix, there are values of particular attributes for particular observed objects.

An example of such a data matrix is the data matrix $\mathcal{M}$ in Fig. 1.1. There are attributes $A_1, \ldots, A_K$ – columns of $\mathcal{M}$. The rows of $\mathcal{M}$ correspond to observed objects $o_1, \ldots, o_n$, see Fig. 1.1. The value of the attribute A_1 for object o_1 is 1, the value of the attribute A_2 for object o_1 is 9 etc.

object i.e. row of $\mathcal{M}$	columns of $\mathcal{M}$ i.e. attributes			examples of Boolean attributes			
				basic Boolean attributes		Boolean attributes	
	A_1	A_2	...	A_K	$A_1(1)$	$A_2(2,14,15)$	$A_1(1) \wedge A_2(2,14,15)$	$\neg A_K(6)$
o_1	1	9	...	4	1	0	0	1
o_2	1	14	...	6	1	1	1	0
o_3	3	15	...	7	0	1	0	1
$\vdots$	$\vdots$	$\vdots$	$\ddots$	$\vdots$	$\vdots$	$\vdots$	$\vdots$	$\vdots$
o_n	2	3	...	6	0	0	0	0

Fig. 1.1 Data matrix $\mathcal{M}$ and examples of Boolean attributes

The association rule is an expression $\varphi \approx \psi$ where φ and ψ are Boolean attributes. φ is called *antecedent* and ψ is called *succedent*. The association rule $\varphi \approx \psi$ means that the Boolean attributes φ and ψ are associated in the way given by the symbol $\approx$. The symbol $\approx$ is called the *4ft-quantifier*. It is related to a condition concerning a contingency table of φ and ψ.

The Boolean attributes φ and ψ are derived from attributes – columns of the analysed data matrix $\mathcal{M}$. *Basic Boolean attributes* are created first. A basic Boolean attribute is an expression of the form $A(\alpha)$ where $\alpha \subset \{a_1, \ldots a_t\}$ and $\{a_1, \ldots a_t\}$ is the set of all categories of the attribute A. The basic Boolean attribute $A(\alpha)$ is true in row o of $\mathcal{M}$ if it is $A[o, \mathcal{M}] \in \alpha$ where $A[o, \mathcal{M}]$ is the value of the attribute A in row o of data matrix $\mathcal{M}$. Boolean attributes φ and ψ are derived from basic Boolean attributes using propositional connectives $\vee$, $\wedge$ and $\neg$ in the usual way. The set α is called a *coefficient* of basic Boolean attribute $A(\alpha)$.

Expressions $A_1(1)$ and $A_2(2,14,15)$ in Fig. 1.1 are examples of basic Boolean attributes and expressions $A_1(1) \wedge A_2(2,14,15)$ and $\neg A_K(6)$ are examples of Boolean attributes. Basic Boolean attribute $A_1(1)$ is true in row o_1 because $A_1[o_1, \mathcal{M}] = 1$ and $1 \in \{1\}$; we write "1" in row o_1 and column A_1. Basic Boolean attribute $A_2(2,14,15)$ is false in row o_1 because $A_2[o_1, \mathcal{M}] = 9$ and $9 \notin \{2,14,15\}$; we write "0" in row o_1 and column $A_2(2,14,15)$. (Pedantically we should write $A_1(\{1\})$ and $A_2(\{2,14,15\})$ etc., however, we will not do it.)

The expression

$$A_1(1) \wedge A_2(2,14,15) \approx A_K(6)$$

is an example of the association rule. An association rule $\varphi \approx \psi$ can be true or false in the analysed data matrix $\mathcal{M}$. The rule $\varphi \approx \psi$ is verified using the *four-fold table* $4ft(\varphi,\psi,\mathcal{M})$ of φ and ψ in $\mathcal{M}$ which is also called *4ft-table of φ and ψ in $\mathcal{M}$*, see Table 1.1. Here a is the number of the objects (i.e. the rows of $\mathcal{M}$) satisfying both

Table 1.1 4ft-table $4ft(\varphi,\psi,\mathcal{M})$ of φ and ψ in $\mathcal{M}$

$\mathcal{M}$	ψ	$\neg\psi$
φ	a	b
$\neg\varphi$	c	d

φ and ψ, b is the number of the objects satisfying φ and not satisfying ψ, etc. We write $4ft(\varphi,\psi,\mathcal{M}) = \langle a,b,c,d \rangle$.

The rule $\varphi \approx \psi$ is *true in data matrix $\mathcal{M}$* if the condition related to the 4ft-quantifier $\approx$ is satisfied in the 4ft-table $4ft(\varphi,\psi,\mathcal{M})$. Various types of dependencies of φ and ψ can be expressed by 4ft-quantifiers including relations corresponding to statistical hypothesis tests. Four examples follow.

The 4ft-quantifier $\Rightarrow_{p,Base}$ of *founded implication* [18] is defined for $0 < p \le 1$ and $Base > 0$ by the condition

$$\frac{a}{a+b} \ge p \wedge a \ge Base .$$

The association rule $\varphi \Rightarrow_{p,Base} \psi$ means that at least $100p$ per cent of rows of $\mathcal{M}$ satisfying φ satisfy also ψ and that there are at least $Base$ rows of $\mathcal{M}$ satisfying both φ and ψ.

The 4ft-quantifier $\rightarrow_{p,s}$ is defined for $0 < p \le 1$ and $0 < s \le 1$ by the condition

$$\frac{a}{a+b} \ge p \wedge \frac{a}{a+b+c+d} \ge s .$$

The 4ft-quantifier $\rightarrow_{p,s}$ is called the *MB-quantifier*. Association rule $\varphi \rightarrow_{p,s} \psi$ can be seen as a generalization of association rule $X \rightarrow Y$ used for analysis of market basket, see above. The generalization consists in using general Boolean attributes φ and ψ instead of itemsets X and Y respectively.

The 4ft-quantifier $\Rightarrow^!_{p,\alpha,Base}$ of *lower critical implication* is defined in [18] for $0 < p \leq 1, 0 < \alpha < 0.5$ and $Base > 0$ by the condition

$$\sum_{i=a}^{a+b} \binom{a+b}{i} p^i (1-p)^{a+b-i} \leq \alpha \wedge a \geq Base .$$

Association rule $\varphi \Rightarrow^!_{p,\alpha,Base} \psi$ can be derived from the statistical binomial test (on the significance level α) of the null hypothesis $H_0 : P(\psi|\varphi) \leq p$ against the alternative $H_1 : P(\psi|\varphi) > p$. The association rule $\varphi \Rightarrow^!_{p,\alpha,Base} \psi$ is true in data matrix $\mathcal{M}$ exactly in those cases when H_0 is rejected by the test in favour of H_1. Hence, it can be interpreted as "the binomial test rejects on the level α the null hypothesis $P(\psi|\varphi) \leq p$ in favour of the alternative $P(\psi|\varphi) > p$".

The 4ft-quantifier $\sim^+_{p,Base}$ of *above average dependence* [80] is defined for $0 < p$ and $Base > 0$ by the condition

$$\frac{a}{a+b} \geq (1+p) \frac{a+c}{a+b+c+d} \wedge a \geq Base .$$

Association rule $\varphi \sim^+_{p,Base} \psi$ means that among the objects satisfying φ there is at least $100p$ per cent more objects satisfying ψ than among all observed objects and that there are at least $Base$ rows satisfying both φ and ψ.

1.2 ASSOC Procedure

The ASSOC procedure is a GUHA procedure, meaning its input consists of an analysed data matrix $\mathcal{M}$ and of several parameters defining a large set Ω of relevant association rules. Its output consists of all prime association rules. An association rule is prime if it is true in the analysed data matrix and if it does not immediately follow from another more simple output association rule.

Particular implementations of the ASSOC procedure [25, 28, 56, 80] differ regarding the possibilities of definition of the set Ω of relevant association rules. The most contemporarily used ASSOC procedure is probably the 4ft-Miner procedure [80]. It is capable of fine-tuning the set of relevant association rules very well. Note that similar possibilities has the procedure 4FT which is a part of the FERDA system [56]. These possibilities are outlined in Sect. 1.2.1. The output prime association rules are described in Sect. 1.2.2. The ASSOC procedure has a very elaborated approach for dealing with missing information. It is outlined in Sect. 1.2.3. A summary of important features of the ASSOC procedure can be found in Sect. 1.2.4.

1.2.1 Set of Relevant Association Rules

The set Ω of relevant association rules is usually defined such that sets Φ and Ψ of relevant antecedents and succedents respectively and a 4ft-quantifier $\approx$ are given. Then Ω is the set of all rules $\varphi \approx \psi$ such that $\varphi \in \Phi$, $\psi \in \Psi$ and φ and ψ have no common attributes. This means that there is no attribute A such that a basic Boolean attribute $A(\alpha)$ is used in φ and another basic Boolean attribute $A(\alpha')$ is used in ψ.

Both the 4ft-Miner and 4FT procedures have more than 15 basic 4ft-quantifiers including the quantifiers mentioned in Sect. 1.1. It is also possible to use conjunctions of particular basic 4ft-quantifiers. Both procedures provide rich possibilities to define the sets Φ and Ψ of relevant antecedents and succedents. It is important that there is a possibility of automatic generation of basic Boolean attributes such as $A(\alpha)$, where α is a general subset of all categories $a_1, \ldots, a_t$ of A. In other words, not only attribute-category pairs $A(a_1), \ldots, A(a_t)$ can be used. There are several possibilities of defining a set $\mathscr{B}(A)$ of basic Boolean attribute to be automatically generated for a given attribute A.

One possibility is to define intervals of categories with given minimal and maximal length, symbolically $\mathscr{B}(A) = Int[min, max](A)$. If we apply this possibility to the attribute Age with categories $1, \ldots, 100$, then the expression $Int[10, 10](Age)$ defines 91 basic Boolean attributes $Age(1 - 10), Age(2 - 11), \ldots, Age(91 - 100)$ with coefficients - intervals of categories with length of 10. Here $Age(1 - 10)$ means $Age(1, 2, \ldots, 10)$, etc.

We can use left cuts, i.e. intervals of categories "cut from left" with given minimal and maximal length, symbolically $\mathscr{B}(A) = Lcut[min, max](A)$. The expression $Lcut[1, 20](Age)$ defines 20 basic Boolean attributes for the above introduced attribute Age: $Age(1), Age(1, 2), Age(1 - 3), \ldots, Age(1 - 20)$. We can say that left cuts applied to attribute A give basic Boolean attributes corresponding to the low values of attribute A.

We can also use right cuts, i.e. intervals of categories "cut from right" with given minimal and maximal length, symbolically $\mathscr{B}(A) = Rcut[min, max](A)$. The expression $Rcut[1, 15](Age)$ defines 15 basic Boolean attributes for the above introduced attribute Age: $Age(100), Age(99, 100), Age(98 - 100), \ldots, Age(86 - 100)$. We can say that right cuts applied to attribute A give basic Boolean attributes corresponding to the high values of attribute A.

Cuts can also be used. We write $\mathscr{B}(A) = Cut[min, max](A)$ and we define $Cut[min, max](A) = Lcut[min, max](A) \bigcup Rcut[min, max](A)$. We can say that cuts applied to attribute A give basic Boolean attributes corresponding to extreme (i.e. low or high) values of attribute A. There are several additional possibilities of defining the set $\mathscr{B}(A)$ of basic Boolean attribute to be automatically generated for a given attribute A, see [56, 80].

The set Φ of relevant antecedents is defined such that a list of possible antecedent attributes is given and for each attribute A from this list a set $\mathscr{B}(A)$ of basic Boolean attributes to be automatically generated is defined. Then, some additional conditions are given on how the basic Boolean attributes can be used to derive particular $\varphi \in \Phi$. The set Ψ of relevant succedents is defined in a similar way. A more detailed description can be found in [56, 80]. It can be said that a lot of important analytical questions can be solved this way in a reasonable time [80, 83].

1.2.2 Prime Association Rules

The input of the ASSOC procedure consists of the analysed data matrix $\mathscr{M}$ and of the definition of the set Ω of relevant association rules. The output of the ASSOC procedure is a set $Pr(\mathscr{M}, \Omega)$ of all association rules from Ω which are prime with respect to Ω and $\mathscr{M}$. The association rule $\varphi \approx \psi$ is prime with respect to Ω and $\mathscr{M}$ if it is $\varphi \approx \psi \in \Omega$, $\varphi \approx \psi$ is true in $\mathscr{M}$ and if there is no association rule $\varphi_0 \approx \psi_0$ simpler than $\varphi \approx \psi$ such that $\varphi_0 \approx \psi_0$ is true in $\mathscr{M}$, $\varphi_0 \approx \psi_0$ already belongs to $Pr(\mathscr{M}, \Omega)$, and $\varphi \approx \psi$ logically follows from $\varphi_0 \approx \psi_0$ in a simple way. It must be carefully defined what it means that association rule $\varphi_0 \approx \psi_0$ is simpler than $\varphi \approx \psi$ and that $\varphi \approx \psi$ logically follows from $\varphi_0 \approx \psi_0$ in a simple way.

We give only an informal example of a rule which is true but not prime. The rule $A_1(1) \Rightarrow_{p,Base} A_2(1)$ is simpler than the rule $A_1(1) \Rightarrow_{p,Base} A_2(1) \vee A_3(1)$ because of the succedent $A_2(1)$ is simpler than the succedent $A_2(1) \vee A_3(1)$. It can be easily verified that if $A_1(1) \Rightarrow_{p,Base} A_2(1)$ is true in data matrix $\mathscr{M}$ then the rule $A_1(1) \Rightarrow_{p,Base} A_2(1) \vee A_3(1)$ is also true in data matrix $\mathscr{M}$. We can see this from 4ft-tables $4ft(A_1(1), A_2(1), \mathscr{M})$ and $4ft(A_1(1), A_2(1) \vee A_3(1), \mathscr{M})$, see Fig. 1.2.

$\mathscr{M}$	$A_2(1)$	$\neg A_2(1)$
$A_1(1)$	a	b
$\neg A_1(1)$	c	d

$\mathscr{M}$	$A_2(1) \vee A_3(1)$	$\neg(A_2(1) \vee A_3(1))$
$A_1(1)$	a'	b'
$\neg A_1(1)$	c'	d'

$4ft(A_1(1), A_2(1), \mathscr{M})$ $4ft(A_1(1), A_2(1) \vee A_3(1), \mathscr{M})$

Fig. 1.2 4ft-tables $4ft(A_1(1), A_2(1), \mathscr{M})$ and $4ft(A_1(1), A_2(1) \vee A_3(1), \mathscr{M})$

It holds $a + b = a' + b' =$ the number of rows satisfying $A_1(1)$. It also holds $a' \geq a$ because of each row satisfying $A_1(1) \wedge A_2(1)$ satisfy also $A_1(1) \wedge (A_2(1) \vee A_3(1))$. This means $b' \leq b$ and thus $\frac{a'}{a'+b'} \geq \frac{a}{a+b}$. We can conclude that if $A_1(1) \Rightarrow_{p,Base} A_2(1)$ is true in $\mathscr{M}$ then $A_1(1) \Rightarrow_{p,Base} A_2(1) \vee A_3(1)$ is also true in $\mathscr{M}$.

We have shown that if $A_1(1) \Rightarrow_{p,Base} A_2(1)$ is true in data matrix $\mathscr{M}$ then $A_1(1) \Rightarrow_{p,Base} A_2(1) \vee A_3(1)$ is also true in data matrix $\mathscr{M}$. We say that $A_1(1) \Rightarrow_{p,Base} A_2(1) \vee A_3(1)$ logically follows from $A_1(1) \Rightarrow_{p,Base} A_2(1)$ or that

$$\frac{A_1(1) \Rightarrow_{p,Base} A_2(1)}{A_1(1) \Rightarrow_{p,Base} A_2(1) \vee A_3(1)}$$

is a sound deduction rule.

This means that if the rule $A_1(1) \Rightarrow_{p,Base} A_2(1)$ belongs to $Pr(\mathscr{M}, \Omega)$ then the rule $A_1(1) \Rightarrow_{p,Base} A_2(1) \vee A_3(1)$ is true but not prime. It is important that only prime association rules are used in the output of the ASSOC procedure, this way the output can be remarkable reduced without loss of true rules. All true rules can be easily deduced by the user from output prime rules.

We are interested in prime rules for all 4ft-quantifiers, which means that we are interested in sound deduction rules

$$\frac{\varphi \approx \psi}{\varphi' \approx \psi'}$$

where $\varphi \approx \psi$ and $\varphi' \approx \psi'$ are association rules. In addition, we are interested in sound deduction rules which are simple enough to be applied by the user of the ASSOC procedure. There are important results [68] making possible to decide if a given deduction rule of the form $\frac{\varphi \approx \psi}{\varphi' \approx \psi'}$ is sound or not. Please note that such deduction rules can be also used to optimize the run of the ASSOC procedure [61].

1.2.3 Missing Information

A serious problem of real data analysis is missing information. A specific approach called *secured X-extension* is developed in [18] and applied for the ASSOC procedure. We informally outline its main features.

We assume there is a special symbol X that we interpret as the fact "the value of the corresponding attribute is not known for the corresponding object". An example of a data matrix with missing information is data matrix $\mathcal{M}^X$ in Fig. 1.3. It is the data matrix from Fig. 1.1 where some values are changed to X. We see that the value of attribute A_1 is not known for object o_1, the value of attribute A_2 is not known for object o_2 etc. We also say that the value of A_1 for object o_1 in data matrix $\mathcal{M}^X$ is X, and write $A_1[o_1, \mathcal{M}^X] = X$, similarly for additional attributes and objects.

There are two basic problems – how to deal with Boolean attributes and how to deal with association rules in data matrices with missing information. It is natural that in some cases the values of Boolean attributes cannot be known, and the same is true for association rules. However, in some cases we can determine the value of a Boolean attribute even in a data matrix with missing information without a danger of a mistake. This means that we can unmistakably say that the Boolean attribute is true or false for row o of a given data matrix. If we cannot do so, then we say that the value of the Boolean attribute for row o of the given data matrix is X. The value of an association rule in the given data matrix is determined similarly. This approach is formulated in [18] as *principle of secured X-extension*.

object	columns of $\mathcal{M}$			examples of Boolean attributes			
i.e. row	i.e. attributes			basic Boolean attributes		Boolean attributes	
of $\mathcal{M}$	A_1	A_2 ...	A_K	$A_1(1)$	$A_2(2,14,15)$	$A_1(1) \wedge A_2(2,14,15)$	$A_1(1) \vee A_2(2,14,15)$
o_1	X	9 ...	4	X	0	0	X
o_2	1	X ...	6	1	X	X	1
o_3	3	15 ...	X	0	1	0	1
$\vdots$	$\vdots$	$\vdots$ $\ddots$	$\vdots$	$\vdots$	$\vdots$	$\vdots$	$\vdots$
o_n	2	3 ...	6	0	0	0	0

Fig. 1.3 Data matrix $\mathcal{M}^X$ with missing information

To introduce this in more details we need the notion of *completion of the data matrix with missing information*. A completion of the data matrix $\mathcal{M}^X$ with missing information is any data matrix $\mathcal{M}$ with the same rows and columns such that each symbol X is replaced by one of possible values of the corresponding attribute. An example of a completion of data matrix $\mathcal{M}^X$ from Fig. 1.3 is the data matrix $\mathcal{M}$ from Fig. 1.1.

We write $\varphi[o, \mathcal{D}] = 1$ or $\varphi[o, \mathcal{D}] = 0$ if a Boolean attribute φ is true or false in row o of data matrix $\mathcal{D}$ respectively and we write $\varphi[o, \mathcal{D}] = X$ if the value of Boolean attribute for row o of the data matrix $\mathcal{D}$ is X. If we apply the principle of secured X-extension for Boolean attributes we get these rules.

- $\varphi[o, \mathcal{M}^X] = 1$ if $\varphi[o, \mathcal{M}] = 1$ in each completion $\mathcal{M}$ of $\mathcal{M}^X$
- $\varphi[o, \mathcal{M}^X] = 0$ if $\varphi[o, \mathcal{M}] = 0$ in each completion $\mathcal{M}$ of $\mathcal{M}^X$
- $\varphi[o, \mathcal{M}^X] = X$ otherwise.

Value $A(\alpha)[o, \mathcal{M}^X]$ of basic Boolean attribute $A(\alpha)$ in the row o of data matrix $\mathcal{M}^X$ is according to the principle of secured X-extension defined such that

- $A(\alpha)[o, \mathcal{M}^X] = 1$ if $A[o, \mathcal{M}^X] \in \alpha$,
- $A(\alpha)[o, \mathcal{M}^X] = 0$ if $A[o, \mathcal{M}^X] \notin \alpha$ and $A[o, \mathcal{M}^X] \neq X$
- $A(\alpha)[o, \mathcal{M}^X] = X$ otherwise.

The values of Boolean attributes derived from basic Boolean attributes using propositional connectives $\neg$, $\wedge$, and $\vee$ are defined by truth tables of these connectives extended by the principle of secured X-extension, see Fig. 1.4. Examples of values of basic and derived Boolean attributes in data matrix $\mathcal{M}^X$ with missing information are located in Fig. 1.3.

The principle of secured X-extension is applied also for association rules. This means:

- $Val(\varphi \approx \psi, \mathcal{M}^X) = 1$ if $Val(\varphi \approx \psi, \mathcal{M}) = 1$ in each completion $\mathcal{M}$ of $\mathcal{M}^X$
- $Val(\varphi \approx \psi, \mathcal{M}^X) = 0$ if $Val(\varphi \approx \psi, \mathcal{M}) = 0$ in each completion $\mathcal{M}$ of $\mathcal{M}^X$
- $Val(\varphi \approx \psi, \mathcal{M}^X) = X$ otherwise.

The values of φ and ψ belong to the set $\{0, X, 1\}$. Thus we can compute a nine-fold table of φ and ψ in $\mathcal{M}^X$ instead of the four-fold table $4ft(\varphi, \psi, \mathcal{M})$ of φ and ψ in data matrix $\mathcal{M}$ without missing information. The nine-fold table of φ and ψ in $\mathcal{M}^X$ is denoted as $9ft(\varphi, \psi, \mathcal{M}^X)$, see Table 1.2. Here $f_{1,1}$ is the number of rows o of $\mathcal{M}^X$ such that both $\varphi(o, \mathcal{M}^X) = 1$ and $\psi(o, \mathcal{M}^X) = 1$, $f_{1,X}$ is the number of rows o of $\mathcal{M}^X$ such that both $\varphi(o, \mathcal{M}^X) = 1$ and $\psi(o, \mathcal{M}^X) = X$, etc. The problem is that

$\neg$		$\wedge$	1	X	0		$\vee$	1	X	0
1	0	1	1	X	0		1	1	1	1
X	X	X	X	X	0		X	1	X	X
0	1	0	0	0	0		0	1	X	0

Fig. 1.4 Extended truth tables of $\vee$, $\wedge$ and $\neg$

Table 1.2 Nine-fold table $9ft(\varphi, \psi, \mathscr{M}^X)$

$\mathscr{M}^X$	ψ	ψ_X	$\neg\psi$
φ	$f_{1,1}$	$f_{1,X}$	$f_{1,0}$
φ_X	$f_{X,1}$	$f_{X,X}$	$f_{X,0}$
$\neg\varphi$	$f_{0,1}$	$f_{0,X}$	$f_{0,0}$

we have to deal with many 4ft-tables corresponding to all possible completions of data matrix $\mathscr{M}^X$. We say that the *4ft-table* $\langle a,b,c,d \rangle$ *is a completion of the nine-fold table* $9ft(\varphi, \psi, \mathscr{M}^X)$ if $\langle a,b,c,d \rangle = 4ft(\varphi, \psi, \mathscr{M})$, where $\mathscr{M}$ is a completion of $\mathscr{M}^X$ [18]. It is, however, practically not possible to find the value $Val(\varphi \approx \psi, \mathscr{M}^X)$ by computing values $Val(\varphi \approx \psi, \mathscr{M})$ for all completions $\mathscr{M}$ of $\mathscr{M}^X$.

For important 4ft-quantifiers it can be proved that for each nine-fold table $9ft(\varphi, \psi, \mathscr{M}^X)$ there are 4ft-tables $\langle a_s, b_s, c_s, d_s \rangle$ and $\langle a_o, b_o, c_o, d_o \rangle$ such that

- $Val(\varphi \approx \psi, \mathscr{M}^X) = 1$ if $\approx (a_s, b_s, c_s, d_s) = 1$
- $Val(\varphi \approx \psi, \mathscr{M}^X) = 0$ if $\approx (a_o, b_o, c_o, d_o) = 0$
- $Val(\varphi \approx \psi, \mathscr{M}^X) = X$ otherwise.

Details are given later in this book, see also previous results [18, 60, 71]. 4ft-table $\langle a_s, b_s, c_s, d_s \rangle$ is called a *secured* or *pessimistic 4ft-table* for nine-fold table $9ft(\varphi, \psi, \mathscr{M}^X)$ and 4ft-quantifier $\approx$. 4ft-table $\langle a_o, b_o, c_o, d_o \rangle$ is called an *optimistic 4ft-table* for nine-fold table $9ft(\varphi, \psi, \mathscr{M}^X)$ and 4ft-quantifier $\approx$.

1.2.4 Summary of Features of the ASSOC Procedure

The above informal and very short introduction of the procedure ASSOC can be summarized as follows:

- The procedure ASSOC was defined in the 1970's [18]. It mines for interesting pairs of Boolean attributes $\varphi \approx \psi$ which are also called association rules. The association rules $X \to Y$ defined and studied in relation with analysis of market basket data are a special case of association rules the procedure ASSOC deals with.
- Boolean attributes φ and ψ are built from basic Boolean attributes of the form $A(\alpha)$ where α is a subset of all possible values of attribute A. The basic Boolean attributes $A(\alpha)$ are automatically generated. Various sets $\mathscr{B}(A)$ of basic Boolean attributes $A(\alpha)$ can be specified in a simple way, e.g. $\mathscr{B}(A) = Int[min, max](A)$ or $\mathscr{B}(A) = Cut[min, max](A)$.
- There are theoretical results making possible to decide if a deduction rule of the form $\frac{\varphi \approx \psi}{\varphi' \approx \psi'}$ is sound or not. Such deduction rules are applied in the ASSOC procedure in various ways.
- There is a special approach for dealing with missing information in the ASSOC procedure. It is based on the principle of secured X-extension. This way, a value *true* or *false* is assigned to a Boolean attribute or to an association rule even in

data matrices with missing information whenever this is possible without danger of a mistake.

1.3 Logic of Association Rules

The above mentioned results on deduction rules of the form $\frac{\varphi \approx \psi}{\varphi' \approx \psi'}$ as well as results concerning dealing with missing information can be naturally formulated using special logical calculi. Formulas of such calculi correspond to association rules. Formulas, i.e. association rules, are interpreted in models – data structures corresponding to data matrices. There is a formal definition of truthfulness of an association rule in a given data matrix.

Such special calculi are defined and studied in [18] as a tool for logic of discovery which is also developed in [18]. There are two types of such calculi – observational predicate calculi and calculi with qualitative values. Observational predicate calculi can be seen as a modification of classical predicate calculi – only finite models are allowed and generalized quantifiers are added. Finite models correspond to analysed data and generalized quantifiers make it possible to express various assertion on the analysed data.

Formulas $(\forall x)P_1(x)$ and $(\exists x)(P_3(x) \wedge P_4(y))$ are examples of formulas of a predicate calculus. They can also be seen as formulas of observational predicate calculus with unary predicates P_1, P_2, P_3, P_4. However, they are then interpreted in $\{0,1\}$-valued data matrices i.e. finite structures. Particular predicates are interpreted by columns of such matrices. Value 1 usually means *true* and 0 means *false*. The first formula is closed, both formulas only contain classical quantifiers $\forall$ and $\exists$. The formula $(\forall x)P_1(x)$ is true in data matrix $\mathcal{M}$ if the column interpreting P_1 contains only 1's.

The formula $(\Rightarrow_{p,Base} x)(P_1(x), P_2(x))$ is an example of a closed formula of an observational predicate calculus with unary predicates. A generalized quantifier is used in this formula, specifically the 4ft-quantifier $\Rightarrow_{p,Base}$. Additional examples of formulas of the observational predicate calculus are formulas $(\Rightarrow_{p,Base} x)(P_1(x), P_2(y))$ and $(\sim^+_{p,Base} x)(P_1(x), P_2(y) \wedge P_5(y,z))$. Here P_5 is a binary predicate. A formal definition of truthfulness of such formulas is given in [18]. Calculi with qualitative values defined in [18] are intended to be a generalization of observational predicate calculus allowing to talk about data matrices with general values, not only about $\{0,1\}$-valued data matrices.

Theoretically interesting and practically important results about association rules – formulas of observational predicate calculi are achieved in [18]. Several classes of 4ft-quantifiers rules are defined and studied. The 4ft-quantifier $\Rightarrow_{p,Base}$ is an example of an implicational quantifier, the 4ft-quantifier $\sim^+_{p,Base}$ is an example of a symmetrical quantifiers. The results concern namely the deduction rules, dealing with missing information, and the definability of association rules in classical predicate calculus (i.e. calculus with only $\forall$ and $\exists$ quantifiers).

A new effort in the study of association rules, i.e. interesting pairs of Boolean attributes $\varphi \approx \psi$, started in the 1990's and new results were achieved, see e.g. [68].

The syntax of used formulas of observational calculi has been significantly simplified, and only calculi with monadic predicates are further studied. Free and bound variables are omitted and basic Boolean attributes are used instead of predicates. The rule $A_1(1) \wedge A_2(3,4) \Rightarrow_{p,Base} A_3(6)$ is an example of such simplified association rules. Please note that such calculi can be defined also by suitable simplification of calculi with qualitative values defined in [18].

New classes of association rules are defined and studied and new theoretically interesting and practically useful results were obtained. Most of the important results are related to classes of association rules. The results can be considered as a logic of association rules.

1.4 Goals and Structure of the Book

The goal of this book is to present results on logic of association rules and to argue for their usefulness for data mining. There are three types of results:

- definitions of crucial notions
- theoretical results in the form of theorems with proofs
- applications and research projects.

Definitions of crucial notions are provided as simple as possible, while still allowing the presentation of theoretical results. The following notions are defined:

- *Logical calculi of association rules.* These are defined in a way informally outlined in Sect. 1.1.
- *4ft-quantifiers.* 4ft-quantifiers studied with relation to the ASSOC procedure as well as additional 4ft-quantifiers derived from well known measures of interestingness of association rules are presented.
- *Classes of association rules.* All classes related to known interesting results are introduced. Classes of association rules are defined by classes of 4ft-quantifiers. We say that the rule $\varphi \approx \psi$ belongs to the class $\mathscr{C}$ if the 4ft-quantifier $\approx$ belongs to the class $\mathscr{C}$ of 4ft-quantifiers. Membership of all defined 4ft-quantifiers to particular classes is presented and proved. Some of defined classes have important subclasses, and such subclasses are also introduced. Mutual relations of defined classes and subclasses are presented and proved when important.

The defined notions are used to present achieved theoretical results. The following theoretical results are presented:

- *Criteria of soundness of deduction rules of the form* $\frac{\varphi \approx \psi}{\varphi' \approx \psi'}$. These criteria are used in various ways in relation to the ASSOC procedure.
- *Logical calculi of association rules with missing information.* These calculi formalize dealing with missing information outlined in Sect. 1.2.3.
- *Tables of critical frequencies* which allow to optimize verification of association rules with some 4ft-quantifiers corresponding to hypothesis tests. Some verifications require complex computation, an example is the 4ft-quantifier $\Rightarrow^!_{p,\alpha,Base}$ of lower critical implication which is defined by the condition

$\sum_{i=a}^{a+b} \binom{a+b}{i} p^i (1-p)^{a+b-i} \le \alpha \land a \ge Base$, see Sect. 1.1. Tables of critical frequencies allow the conversion of such a verification into a verification of a simple inequality. Tables of critical frequencies can be also used to set a criterion of definability of 4ft-quantifiers in classical predicate calculus.

- *Definability of 4ft-quantifiers*. It can be proved that in some cases it is possible to express association rules with 4ft-quantifiers by formulas of classical predicate calculus with equality, i.e. predicate calculus with equality and only classical quantifiers $\exists$ and $\forall$. We say that such a 4ft-quantifier is definable in classical predicate calculus. There are relatively simple criteria of definability of 4ft-quantifiers which can be seen as an intuitive characterization of definable 4ft-quantifiers.

All the presented theoretical results are related to classes of association rules. The results are used in various ways related to the GUHA procedure ASSOC.

The GUHA procedure 4ft-Miner is probably the currently most used implementation of the ASSOC procedure. The 4ft-Miner procedure together with the presented theoretical results inspired several research projects which have led to new challenges in the research of observational calculi and hopefully can contribute to solution of current research challenges related to data mining research [96]. Main features of the 4ft-Miner procedure and of the related research projects are also presented in the book.

The book is divided into four parts:

I   Logical calculi of association rules
II   Classes of association rules
III Results on Classes of Association Rules
IV   Applications and related research.

The contents of particular parts correspond to their names. A formal definition of logical calculi of association rules is located in part I. All classes of association rules related to important results are introduced in part II together with interesting subclasses and their mutual relations. Tables of critical frequencies are also presented in part II. This is due to the tables of critical frequencies being used to define some important subclasses of association rules.

Additional theoretical results, i.e. deduction rules, approaches to dealing with missing information and results on definability of 4ft-quantifiers in classical predicate calculi are presented in part III. The possibilities of applications of presented theoretical results are introduced in part IV together with several research projects and open problems.

Part I
Logical Calculi of Association Rules

The goal of this part is to give a formal definition of logical calculi of association rules and to introduce important 4ft-quantifiers. We are going to define calculus of association rules as a semantic system in the sense of [18]. This means that we first have to clarify what are the formal structures our calculi speak about.

We introduced an association rule as a pair of Boolean attributes derived from columns of an analysed data matrix. Thus calculi of association rules deal with data matrices. However, to keep things simple we limit ourselves to data matrices with values – positive integer numbers only. The data matrices we are going to deal with are defined in Chap. 2.

Having clarified data matrices, we can define calculus of association rules. Specifically, we define a set of formulas which correspond to association rules and then we specify how truthfulness of a particular association rule is determined for a particular data matrix. This is done in Chap. 3.

In the definition of calculus of association rules a notion of general 4ft-quantifier is used. We are, however, interested in concrete 4ft-quantifiers which are defined to express particular practically important relations between two Boolean attributes. There are many such 4ft-quantifiers. All 4ft-quantifiers studied in this book are introduced in Chap. 4. Chap. 5 contains several results on particular 4ft-quantifiers which are used in the rest of the book.

Chapter 2
Data Matrices

We consider an association rule to be an interesting pair of Boolean attributes. These Boolean attributes are derived from columns of an analysed data matrix. An example of such a data matrix with general values is given in Sect. 2.1. However, we are interested only in a simple form of data matrices. This means that we consider data matrices with positive integer numbers as their only values, see Sect. 2.2. Association rules were defined in [1] as a tool of analysing market baskets data. This is the reason why the market basket data is introduced in Sect. 2.3. In addition, it is shown that market basket data can also be simply considered as a data matrix.

2.1 Data Matrix - An Example

An example of a data matrix is data matrix *LoanDetails* concerning financial data. It is derived from data on a fictitious bank [80]. There are 6181 rows in the *LoanDetails* data matrix. Each row corresponds to a particular loan of a client, see Fig. 2.1.

loan	Sex	Age	Salary	District	Amount	Repayment	Months	Quality
o_1	male	$\langle 20,30 \rangle$	average	Prague	$\langle 50;100 \rangle$	$\langle 2,3 \rangle$	37–48	good
o_2	female	$\langle 40,50 \rangle$	high	Brno	$\langle 250;500 \rangle$	> 9	13–24	bad
o_3	female	$\langle 60,70 \rangle$	very low	Ostrava	$\langle 1;20 \rangle$	$\langle 1,2 \rangle$	1–12	good
o_4	male	$\langle 40,50 \rangle$	very high	Prague	≥ 500	> 9	49–60	good
$\vdots$	$\vdots$	$\vdots$	$\vdots$	$\vdots$	$\vdots$	$\vdots$	$\vdots$	$\vdots$
o_{6180}	male	$\langle 50,60 \rangle$	low	Pilsen	$\langle 20;50 \rangle$	$\langle 0,1 \rangle$	25–36	bad
o_{6181}	female	$\langle 30,40 \rangle$	high	Brno	$\langle 100;250 \rangle$	$\langle 2,3 \rangle$	37–48	good

Fig. 2.1 Data matrix *LoanDetails*

There are 8 columns in the data matrix. Each column corresponds to one attribute describing particular loans. There are several possible values for each attribute, these values are called *categories*. It is important that each attribute can have only a finite

J. Rauch: *Observational Calculi and Association Rules*, SCI 469, pp. 17–20.
DOI: 10.1007/978-3-642-11737-4_2   © Springer-Verlag Berlin Heidelberg 2013

number of categories. Particular values can be e.g. numbers, intervals of numbers or strings of symbols.

The attributes of data matrix *LoanDetails* can be divided into two groups. The group *Client* has four attributes:

- *Sex* with 2 categories *female* and *male*
- *Age* with 5 categories – intervals $\langle 20,30 \rangle, \ldots, \langle 60,70 \rangle$
- *Salary* with 5 categories *very low, low, average, high* and *very high*
- *District* with 77 categories i.e. districts where the clients live, e.g. *Brno, Ostrava, Pilsen, Prague*.

The group *Loan* has four attributes:

- *Amount* – i.e. the amount of borrowed money in thousands of Czech crowns divided into 6 categories - intervals $\langle 1;20 \rangle$, $\langle 20;50 \rangle$, $\langle 50;100 \rangle$, $\langle 100;250 \rangle$, $\langle 250;500 \rangle$ and ≥ 500
- *Repayment* – i.e. repayment in thousands of Czech crowns with 10 categories – intervals $(0,1\rangle, \ldots, (8,9\rangle$ and > 9
- *Months* – i.e. the number of months to repay the whole amount with values divided into 5 categories – intervals 1–12, 13–24, 25–36, 37–48 and 49–60
- *Quality* – with 2 categories *good* (i.e. the loan is already paid up or it is repaid without problems) and *bad* (i.e. the loan finished and it is not paid up or it is repaid with problems).

2.2   Data Matrices for Calculi of Association Rules

We are interested in logical calculi of association rules. The association rules talk about data matrices that represent results of observations of objects of our interest. An example is data matrix *LoanDetails* in Fig. 2.1. To keep the data matrices we deal with simple, we consider data matrices with values – positive integer numbers only. The numbers represent the real possible values such as *male*, $\langle 20,30 \rangle$, *average*, *Prague* etc.

There is only a finite number of possible values for each column. Let us assume that the number of possible values of a column is t and the possible values in this column are integer numbers $1, \ldots, t$. Then all the possible values in the data matrix are described by the number of its columns and by the numbers of possible values for each column. These numbers determine a type of a data matrix and also a type of a logical calculus of association rules. We define:

Definition 2.1. A *type of a logical calculus of association rules* is a K-tuple $\mathscr{T} = \langle t_1, \ldots, t_K \rangle$ where $K \geq 2$ is an integer number and $t_i \geq 2$ are integer numbers for $i = 1, \ldots, K$.

We are going to define a data matrix of the type $\mathscr{T} = \langle t_1, \ldots, t_K \rangle$. The rows of the data matrix correspond to observed objects, the columns contains values of attributes for particular objects. Let us assume that a data matrix has n rows. We can then see the data matrix as a K+1-tuple according to definition 2.2.

Definition 2.2. Let $\mathscr{T} = \langle t_1, \ldots, t_K \rangle$ be the type of a logical calculus of association rules. Then a *data matrix of the type* $\mathscr{T}$ is a $K+1$-tuple $\mathscr{M} = \langle M, f_1, \ldots, f_K \rangle$, where M is a non-empty finite set and f_i is an unary function from M to $\{1, \ldots, t_i\}$ for $i = 1, \ldots, K$. Elements of M are called rows of $\mathscr{M}$, and the set M is a *set of rows* of data matrix $\mathscr{M}$.

Figure 2.2 contains an example of a data matrix $\mathscr{M} = \langle M, f_1, \ldots, f_K \rangle$. We assume that $M = \{o_1, \ldots, o_n\}$.

row	f_1	f_2	$\cdots$	f_K
o_1	$f_1(o_1)$	$f_2(o_1)$	$\cdots$	$f_K(o_1)$
o_2	$f_1(o_2)$	$f_2(o_2)$	$\cdots$	$f_K(o_2)$
$\vdots$	$\vdots$	$\vdots$	$\ddots$	$\vdots$
o_{n-1}	$f_1(o_{n-1})$	$f_2(o_{n-1})$	$\cdots$	$f_K(o_{n-1})$
o_n	$f_1(o_n)$	$f_2(o_n)$	$\cdots$	$f_K(o_n)$

Fig. 2.2 Data matrix $\mathscr{M}$

2.3 Data Matrix and Market Baskets

Association rules defined in [1] concern data on market baskets. This data describes particular transactions in supermarkets, with each transaction corresponding to one market basket. Let us assume that the supermarket sells items $I_1, I_2, \ldots, I_K$ and that there is a database $\mathscr{D}$ of transactions, i.e. market baskets $b_1, b_2, \ldots, b_n$. An example of such a database of transactions is located in Fig. 2.3.

basket	items
b_1	I_1, I_2
b_2	I_2, I_7, I_K
$\vdots$	$\vdots$
b_{n-1}	I_4, I_{K-1}, I_K
b_n	I_2, I_3, I_K

Fig. 2.3 Example of the database $\mathscr{D}$ of transactions

This means that market basket b_1 contains items I_1, I_2, market basket b_2 contains items I_2, I_7, I_K, etc. The database $\mathscr{D}$ of transactions can thus be seen as a data matrix $\mathscr{M}_{\mathscr{D}} = \langle M_{\mathscr{D}}, g_1, \ldots, g_K \rangle$ of type $\mathscr{T}_{\mathscr{D}} = \langle \overbrace{2, \ldots, 2}^{K\times} \rangle$, see Fig. 2.4.

Functions $g_1, \ldots, g_K$ attain values $\{1, 2\}$, $g_i(b_j) = 1$ if item I_i is in market basket b_j and $g_i(b_j) = 2$ if item I_i is not in market basket b_j for $i = 1, \ldots, K$ and $j = 1, \ldots, n$.

basket	g_1	g_2	$\cdots$	g_{K-1}	g_K
b_1	1	1	$\cdots$	2	2
b_2	2	1	$\cdots$	2	1
$\vdots$	$\vdots$	$\vdots$	$\ddots$	$\vdots$	$\vdots$
b_{n-1}	2	2	$\cdots$	1	1
b_n	2	1	$\cdots$	2	1

Fig. 2.4 Data matrix $\mathcal{M}_{\mathcal{D}} = \langle M_{\mathcal{D}}, g_1, \ldots, g_K \rangle$

Chapter 3
Logical Calculus of Association Rules

We are going to define logical calculus of association rules as a semantic system $\langle \text{Sent}, \text{M}, V, Val \rangle$ introduced in [18]. The semantic system consists of

- a non-empty set Sent of *sentences*
- a non-empty set M of *models*
- a non-empty set V of *abstract values*
- an *evaluating function Val* : Sent $\times$ M $\rightarrow V$.

If $\Omega \in$ Sent and $\mathcal{M} \in$ M then $Val(\Omega, \mathcal{M})$ is the value of Ω in $\mathcal{M}$.

We define sentences as association rules $\varphi \approx \psi$ introduced in Sect. 1.1. Association rules are evaluated in data matrices. We have decided to deal with data matrices with values – positive integer numbers only. It means that the set M of models will be a set of such data matrices. The set Sent of sentences will be a set of association rules which can be formulated for a given set of data matrices. Boolean attributes we can derive from columns of a given data matrix are determined by the type of data matrix. This leads to a definition of the set Sent as a set of formulas of a language of association rules of type $\mathcal{T}$, see Sect. 3.1.

The set V of abstract values is the set $\{0, 1\}$, where 1 means *true* and 0 means *false*. The function *Val* assigns a value $Val(\varphi \approx \psi, \mathcal{M})$ of $\varphi \approx \psi$ in $\mathcal{M}$ to each association rule $\varphi \approx \psi \in$ Sent and to each $\mathcal{M} \in$ M. If $Val(\varphi \approx \psi, \mathcal{M}) = 1$ then we say that $\varphi \approx \psi$ is true in $\mathcal{M}$. If $Val(\varphi \approx \psi, \mathcal{M}) = 0$ then we say that $\varphi \approx \psi$ is false in $\mathcal{M}$.

The function $Val(\varphi \approx \psi, \mathcal{M})$ is evaluated in several steps. In the first step we need to interpret Boolean attributes φ and ψ in the data matrix $\mathcal{M}$. This means that we decide for each row o of $\mathcal{M}$ if the Boolean attribute φ is true or false for this row, and we do the same for ψ. Then we can compute a 4ft-table $4ft(\varphi, \psi, \mathcal{M})$ of φ and ψ in $\mathcal{M}$. The interpretation of Boolean attributes is described in Sect. 3.2, and the computation of the 4ft-table in Sect. 3.3. Finally, we use the associated function $F_\approx$ which corresponds to the condition associated to 4ft-quantifier $\approx$. Thus, the definition of logical calculus of association rules is finished, see Sect. 3.4.

J. Rauch: *Observational Calculi and Association Rules*, SCI 469, pp. 21–26.
DOI: 10.1007/978-3-642-11737-4_3 © Springer-Verlag Berlin Heidelberg 2013

3.1   Language of Association Rules

We define association rules as formulas of a formal language. This means that we
have to give a list of symbols from which the formulas (i.e. association rules) will be
created as well as rules how to create formulas from symbols. Our formal language
is called language of association rules.

An example of an association rule informally introduced in Sect. 1.1 is the rule
$A_1(1) \wedge A_2(2,14,15) \approx A_K(6)$. Each association rule is composed from two Boolean
attributes derived from columns of an analysed data matrix and from a 4ft-quantifier.
The expressions $A_1(1) \wedge A_2(2,14,15)$ and $A_K(6)$ are examples of Boolean attributes.
The Boolean attributes we deal with are created from basic Boolean attributes.

Expression $A_1(1)$ is an example of a basic Boolean attribute. It is comprised of
the symbol A_1 and digit 1. A_1 is an attribute, and is interpreted by a column of an
analysed data matrix. The digit "1" in $A_1(1)$ denotes a category, i.e. a positive integer
number. The digit "2" and strings of digits "14" and "15" in the basic Boolean
attribute $A_2(2,14,15)$ also denote categories.

We will distinguish between integer numbers and their digital representation. If
k is an integer number then $\diamond(k)$ denotes its decimal representation. Each decimal
representation of an integer number will be considered as one symbol. Thus, 100
is the number "one hundred", while $\diamond(100)$ is the 3-digit string "100", which is
considered to be one symbol. However, if there is no danger of misunderstanding
we will use only k instead of $\diamond(k)$. This means that usually we write $A_2(2,14,15)$
instead of $A_1(\diamond(2),\diamond(14),\diamond(15))$ etc.

Based on these considerations we can define a language $\mathscr{L}$ of association rules.
We are going to define:

- basic symbols, i. e. attributes, categories, propositional connectives and 4ft-quantifiers, see definition 3.1
- rules for creating Boolean attributes, see definition 3.2
- the language $\mathscr{L}$ of association rules, see definition 3.3.

We also use additional symbols such as parentheses and commas in the usual way.

We sometimes use the abbreviation LAR instead of the expression "language of
association rules". Remember that type $\mathscr{T} = \langle t_1,\dots,t_K \rangle$ is a K-tuple of integer numbers $t_1,\dots,t_K$, see definition 2.1. This determines the number of columns of a data
matrix and possible values of particular columns. It also determines the number of
attributes and the numbers of categories of particular attributes of a logical calculus
of association rules. Thus we speak about LAR of type $\mathscr{T}$.

Definition 3.1. The *basic symbols* $\mathscr{BS}(\mathscr{L}_\mathscr{T})$ of a language $\mathscr{L}_\mathscr{T}$ of association
rules of type $\mathscr{T} = \langle t_1,\dots,t_K \rangle$ are:

1. *basic attributes* $A_1,\dots,A_K$, we say that A_i is of type t_i for $i = 1,\dots,K$
2. *categories* $\diamond(1),\dots,\diamond(u)$, where $u = \max\{t_1,\dots,t_K\}$
3. *propositional connectives* $\wedge$, $\vee$, and $\neg$
4. *4ft-quantifiers* $\approx_1,\dots\approx_Q$.

Definition 3.2. The *Boolean attributes of a language* $\mathscr{L}_{\mathscr{T}}$ of association rules of type $\mathscr{T} = \langle t_1,\ldots,t_K \rangle$ with basic symbols $\mathscr{BS}(\mathscr{L}_{\mathscr{T}})$ according to definition 3.1 are defined this way:

1. An expression $A_i(\diamond(v_1), \ldots, \diamond(v_k))$ such that $i = 1,\ldots,K$ and $\{v_1,\ldots,v_k\} \subset \{1,\ldots,t_i\}$ is called a *basic Boolean attribute* of $\mathscr{L}_{\mathscr{T}}$.
2. Each basic Boolean attribute of $\mathscr{L}_{\mathscr{T}}$ is a *Boolean attribute* of $\mathscr{L}_{\mathscr{T}}$.
3. If φ and ψ are Boolean attributes of $\mathscr{L}_{\mathscr{T}}$ then $\neg\varphi$, $\varphi \vee \psi$ and $\varphi \wedge \psi$ are *Boolean attributes* of $\mathscr{L}_{\mathscr{T}}$, and the usual conventions concerning parentheses are valid.

We define also a *coefficient of a basic Boolean attribute*: If $A_i(\diamond(v_1), \ldots, \diamond(v_k))$ is a basic Boolean attribute then the expression $\diamond(v_1), \ldots, \diamond(v_k)$ is a *coefficient* of $A_i(\diamond(v_1), \ldots, \diamond(v_k))$.

Definition 3.3. The *language* $\mathscr{L}_{\mathscr{T}}$ *of association rules of type* $\mathscr{T} = \langle t_1,\ldots,t_K \rangle$ consists of the following:

1. basic symbols $\mathscr{BS}(\mathscr{L}_{\mathscr{T}})$ of language $\mathscr{L}_{\mathscr{T}}$ of type $\mathscr{T}$ according to definition 3.1
2. set $\mathscr{BA}(\mathscr{L}_{\mathscr{T}})$ of Boolean attributes of language $\mathscr{L}_{\mathscr{T}}$ of type $\mathscr{T}$ defined according according to definition 3.2
3. set of association rules $\mathscr{AR}(\mathscr{L}_{\mathscr{T}})$ of language $\mathscr{L}_{\mathscr{T}}$, with each association rule in the form

$$\varphi \approx \psi$$

where φ and ψ are Boolean attributes of language $\mathscr{L}_{\mathscr{T}}$ and $\approx$ is 4ft-quantifier belonging to $\mathscr{BS}(\mathscr{L})$.

3.2   Interpretation of Boolean Attributes

The truthfulness of an association rule $\varphi \approx \psi$ in data matrix $\mathscr{M} = \langle M, f_1,\ldots,f_K \rangle$ is defined using a 4ft-table of φ and ψ. To compute the 4ft-table we need to know the values of Boolean attributes φ and ψ for each $o \in M$.

We show how the values of a given Boolean attribute φ are defined for each $o \in M$. Values of attributes $A_1,\ldots A_K$ are given by functions $f_1,\ldots,f_K$. Values of a Boolean attribute φ derived from attributes $A_1,\ldots A_K$ are given by a $\{0,1\}$-valued function f_φ derived from functions $f_1,\ldots,f_K$.

The function f_φ is called an *interpretation of φ in* $\mathscr{M} = \langle M, f_1,\ldots,f_K \rangle$. It is specified in definitions 3.4 and 3.5.

Definition 3.4. Let $\mathscr{L}_{\mathscr{T}}$ be a LAR of type $\mathscr{T} = \langle t_1,\ldots,t_K \rangle$ with basic symbols according to definition 3.1 and let $\mathscr{M} = \langle M, f_1,\ldots,f_K \rangle$ be a data matrix of the type $\mathscr{T}$. Then we define:

1. The *interpretation* $\Im(A_i)$ *of attribute* A_i in $\mathscr{M}$ is the function f_i for $i = 1,\ldots,K$.
2. If $A_i(\diamond(v_1), \ldots, \diamond(v_k))$ is a basic Boolean attribute of $\mathscr{L}_{\mathscr{T}}$ then the *interpretation* $\Im(\diamond(v_1), \ldots, \diamond(v_k))$ *of coefficient* $\diamond(v_1), \ldots, \diamond(v_k)$ in $\mathscr{M}$ is the set $\{v_1,\ldots,v_k\}$.

Definition 3.5. Let $\mathscr{L}_{\mathscr{T}}$ be a LAR of type $\mathscr{T} = \langle t_1, \ldots, t_K \rangle$ with basic symbols according to definition 3.1. Let $\mathscr{M} = \langle M, f_1, \ldots, f_K \rangle$ be a data matrix of the type $\mathscr{T}$. Then we define:

1. The *interpretation of a basic Boolean attribute* $A_i(\alpha)$ in $\mathscr{M}$ is a $\{0,1\}$-valued function $f_{A_i(\alpha)}$ defined for each $o \in M$ such that

$$f_{A_i(\alpha)}(o) = \begin{cases} 1 \text{ if } f_i(o) \in \mathfrak{I}(\alpha) \\ 0 \text{ otherwise}, \end{cases}$$

   where $\mathfrak{I}(\alpha)$ is the interpretation of the coefficient α in $\mathscr{M}$.
2. Let φ be a Boolean attribute of $\mathscr{L}_{\mathscr{T}}$ and let f_φ be the interpretation of φ in $\mathscr{M}$. Then the *interpretation of the Boolean attribute* $\neg\varphi$ in $\mathscr{M}$ is the function $f_{\neg\varphi}$ defined for each $o \in M$ such that

$$f_{\neg\varphi}(o) = 1 - f_\varphi(o) \, .$$

3. Let φ and ψ be Boolean attributes of $\mathscr{L}_{\mathscr{T}}$. In addition, let f_φ be the interpretation of φ in $\mathscr{M}$ and let f_ψ be the interpretation of ψ in $\mathscr{M}$. Then the *interpretation of the Boolean attribute* $\varphi \wedge \psi$ in $\mathscr{M}$ is the function $f_{\varphi\wedge\psi}$ defined for each $o \in M$ such that

$$f_{\varphi\wedge\psi}(o) = \min(f_\varphi(o), f_\psi(o)) \, .$$

4. Let φ and ψ be Boolean attributes of $\mathscr{L}_{\mathscr{T}}$. In addition, let f_φ be the interpretation of φ in $\mathscr{M}$ and let f_ψ be the interpretation of ψ in $\mathscr{M}$. Then the *interpretation of the Boolean attribute* $\varphi \vee \psi$ in $\mathscr{M}$ is the function $f_{\varphi\vee\psi}$ defined for each $o \in M$ such that

$$f_{\varphi\vee\psi}(o) = \max(f_\varphi(o), f_\psi(o)) \, .$$

An example of an interpretation of attributes $A_1, \ldots, A_K, A_1(\alpha)$, φ, $\neg\varphi$, $\varphi \wedge \psi$, and $\varphi \vee \psi$ in data matrix $\mathscr{M} = \langle M, f_1, \ldots, f_K \rangle$ is located in Fig. 3.1. We assume that the rows of $\mathscr{M}$ are $o_1, \ldots, o_n$ i.e. $M = \{o_1, \ldots, o_n\}$.

	A_1	$\ldots$	A_K	$A_1(\alpha)$	φ	$\neg\varphi$	$\varphi \wedge \psi$	$\varphi \vee \psi$
row	f_1	$\ldots$	f_K	$f_{A_1(\alpha)}$	f_φ	$f_{\neg\varphi}$	$f_{\varphi\wedge\psi}$	$f_{\varphi\vee\psi}$
o_1	$f_1(o_1)$	$\ldots$	$f_K(o_1)$	$f_{A_1(\alpha)}(o_1)$	$f_\varphi(o_1)$	$f_{\neg\varphi}(o_1)$	$f_{\varphi\wedge\psi}(o_1)$	$f_{\varphi\vee\psi}(o_1)$
o_2	$f_1(o_2)$	$\ldots$	$f_K(o_2)$	$f_{A_1(\alpha)}(o_2)$	$f_\varphi(o_2)$	$f_{\neg\varphi}(o_2)$	$f_{\varphi\wedge\psi}(o_2)$	$f_{\varphi\vee\psi}(o_2)$
$\vdots$	$\vdots$	$\ddots$	$\vdots$	$\vdots$	$\vdots$	$\vdots$	$\vdots$	$\vdots$
o_n	$f_1(o_n)$	$\ldots$	$f_K(o_n)$	$f_{A_1(\alpha)}(o_n)$	$f_\varphi(o_n)$	$f_{\neg\varphi}(o_n)$	$f_{\varphi\wedge\psi}(o_n)$	$f_{\varphi\vee\psi}(o_n)$

Fig. 3.1 Example of interpretation of attributes $A_1, \ldots, A_K, A_1(\alpha)$, φ, $\neg\varphi$, $\varphi \wedge \psi$, and $\varphi \vee \psi$

An example of an interpretation of concrete Boolean attributes in a data matrix $\mathscr{M} = \langle M, f_1, f_2, f_3 \rangle$ with concrete values is located in Fig. 3.2. We assume that the type $\mathscr{T}$ of this data matrix is $\mathscr{T} = \langle 3, 5, 4 \rangle$ and rows of $\mathscr{M}$ are $o_1, \ldots, o_{1000}$.

row	A_1 f_1	A_2 f_2	A_3 f_3	$A_1(1)$ $f_{A_1(1)}$	$A_2(4,5)$ $f_{A_2(4,5)}$	$A_1(1) \wedge A_2(4,5)$ $f_{A_1(1) \wedge A_2(4,5)}$	$A_1(1) \vee A_2(4,5)$ $f_{A_1(1) \vee A_2(4,5)}$
o_1	1	1	1	1	0	0	1
o_2	1	2	2	1	0	0	1
o_3	2	3	3	0	0	0	0
o_4	1	4	4	1	1	1	1
$\vdots$	$\vdots$	$\vdots$	$\vdots$	$\vdots$	$\vdots$	$\vdots$	$\vdots$
o_{1000}	3	5	1	0	1	0	1

Fig. 3.2 Interpretation of attributes in a concrete data matrix $\mathcal{M} = \langle M, f_1, f_2, f_3 \rangle$

In addition we define:

Definition 3.6. Let $\mathcal{L}_{\mathcal{T}}$ be a LAR of type $\mathcal{T} = \langle t_1, \ldots, t_K \rangle$ and $\mathcal{M} = \langle M, f_1, \ldots, f_K \rangle$ be a data matrix of the type $\mathcal{T}$. Let φ be a Boolean attributes of $\mathcal{L}_{\mathcal{T}}$ and let f_φ be the interpretation of φ in M. Then:

- If $o \in M$ and $f_\varphi(o) = 1$ then we say that o *satisfies* φ in $\mathcal{M}$ or that φ *is true for o* in $\mathcal{M}$ or that *a value of φ for o* in $\mathcal{M}$ is 1.
- If $o \in M$ and $f_\varphi(o) = 0$ then we say that o *does not satisfy* φ in $\mathcal{M}$ or that φ is *false for o* in $\mathcal{M}$ or that *a value of φ for o* in $\mathcal{M}$ is 0.

3.3 4ft-table

A 4ft-table is computed using the $Fr(\varphi, \mathcal{M})$ function which determines the number of rows $o \in M$ satisfying φ in data matrix $\mathcal{M} = \langle M, f_1, \ldots, f_K \rangle$. It is described in the following definition.

Definition 3.7. Let $\mathcal{L}_{\mathcal{T}}$ be a LAR of type $\mathcal{T} = \langle t_1, \ldots, t_K \rangle$ with basic symbols according to definition 3.1. Let $\mathcal{M} = \langle M, f_1, \ldots, f_K \rangle$ be a data matrix of the type $\mathcal{T}$ and φ and ψ be Boolean attributes of $\mathcal{L}_{\mathcal{T}}$. Then we define:

1. Let ω be a Boolean attribute of $\mathcal{L}_{\mathcal{T}}$ and let f_ω be the interpretation of ω in M. Then the *frequency of a Boolean attribute* ω in $\mathcal{M}$ is the number $Fr(\omega, \mathcal{M})$ defined such that

$$Fr(\omega, \mathcal{M}) = \sum_{o \in M} f_\omega(o) .$$

2. The *4ft-table* $4ft(\varphi, \psi, \mathcal{M})$ of φ and ψ in $\mathcal{M}$, is a quadruple of non-negative integer numbers

$$4ft(\varphi, \psi, \mathcal{M}) = \langle a, b, c, d \rangle$$

where $a = Fr(\varphi \wedge \psi, \mathcal{M})$, $b = Fr(\varphi \wedge \neg\psi, \mathcal{M})$, $c = Fr(\neg\varphi \wedge \psi, \mathcal{M})$, and $d = Fr(\neg\varphi \wedge \neg\psi, \mathcal{M})$.

Please note that 4ft-table $4ft(\varphi, \psi, \mathcal{M})$ of φ and ψ in $\mathcal{M}$ can be written in the form introduced in Table 1.1. Table 1.1 is shown in the left part of Fig. 3.3 together with the form of $4ft(\varphi, \psi, \mathcal{M})$ corresponding to definition 3.7.

$\mathcal{M}$	ψ	$\neg\psi$
φ	a	b
$\neg\varphi$	c	d

$\mathcal{M}$	ψ	$\neg\psi$
φ	$Fr(\varphi \wedge \psi, \mathcal{M})$	$Fr(\varphi \wedge \neg\psi, \mathcal{M})$
$\neg\varphi$	$Fr(\neg\varphi \wedge \psi, \mathcal{M})$	$Fr(\neg\varphi \wedge \neg\psi, \mathcal{M})$

Fig. 3.3 4ft-table of φ and ψ in $\mathcal{M}$: 4ft($\varphi, \psi, \mathcal{M}$) = $\langle a,b,c,d \rangle$

3.4   Calculus of Association Rules

We have defined both the language of association rule formulas which correspond to association rules and the data matrices the association rules talk about. The language of association rules and the data matrix are related through the type $\mathcal{T} = \langle t_1, \ldots, t_K \rangle$ of a logical calculus, see definition 2.1.

To finish the definition of calculus of association rules, we need to define truthfulness of an association rule $\varphi \approx \psi$ in a data matrix $\mathcal{M}$. The association rule $\varphi \approx \psi$ is verified in data matrix $\mathcal{M}$ using a 4ft-table $4ft(\varphi, \psi, \mathcal{M})$. The rule is true if a condition associated to the 4ft-quantifier $\approx$ is satisfied for $4ft(\varphi, \psi, \mathcal{M})$. The way of computing of 4ft-table $4ft(\varphi, \psi, \mathcal{M})$ is given by definition 3.7. The condition associated to 4ft-quantifier $\approx$ is formalized by function $F_\approx$ associated to 4ft-quantifier $\approx$.

Definition 3.8. The *associated function $F_\approx$ of the 4ft-quantifier* $\approx$ is a $\{0,1\}$ - valued function defined for all quadruples $\langle a,b,c,d \rangle$ of non-negative integer numbers satisfying $a + b + c + d > 0$.

The associated functions of 4ft-quantifiers are used to finish the definition of logical calculus of association rules.

Definition 3.9. The *logical calculus* $\mathcal{LC}_\mathcal{T}$ *of association rules* of type $\mathcal{T} = \langle t_1, \ldots, t_K \rangle$ is given by:

1. the language $\mathcal{L}_\mathcal{T}$ of association rules of type $\mathcal{T} = \langle t_1, \ldots, t_K \rangle$
2. the set $\mathrm{M}_\mathcal{T}$ of all data matrices $\mathcal{M} = \langle M, f_1, \ldots, f_K \rangle$ of type $\mathcal{T}$
3. the associated functions $F_{\approx_1}, \ldots, F_{\approx_Q}$ of 4ft-quantifiers $\approx_1, \ldots \approx_Q$ of language $\mathcal{L}_\mathcal{T}$ respectively.

The *value Val*($\varphi \approx \psi, \mathcal{M}$) *of association rule* $\varphi \approx \psi$ of $\mathcal{L}_\mathcal{T}$ *in data matrix* $\mathcal{M} \in \mathrm{M}_\mathcal{T}$ is defined such that

$$Val(\varphi \approx \psi, \mathcal{M}) = F_\approx(a,b,c,d)$$

where $\langle a,b,c,d \rangle = 4ft(\varphi, \psi, \mathcal{M})$ is the 4ft-table of φ and ψ in data matrix $\mathcal{M}$ and $F_\approx$ is an associated function of 4ft-quantifier $\approx$.

The *association rule $\varphi \approx \psi$ is true in data matrix $\mathcal{M}$* if $Val(\varphi \approx \psi, \mathcal{M}) = 1$, otherwise the *association rule $\varphi \approx \psi$ is false in data matrix $\mathcal{M}$*.

This finishes the definition of logical calculi of association rules.

Note 3.1. Please note that the associated function $F_\approx$ of the 4ft-quantifier $\approx$ is defined for all quadruples $\langle a,b,c,d \rangle$ of non-negative integer numbers satisfying $a + b + c + d > 0$, see definition 3.8. This is because the set M of rows of data matrix $\mathcal{M} = \langle M, f_1, \ldots, f_K \rangle$ is non-empty, see definition 2.2.

Chapter 4
4ft-quantifiers

We have defined the association rule $\varphi \approx \psi$ with general 4ft-quantifier $\approx$. There exist various practically important 4ft-quantifiers defined in connection to the GUHA procedure ASSOC. These are introduced in Sect. 4.1. There is also a possibility to use various measures of interestingness of association rules together with suitable thresholds for their values to define additional 4ft-quantifiers. An overview of well known measures of interestingness is given in Sect. 4.2. In Sect. 4.3, these measures of interestingness are used to define additional 4ft-quantifiers. Practical experience has led to the definition of so called founded and supported versions of 4ft-quantifiers and also to the use of combinations of particular 4ft-quantifiers. We call such 4ft-quantifiers as compound 4ft-quantifiers. Founded, supported and compound 4ft-quantifiers are introduced in Sect. 4.4.

4.1 4ft-quantifiers and the GUHA Procedure ASSOC

There are several 4ft-quantifiers defined in [18] and implemented in the GUHA procedures ASSOC described in [24]. There are additional 4ft-quantifiers implemented in the 4ft-Miner procedure [80]. The same 4ft-quantifiers are implemented in the 4FT procedure [56]. Both 4ft-Miner and 4FT are implementations of the ASSOC procedure. Additional 4ft-quantifiers are studied in connection to the ASSOC procedure in the master thesis [48].

Results presented in this book are closely related to classes of 4ft-quantifiers which are introduced and studied in Part II. Thus, information to which class of 4ft-quantifiers each below introduced 4ft-quantifier belongs to is associated to particular 4ft-quantifiers. However, this information is indicative only. There are many classes and subclasses of 4ft-quantifiers defined in Chaps. 6–10 and it is necessary to study these chapters to better understand the mutual relations of particular 4ft-quantifiers and classes of 4ft-quantifiers. In addition, information on classes of 4ft-quantifiers which particular 4ft-quantifiers belong to is given only when this is known and proved.

J. Rauch: *Observational Calculi and Association Rules*, SCI 469, pp. 27–38.
DOI: 10.1007/978-3-642-11737-4_4 © Springer-Verlag Berlin Heidelberg 2013

The classes of 4ft-quantifiers used in this chapter together with their abbreviations and information on chapters where they are described are listed in Table 4.1.

Table 4.1 Classes of 4ft-quantifiers used in Chap. 4

Class of association rules	Abbreviation	Details in
Implicational	*IM*	Chap. 7
$\mathcal{M}$-dependent Implicational	$\mathcal{M}$-IM	Chap. 7
Σ-Double Implicational	ΣDI	Chap. 8
Equivalence	*EQ*	Chap. 9
Σ-Equivalence	ΣEQ	Chap. 9
with the *F*-property	*FPR*	Chap. 10

An overview of all 4ft-quantifiers defined and implemented or studied in relation to the GUHA procedure ASSOC is located in Table 4.2. Each 4ft-quantifier has one or two parameters which are written in the lower index; the 4ft-quantifier $\Rightarrow_p$ of p-implication has parameter p, the 4ft-quantifier $\Rightarrow_{p,\alpha}^!$ of likely p-implication has parameters p and α etc. Some of 4ft-quantifiers defined in relation to the GUHA procedure ASSOC can be also defined using various measures of interestingness of association rules. 4ft-quantifiers defined using measures of interestingness are introduced in Table 4.6.

The following information is given for each 4ft-quantifier in Table 4.2:

- *No.* - used for reference in Table 4.4
- *Name* - some quantifiers have several names
- *Symbol* - also denotes the parameters as mentioned above
- *Defined in* - literature where the 4ft-quantifier is defined, two different references may be used when the 4ft-quantifier is defined under different names; in addition the number of the respective section in the book may be given
- *No. in Table 4.6* - number of the corresponding entry in Table 4.6.

Please note that quantifiers $\sim_p^-$ of below average (No. 14 in Table 4.2), $\odot_s$ of support (No. 19 in Table 4.2), $\overline{\oplus}_{Base}$ of ceil (No. 20 in Table 4.2), and $\overline{\odot}_s$ of ceil support (No. 21 in Table 4.2) are implemented in the 4ft-Miner procedure. Quantifier $\odot_s$ is defined using measure support, see also Sect. 4.3.

Associated functions of 4ft-quantifiers listed in Table 4.2 are given in definition 4.1. Please note that definition 4.1 is rather long. It finishes with points G_1–G_{10}. Please note also that there are two parameters p and s in definition 4.1 with the same range $(0;1\rangle$, this is due to the fact that they are both used in one compound quantifier, see also Sect. 4.4 and theorem 7.4.

Table 4.2 Overview of 4ft-quantifiers related to the GUHA procedure ASSOC

No.	Name	Symbol	Defined in	No. in Table 4.6
1	p-implication	$\Rightarrow_p$	4.4.27 in [18]	2
2	Likely p-implication Lower critical implication	$\Rightarrow^!_{p,\alpha}$	4.4.12 in [18] 1.2.25 in [24]	
3	Suspicious critical implication Upper critical implication	$\Rightarrow^?_{p,\alpha}$	4.4.12 in [18] 1.2.25 in [24]	
4	Double p-implication	$\Leftrightarrow_p$	3.3.2 in [24]	12
5	Lower critical double implication	$\Leftrightarrow^!_{p,\alpha}$	3.3.2 in [24]	
6	Upper critical double implication	$\Leftrightarrow^?_{p,\alpha}$	3.3.2 in [24]	
7	p-equivalence	$\equiv_p$	3.3.2 in [24]	7
8	Lower critical equivalence	$\equiv^!_{p,\alpha}$	3.3.2 in [24]	
9	Upper critical equivalence	$\equiv^?_{p,\alpha}$	3.3.2 in [24]	
10	Simple deviation	$\sim^!_q$	1.2.25 in [24]	14
11	Fisher's quantifier	$\sim^!_\alpha$	4.4.20 in [18]	
12	χ^2-quantifier	$\sim^2_\alpha$	4.4.23 in [18]	
13	Above average	$\sim^+_q$	[68]	
14	Below average	$\sim^-_p$		
15	E-quantifier	$\equiv^E_p$	[97]	
16	Above negation	$\Rightarrow^{\neg}_{p,q}$	[48]	
17	Pairing quantifier	$\sim^{0,\bullet}_p$	[48]	
18	Base	$\oplus_{Base}$	[18]	
19	Support	$\odot_s$		1
20	Ceil	$\overline{\oplus_{Base}}$		
21	Ceil support	$\overline{\odot_s}$		

Definition 4.1. Associated functions of the 4ft-quantifiers introduced in Table 4.2 are defined in Table 4.4. The 4ft-quantifiers listed in Table 4.2 have the following parameters:

- real number p satisfying $0 < p \leq 1$
- real number s satisfying $0 < s \leq 1$
- real number α satisfying $0 < \alpha \leq 0.5$
- real number q satisfying $q > 0$
- integer number $Base$ satisfying $Base > 0$.

The associated functions are evaluated using quadruples $\langle a,b,c,d \rangle$ which corresponds to 4ft-table $4ft(\varphi,\psi,\mathscr{M})$ of association rule $\varphi \approx \psi$ in data matrix $\mathscr{M}$. In some definitions we use also $r = a+b$, $k = a+c$, and $n = a+b+c+d$, see Table 4.3.

The sentence "$F_{\approx^*} = 1$ if and only if" in the third column of Table 4.4 means that it holds $F_{\approx^*}(a,b,c,d) = 1$ if and only if the condition given in the corresponding row is satisfied. It is assumed $0 < p < 1$ for 4ft-quantifiers $\Rightarrow^!_{p,\alpha}$ (row 2), $\Rightarrow^?_{p,\alpha}$ (row 3), $\Leftrightarrow^!_{p,\alpha}$ (row 5), $\Leftrightarrow^?_{p,\alpha}$ (row 6), $\equiv^!_{p,\alpha}$ (row 8), $\equiv^?_{p,\alpha}$ (row 9), and $\sim^-_p$ (row 14). Definitions of $F_{\approx^*}$ in some rows require additional information or a note. These are

Table 4.3 4ft-table 4ft($\varphi,\psi,\mathscr{M}$) of φ and ψ in $\mathscr{M}$

$\mathscr{M}$	ψ	$\neg\psi$	
φ	a	b	r
$\neg\varphi$	c	d	
	k		n

provided in points $G_1 - G_{10}$ as mentioned in Table 4.4 in the fourth column and the corresponding rows.

Table 4.4 Associated functions of 4ft-quantifiers related to the GUHA procedure ASSOC

No.	$\approx^*$	$F_{\approx^*} = 1$ if and only if	See point	Class
1	$\Rightarrow_p$	$\frac{a}{a+b} \geq p$	G_1	IM
2	$\Rightarrow_{p,\alpha}^!$	$\sum_{i=a}^{a+b} \binom{a+b}{i} p^i (1-p)^{a+b-i} \leq \alpha$	G_2	IM
3	$\Rightarrow_{p,\alpha}^?$	$\sum_{i=0}^{a} \binom{a+b}{i} p^i (1-p)^{a+b-i} > \alpha$		IM
4	$\Leftrightarrow_p$	$\frac{a}{a+b+c} \geq p$	G_3	ΣDI
5	$\Leftrightarrow_{p,\alpha}^!$	$\sum_{i=a}^{a+b+c} \binom{a+b+c}{i} p^i (1-p)^{a+b+c-i} \leq \alpha$	G_4	ΣDI
6	$\Leftrightarrow_{p,\alpha}^?$	$\sum_{i=0}^{a} \binom{a+b+c}{i} p^i (1-p)^{a+b+c-i} > \alpha$		ΣDI
7	$\equiv_p$	$\frac{a+d}{a+b+c+d} \geq p$		ΣEQ
8	$\equiv_{p,\alpha}^!$	$\sum_{i=a+d}^{n} \binom{n}{i} p^i (1-p)^{n-i} \leq \alpha$		ΣEQ
9	$\equiv_{p,\alpha}^?$	$\sum_{i=0}^{a+d} \binom{n}{i} p^i (1-p)^{n-i} > \alpha$		ΣEQ
10	$\sim_q$	$ad > e^q bc$		EQ, FPR
11	$\sim_\alpha^1$	$\sum_{i=a}^{\min(r,k)} \frac{\binom{k}{i}\binom{n-k}{r-i}}{\binom{n}{r}} \leq \alpha \wedge ad > bc$		EQ, FPR
12	$\sim_\alpha^2$	$\frac{(ad-bc)^2}{rk(n-k)(n-r)} n \geq \chi_\alpha^2 \wedge ad > bc$	G_5	EQ, FPR
13	$\sim_q^+$	$\frac{a}{a+b} \geq (1+q)\frac{a+c}{a+b+c+d}$	G_6	FPR
14	$\sim_p^-$	$\frac{a}{a+b} \leq (1-p)\frac{a+c}{a+b+c+d}$	G_7	
15	$\equiv_p^E$	$max(\frac{b}{a+b}, \frac{c}{d+c}) < p$	G_8	EQ
16	$\Rightarrow_{p,q}^-$	$\frac{a}{a+b} \geq p \wedge \frac{ac+ad}{ac+bc} \geq q$	G_9	
17	$\sim_p^{\circ,\bullet}$	$2\frac{a^2 d^2}{a^4+d^4} \geq p$	G_{10}	
18	$\oplus_{Base}$	$a \geq Base$		IM
19	$\odot_s$	$\frac{a}{a+b+c+d} \geq s$		$\mathscr{M}$-IM
20	$\oplus_{Base}$	$a \leq Base$		
21	$\odot_s$	$\frac{a}{a+b+c+d} \leq s$		

G_1) If $a+b=0$ then we define $F_{\Rightarrow_p}(a,b,c,d)=0$, which means $F_{\Rightarrow_p}(0,0,c,d)=0$ for all non-negative integer c,d satisfying $c+d>0$.

G_2) It holds $F_{\Rightarrow_{p,\alpha}^!}(0,0,c,d)=0$ for all non-negative integer c,d satisfying $c+d>0$.

G_3) If $a+b+c=0$ then we define $F_{\Leftrightarrow_p}(a,b,c,d)=0$, thus $F_{\Leftrightarrow_p}(0,0,0,d)=0$ for all integer $d > 0$.

G_4) It holds $F_{\Leftrightarrow_{p,\alpha}^!}(0,0,0,d)=0$ for all integer $d > 0$.

G_5) χ_α^2 is the $1-\alpha$ quantile of the χ^2-distribution function. If $rk(n-k)(n-r)=0$ then we define $F_{\sim_\alpha^2}(a,b,c,d)=0$.

G_6)   If $a+b=0$ then we define $F_{\sim^+_q}(a,b,c,d)=0$.

G_7)   If $a+b=0$ then we define $F_{\sim^-_p}(a,b,c,d)=0$.

G_8)   We distinguish two cases:

- if $a+b=0$ then it must hold $c+d>0$ and we define $F_{\equiv^E_p}(a,b,c,d)=1$ if and only if $\frac{c}{d+c}<p$
- if $c+d=0$ then it must hold $a+b>0$ and we define $F_{\equiv^E_p}(a,b,c,d)=1$ if and only if $\frac{b}{a+b}<p$.

G_9)   If $a+b=0$ or $ac+bc=0$ then we define $F_{\Rightarrow^-_{p,q}}(a,b,c,d)=0$

G_{10})   If $a^4+d^4=0$ then we define $F_{\sim^{\circ\bullet}_p}(a,b,c,d)=0$.

Please note that Table 4.4 also includes the class particular 4ft-quantifiers belong to. The abbreviations of classes of 4ft-quantifiers come from Table 4.1. However, this information is not available for all 4ft-quantifiers. In some cases this is trivial to decide, while in others it would require more effort to find the corresponding class of 4ft-quantifiers – also see Sect. 16.2.

We present additional information to some of the above given definitions. We assume to have data matrix $\mathcal{M}$ and we try to explain the meaning of association rule $\varphi \approx \psi$ for particular 4ft-quantifiers $\approx$.

1. Rule $\varphi \Rightarrow_p \psi$ means that at least $100p$ per cent of objects satisfying φ satisfy also ψ. The quantifier $\Rightarrow_p$ of p-implication corresponds to confidence as defined in [1].

2. Rule $\varphi \Rightarrow^!_{p,\alpha} \psi$ corresponds to the statistical test (on the level α) of the null hypothesis $H_0 : P(\psi|\varphi) \leq p$ against the alternative one $H_1 : P(\psi|\varphi) > p$. Here $P(\psi|\varphi)$ is the conditional probability of the validity of ψ under the condition φ.

3. Rule $\varphi \Rightarrow^?_{p,\alpha} \psi$ corresponds to the statistical test (on the level α) of the null hypothesis $H_0 : P(\psi|\varphi) \geq p$ against the alternative one $H_1 : P(\psi|\varphi) < p$. Here $P(\psi|\varphi)$ is the conditional probability of the validity of ψ under the condition φ.

4. Rule $\varphi \Leftrightarrow_p \psi$ means that at least $100p$ per cent of objects satisfying φ or ψ satisfy both φ and ψ.

5. Rule $\varphi \Leftrightarrow^!_{p,\alpha} \psi$ corresponds to the statistical test (on the level α) of the null hypothesis $H_0 : P(\psi \wedge \varphi \mid \psi \vee \varphi) \leq p$ against the alternative hypothesis $H_1 : P(\psi \wedge \varphi \mid \psi \vee \varphi) > p$. Here $P(\psi \wedge \varphi \mid \psi \vee \varphi)$ is the conditional probability of the validity of $\psi \wedge \varphi$ under the condition $\psi \vee \varphi$.

6. Rule $\varphi \Leftrightarrow^?_{p,\alpha} \psi$ corresponds to the statistical test (on the level α) of the null hypothesis $H_0 : P(\psi \wedge \varphi \mid \psi \vee \varphi) \geq p$ against the alternative hypothesis $H_1 : P(\psi \wedge \varphi \mid \psi \vee \varphi) < p$. Here $P(\psi \wedge \varphi \mid \psi \vee \varphi)$ is the conditional probability of the validity of $\psi \wedge \varphi$ under the condition $\psi \vee \varphi$.

7. Rule $\varphi \equiv_p \psi$ means that at least $100p$ per cent of objects have the same truth value for φ and ψ.

8. Rule $\varphi \equiv^!_{p,\alpha} \psi$ corresponds to the statistical test (on the level α) of the null hypothesis $H_0 : P(\psi$ and φ have the same truth value$) \leq p$ against the alternative one $H_1 : P(\psi$ and φ have the same truth value$) > p$.

9. Rule $\varphi \equiv^?_{p,\alpha} \psi$ corresponds to the statistical test (on the level α) of the null hypothesis $H_0 : P(\psi$ and φ have the same truth value$) \geq p$ against the alternative one $H_1 : P(\psi$ and φ have the same truth value$) < p$.

10. Rule $\varphi \sim_q \psi$ can be interpreted as *the logarithmic interaction of φ and ψ is estimated to be greater than q*.

11. Rule $\varphi \sim^1_\alpha \psi$ can be derived from the statistical one-sided Fisher test (on the level α) of the null hypothesis $H_0(\varphi$ and ψ are independent) against the alternative one $H_1(\,$the logarithmic interaction of φ and ψ is positive).

12. Rule $\varphi \sim^2_\alpha \psi$ can be derived from the statistical χ^2 test (on the level α) of the null hypothesis $H_0(\varphi$ and ψ are independent) against the alternative one $H_1(\,$the logarithmic interaction of φ and ψ is positive).

13. Rule $\varphi \sim^+_q \psi$ means that among the objects satisfying φ there is at least $100q$ per cent more objects satisfying ψ than among all objects.

14. Rule $\varphi \sim^-_p \psi$ means that among the objects satisfying φ there is at least $100p$ per cent less objects satisfying ψ than among all objects.

Let us emphasize that the GUHA procedure ASSOC was developed approximately 45 years ago to mine for interesting pairs of derived Boolean attributes. Such pairs are called association rules since the 1990's. The definition of interestingness is based on statistical approaches to data analysis. This resulted into 4ft-quantifiers corresponding to conditions concerning estimations of used parameters or to statements given by statistical hypothesis tests. The statistical nature of these quantifiers is described in [18, 24]. This book does not deal with the statistical aspects of association rules.

4.2   Measures of Interestingness of Association Rules

There are many various measures of interestingness of association rules. By applying suitable thresholds for their values we can get additional 4ft-quantifiers [74]. We are going to use this way some of the measures of interestingness introduced in [9]. In [9], measures of interestingness are defined for a rule $A \rightarrow B$ instead of $\varphi \approx \psi$. This means that 4ft-table $4ft(A,B,\mathcal{M})$ is used instead of $4ft(\varphi,\psi,\mathcal{M})$, see Fig. 4.1.

Here $n(AB)$ corresponds to a, $n(A\overline{B})$ corresponds to b, etc, see Fig. 4.1. Frequencies (e.g. $n(AB)$, $n(A)$) or probabilities and conditional probabilities are usually used in definitions of the interestingness measures [9]. An example of probability is $P(A) = \frac{n(A)}{N}$, which denotes the probability of A. An example of conditional probability is $P(B|A) = \frac{n(AB)}{n(A)}$, which denotes the conditional probability of B, given A.

Table 4.5 lists 29 measures of interestingness from measures introduced in [9]. The names of these measures are located in column *Name*. Several measures are

$\mathcal{M}$	B	$\bar{B}$	
A	$n(AB)$	$n(A\bar{B})$	$n(A)$
$\bar{A}$	$n(\bar{A}B)$	$n(\bar{A}\bar{B})$	$n(\bar{A})$
	$n(B)$	$n(\bar{B})$	N

$$4ft(A,B,\mathcal{M})$$

$\mathcal{M}$	ψ	$\neg\psi$	
φ	a	b	r
$\neg\varphi$	c	d	
	k		n

$$4ft(\varphi,\psi,\mathcal{M})$$

Fig. 4.1 $4ft(A,B,\mathcal{M})$ and $4ft(\varphi,\psi,\mathcal{M})$

defined under various names in additional papers. The additional names are also listed in column *Name* together with a citation of the corresponding paper.

Each measure of interestingness is in [9] defined using probabilities, see column *Probabilities*. Each such measure of interestingness can also be expressed as the function $\mathscr{I}(a,b,c,d)$ of frequencies a,b,c,d from the 4ft-table $4ft(\varphi,\psi,\mathcal{M})$ corresponding to $4ft(A,B,\mathcal{M})$, see Fig. 4.1. This is done in column $\mathscr{I}(a,b,c,d)$ of table 4.5.

4.3 4ft-quantifiers Based on Measures of Interestingness

Measures of interestingness can be used to define additional 4ft-quantifiers by applying suitable thresholds to their values, see definition 4.2. Please note that definition 4.2 is rather long. It ends with points $M_1 - M_{22}$. Please note again that there are two parameters p and s in definition 4.2 with the same range $(0;1\rangle$. This is because of these both parameters are used in one compound quantifier, see also Sect. 4.4 and theorem 7.4.

Definition 4.2. Associated functions of 4ft-quantifiers defined using measures of interestingness introduced in Table 4.5 are defined in Table 4.6. The associated functions are evaluated using quadruples $\langle a,b,c,d \rangle$ which corresponds to 4ft-table $4ft(\varphi,\psi,\mathcal{M})$ of association rule $\varphi \approx \psi$ in data matrix $\mathcal{M}$, see Fig. 3.3 and Fig. 4.1.

Each 4ft-quantifier listed in Table 4.6 has a parameter which is written in the lower index. The admissible values for particular parameters are:

- real number p satisfying $0 < p \leq 1$
- real number s satisfying $0 < s \leq 1$
- real number q satisfying $q > 0$
- real number u satisfying $-1 \leq u \leq 1$
- real number t
- real number v satisfying $v \leq 1$.

The names of 4ft-quantifiers derived from the measures of interestingness are the same as the names of corresponding measures of interestingness. If a measure of interestingness has more names, then each name can be used for the 4ft-quantifier. Correspondence between measures of interestingness listed in Table 4.5 and 4ft-quantifiers listed in Table 4.6 is ensured by numbers of measures and numbers of derived 4ft-quantifiers which are given in the first columns of both tables.

Table 4.5 Measures of interestingness of association rules

#	Name	Probabilities	$\mathscr{I}(a,b,c,d)$
1	Support	$P(AB)$	$\frac{a}{a+b+c+d}$
2	Confidence / Precision p-implication [18]	$P(B\|A)$	$\frac{a}{a+b}$
3	Coverage	$P(A)$	$\frac{a+b}{a+b+c+d}$
4	Prevalence	$P(B)$	$\frac{a+c}{a+b+c+d}$
5	Recall	$P(A\|B)$	$\frac{a}{a+c}$
6	Specificity	$P(\neg B\|\neg A)$	$\frac{d}{b+d}$
7	Accuracy p-equivalence [24] Success Rate [29]	$P(AB)+P(\neg A\neg B)$	$\frac{a+d}{a+b+c+d}$
8	Lift / Interest	$P(B\|A)/P(B)$	$\frac{a(a+b+c+d)}{(a+b)(a+c)}$
9	Leverage	$P(B\|A)-P(A)P(B)$	$\frac{a}{a+b}-\frac{(a+b)(a+c)}{(a+b+c+d)^2}$
10	Added Value / Change of Support	$P(B\|A)-P(B)$	$\frac{ad-bc}{(a+b)(a+b+c+d)}$
11	Relative Risk	$P(B\|A)/P(B\|\neg A)$	$\frac{a(c+d)}{c(a+b)}$
12	Jaccard Double p-implication [24]	$P(A\|B)/(P(A)+P(B)-P(AB))$	$\frac{a}{a+b+c}$
13	Certainty Factor	$(P(B\|A)-P(B))/(1-P(B))$	$\frac{ad-bc}{(a+c)(b+d)}$
14	Odds Ratio Simple Deviation [24]	$\frac{P(AB)P(\neg A\neg B)}{P(\neg AB)P(A\neg B)}$	$\frac{ad}{bc}$
15	Yule's Q	$\frac{P(AB)P(\neg A\neg B)-P(A\neg B)P(\neg AB)}{P(AB)P(\neg A\neg B)+P(A\neg B)P(\neg AB)}$	$\frac{ad-bc}{ad+bc}$
16	Yule's Y	$\frac{\sqrt{P(AB)P(\neg A\neg B)}-\sqrt{P(A\neg B)P(\neg AB)}}{\sqrt{P(AB)P(\neg A\neg B)}+\sqrt{P(A\neg B)P(\neg AB)}}$	$\frac{\sqrt{ad}-\sqrt{bc}}{\sqrt{ad}+\sqrt{bc}}$
17	Conviction	$\frac{P(A)P(\neg B)}{P(A\neg B)}$	$\frac{(a+b)(b+d)}{b(a+b+c+d)}$
18	Laplace Correction	$\frac{n(AB)+1}{n(A)+2}$	$\frac{a+1}{a+b+2}$
19	One-Way Support	$P(B\|A)\log_2\frac{P(AB)}{P(A)P(B)}$	$\frac{a}{a+b}\log_2\left(\frac{a(a+b+c+d)}{(a+b)(a+c)}\right)$
20	Two-Way Support	$P(AB)\log_2\frac{P(AB)}{P(A)P(B)}$	$\frac{a}{a+b+c+d}\log_2\left(\frac{a(a+b+c+d)}{(a+b)(a+c)}\right)$
21	Piatetski-Shapiro	$P(AB)-P(A)P(B)$	$\frac{ad-bc}{(a+b+c+d)^2}$
22	Cosine	$\frac{P(AB)}{\sqrt{P(A)P(B)}}$	$\frac{a}{\sqrt{(a+b)(a+c)}}$
23	Loevinger	$1-\frac{P(A)P(\neg B}{P(A\neg B}$	$\frac{bc-ad}{b(a+b+c+d)}$
24	Information Gain	$\log\frac{P(AB)}{P(A)P(B)}$	$\log\left(\frac{a(a+b+c+d)}{(a+b)(a+c)}\right)$
25	Sebag-Schoenauer	$\frac{P(AB)}{P(A\neg B)}$	$\frac{a}{b}$
26	Least Contradiction	$\frac{P(AB)-P(A\neg B)}{P(B)}$	$\frac{a-b}{a+b}$
27	Odd Multiplier	$\frac{P(AB)P(\neg B)}{P(B)P(A\neg B)}$	$\frac{a(b+d)}{b(a+c)}$
28	Example and Counterexample Rate	$1-\frac{P(A\neg B)}{P(AB)}$	$\frac{a-b}{a}$
29	Zhang	$\frac{P(AB)-P(A)P(B)}{\max P(AB)P(\neg B)P(B)P(A\neg B)}$	$\frac{ad-bc}{\max(a(b+d),b(a+c))}$

Please note that some of the quantifiers defined using measures of interesting-ness were defined also in relation to the GUHA procedure ASSOC and are listed in Tables 4.2 and 4.4.

The following information is given for each 4ft-quantifier in Table 4.6:

- *No.* - number of the corresponding entry in Table 4.5
- $\approx^*$ - the symbol which is used for the particular 4ft-quantifier, with parameters as given above
- $F_{\approx^*} = 1$ *if and only if* - this means that $F_{\approx^*}(a,b,c,d) = 1$ if and only if the condition given in the corresponding row is satisfied
- *See point* - refers to one of points M_1 - M_{22} where necessary additional informa-tion is given
- *Class* - information on the class to which the 4ft-quantifier belongs to, see Table 4.1
- *No. in Tables 4.2 and 4.4* – numbers of the corresponding entries in Tables 4.2 and 4.4.

It is assumed $0 < p < 1$ for 4ft-quantifier $\Rightarrow_p^L$ (row 18). Some definitions require ad-ditional information. This information is provided in points $M_1 - M_{22}$ as mentioned in Table 4.6 in the fourth column and the corresponding row.

M_1)   If $a + b = 0$ then   we   define $F_{\Rightarrow_p}(a,b,c,d) = 0$,   which   implies $F_{\Rightarrow_p}(0,0,c,d) = 0$ for all non-negative integer c,d satisfying $c + d > 0$, see also point G_1 in definition 4.1.

M_2)   If $a + c = 0$ then   we   define $F_{\Rightarrow_p^R}(a,b,c,d) = 0$,   which   implies $F_{\Rightarrow_p^R}(0,b,0,d) = 0$ for all non-negative integer b,d satisfying $b + d > 0$.

M_3)   If $b + d = 0$ then   we   define $F_{\Rightarrow_p^S}(a,b,c,d) = 0$,   which   implies $F_{\Rightarrow_p^S}(a,0,c,0) = 0$ for all non-negative integer a,c satisfying $a + c > 0$.

M_4)   If $a + b = 0$ or $a + c = 0$ then we define $F_{\approx_q^{L+}}(a,b,c,d) = 0$, which im-plies $F_{\approx_q^{L+}}(0,0,c,d) = 0$ for all non-negative integer c,d satisfying $c + d > 0$ and $F_{\approx_q^{L+}}(0,b,0,d) = 0$ for all non-negative integer b,d satisfying $b + d > 0$.

M_5)   If $a + b = 0$ then   we   define $F_{\approx_u^L}(a,b,c,d) = 0$,   which   implies $F_{\approx_u^L}(0,0,c,d) = 0$ for all integer c,d satisfying $c + d > 0$.

M_6)   If $a + b = 0$ then   we   define $F_{\approx_u^A}(a,b,c,d) = 0$,   which   implies $F_{\approx_u^A}(0,0,c,d) = 0$ for all integer c,d satisfying $c + d > 0$.

M_7)   If $a + b = 0$ or $c = 0$ then we define $F_{\approx_s^R}(a,b,c,d) = 0$, which implies $F_{\approx_s^R}(0,0,c,d) = 0$ for all integer c,d satisfying $c + d > 0$.

M_8)   If $a + b + c = 0$ then   we   define $F_{\Leftrightarrow_p}(a,b,c,d) = 0$,   which   implies $F_{\Leftrightarrow_p}(0,0,0,d) = 0$ for all integer $d > 0$, see also point G_3 in definition 4.1.

M_9)   If $a + c = 0$ or $b + d = 0$ then we define $F_{\approx_u^C}(a,b,c,d) = 0$. This means that $F_{\approx_u^C}(0,b,0,d) = 0$ for all non-negative integer b,d satisfying $b + d > 0$ and $F_{\approx_u^C}(a,0,c,0) = 0$ for all non-negative integer b,d satisfying $a + c > 0$.

M_{10})   If $ad + bc = 0$ then we define $F_{\approx\varrho}(a,b,c,d) = 0$.

M_{11})   If $ad + bc = 0$ then we define $F_{\approx_u^Y}(a,b,c,d) = 0$.

M_{12})   If $b = 0$ then we define $F_{\approx_p^V}(a,b,c,d) = 0$.

Table 4.6 4ft-quantifiers defined using measures of interestingness

No.	$\approx^*$	$F_{\approx^*}=1$ if and only if	See point	Class	No. in Tables 4.2 and 4.4
1	$\odot_s$	$\frac{a}{a+b+c+d} \geq s$			19
2	$\Rightarrow_p$	$\frac{a}{a+b} \geq p$	M_1	IM	1
3	$\rightarrow_p^C$	$\frac{a+b}{a+b+c+d} \geq p$			
4	$\rightarrow_p^P$	$\frac{a+c}{a+b+c+d} \geq p$			
5	$\Rightarrow_p^R$	$\frac{a}{a+c} \geq p$	M_2		
6	$\Rightarrow_p^S$	$\frac{d}{b+d} \geq p$	M_3		
7	$\equiv_p$	$\frac{a+d}{a+b+c+d} \geq p$		ΣEQ	7
8	$\approx_q^{L+}$	$\frac{a(a+b+c+d)}{(a+b)(a+c)} \geq q$	M_4		
9	$\approx_u^L$	$\frac{a}{a+b} - \frac{(a+b)(a+c)}{(a+b+c+d)^2} \geq u$	M_5		
10	$\approx_u^A$	$\frac{ad-bc}{(a+b)(a+b+c+d)} \geq u$	M_6		
11	$\approx_s^R$	$\frac{a(c+d)}{c(a+b)} \geq s$	M_7		
12	$\Leftrightarrow_p$	$\frac{a}{a+b+c} \geq p$	M_8	ΣDI	4
13	$\approx_u^C$	$\frac{ad-bc}{(a+c)(b+d)} \geq u$	M_9		
14	$\sim_q$	$ad > e^q cd$		EQ, FPR	10
15	$\approx_u^Q$	$\frac{ad-bc}{ad+bc} \geq u$	M_{10}		
16	$\approx_u^Y$	$\frac{\sqrt{ad}-\sqrt{bc}}{\sqrt{ad}+\sqrt{bc}} \geq u$	M_{11}		
17	$\approx_p^V$	$\frac{(a+b)(b+d)}{b(a+b+c+d)} \geq p$	M_{12}		
18	$\Rightarrow_p^L$	$\frac{a+1}{a+b+2} \geq p$		IM	
19	$\approx_t^{V1}$	$\frac{a}{a+b} \log_2\left(\frac{a(a+b+c+d)}{(a+b)(a+c)}\right) \geq t$	M_{13}		
20	$\approx_t^{V2}$	$\frac{a}{a+b+c+d} \log_2\left(\frac{a(a+b+c+d)}{(a+b)(a+c)}\right) \geq t$	M_{14}		
21	$\approx_u^P$	$\frac{ad-bc}{(a+b+c+d)^2} \geq u$			
22	$\Leftrightarrow_p^C$	$\frac{a}{\sqrt{(a+b)(a+c)}} \geq p$	M_{15}		
23	$\approx_u^G$	$\frac{bc-ad}{b(a+b+c+d)} \geq u$	M_{16}		
24	$\approx_t^I$	$\log\left(\frac{a(a+b+c+d)}{(a+b)(a+c)}\right) \geq t$	M_{17}		
25	$\Rightarrow_q^S$	$\frac{a}{b} \geq q$	M_{18}	IM	
26	$\Rightarrow_u^C$	$\frac{a-b}{a+b} \geq u$	M_{19}	IM	
27	$\approx_q^O$	$\frac{a(b+d)}{b(a+c)} \geq s$	M_{20}		
28	$\Rightarrow_v^E$	$\frac{a-b}{a} \geq v$	M_{21}	IM	
29	$\approx_t^Z$	$\frac{ad-bc}{\max(a(b+d),b(a+c))} \geq t$	M_{22}		

$M_{13})$ If $a=0$ or $a+b=0$ or $a+c=0$ then we define $F_{\approx_t^{V1}}(a,b,c,d)=0$.

$M_{14})$ If $a=0$ or $a+b=0$ or $a+c=0$ then we define $F_{\approx_t^{V2}}(a,b,c,d)=0$.

$M_{15})$ If $a+b=0$ or $a+c=0$ then we define $F_{\Leftrightarrow_p^C}(a,b,c,d)=0$, which implies $F_{\Leftrightarrow_p^C}(0,0,c,d)=0$ for all non-negative integer c,d satisfying $c+d>0$ and $F_{\Leftrightarrow_p^C}(0,b,0,d)=0$ for all non-negative integer b,d satisfying $b+d>0$.

$M_{16})$ If $b=0$ then we define $F_{\approx_u^G}(a,b,c,d)=0$.

$M_{17})$ If $a=0$ or $a+b=0$ or $a+c=0$ then we define $F_{\approx_t^I}(a,b,c,d)=0$.

M_{18})   If $a = b = 0$ then we define $F_{\Rightarrow_q^S}(a,b,c,d) = 0$ and if $a > 0$ and $b = 0$ then we define $F_{\Rightarrow_q^S}(a,b,c,d) = 1$.

M_{19})   If $a+b = 0$ then we define $F_{\Rightarrow_u^C}(a,b,c,d) = 0$. This means $F_{\Rightarrow_u^C}(0,0,c,d) = 0$ for all non-negative integer c,d satisfying $c+d > 0$.

M_{20})   If $b(a+c) = 0$ then we define $F_{\approx_q^O}(a,b,c,d) = 0$.

M_{21})   If $a = 0$ then we define $F_{\Rightarrow_v^E}(a,b,c,d) = 0$.

M_{22})   If $\max(a(b+d), b(a+c)) = 0$ then we define $F_{\approx_t^Z}(a,b,c,d) = 0$.

Note 4.1. Please note that there is a close relation between 4ft-quantifier $\sim_q^+$ of above average (see row 13 in Table 4.2) and 4ft-quantifier $\approx_q^{L+}$ defined by lift (see row 8 in Table 4.6). It is

$$F_{\sim_q^+} = 1 \text{ if and only if } \frac{a}{a+b} \geq (1+q)\frac{a+c}{a+b+c+d}$$

i.e.

$$F_{\sim_q^+} = 1 \text{ if and only if } \frac{a(a+b+c+d)}{(a+b)(a+c)} \geq (1+q)$$

and

$$F_{\approx_q^{L+}} = 1 \text{ if and only if } \frac{a(a+b+c+d)}{(a+b)(a+c)} \geq q ,$$

see also G_6 and M_4.

This means that if $a+b > 0$ and $a+c > 0$ then $F_{\sim_q^+} = F_{\approx_{q+1}^{L+}}$. For example, it holds

$$F_{\sim_{0.5}^+} = 1 \text{ if and only if } \frac{a(a+b+c+d)}{(a+b)(a+c)} \geq (1+0.5)$$

and

$$F_{\approx_{1.5}^{L+}} = 1 \text{ if and only if } \frac{a(a+b+c+d)}{(a+b)(a+c)} \geq 1.5 .$$

4.4   Founded, Supported and Compound 4ft-quantifiers

The quantifiers introduced in Table 4.2 are defined for all 4ft-tables. However, when solving practical tasks concerning association rules $\varphi \approx \psi$ we often have to ensure that there are enough objects satisfying both φ and ψ. This has led to definition of founded and supported 4ft-quantifiers . An example is the quantifier $\Rightarrow_{p,Base}$ of *founded p-implication* introduced in [18]. Its associated function $F_{\Rightarrow_{p,Base}}$ is defined such that

$$F_{\Rightarrow_{p,Base}}(a,b,c,d) = \begin{cases} 1 & \text{if } \frac{a}{a+b} \geq p \wedge a \geq Base \\ 0 & \text{otherwise,} \end{cases}$$

see also theorem 7.3. All 4ft-quantifiers used in the GUHA procedure ASSOC have their founded versions. This concerns quantifiers 1 - 17 in table 4.2.

The 4ft-quantifier $\Rightarrow_{p,Base}$ of founded p-implication can be seen as compound from quantifiers $\Rightarrow_p$ of p-implication and $\oplus_{Base}$ i.e. *Base*, see table 4.2. The associated function $F_{\Rightarrow_{p,Base}}$ of 4ft-quantifier $\Rightarrow_{p,Base}$ is a product of associated functions $F_{\Rightarrow_p}$ and $F_{\oplus_{Base}}$, $F_{\Rightarrow_{p,Base}} = F_{\Rightarrow_p} \times F_{\oplus_{Base}}$. The additional founded 4ft-quantifiers can be defined in a similar way.

Let us note that it is possible to define general compound 4ft-quantifiers as the conjunction of several 4ft-quantifiers.

Definition 4.3. Let $\mathscr{LC}$ be a logical calculus of association rules with 4ft-quantifiers $\approx_1, \ldots \approx_Q$ with associated functions $F_{\approx_1}, \ldots, F_{\approx_Q}$ respectively and let $\approx_{i_1}, \ldots, \approx_{i_k}$ be some of 4ft-quantifiers of $\mathscr{LC}$. Then the expression

$$\approx_{i_1} \wedge \ldots \wedge \approx_{i_k}$$

is *a compound 4ft-quantifier of* $\mathscr{LC}$. We also say that $\approx_{i_1} \wedge \ldots \wedge \approx_{i_k}$ *is a composition of* $\approx_{i_1}, \ldots, \approx_{i_k}$. Its associated function $F_{\approx_{i_1} \wedge \ldots \wedge \approx_{i_k}}$ is defined as a product of functions $F_{\approx_{i_1}}, \ldots, F_{\approx_{i_k}}$. This means that

$$F_{\approx_{i_1} \wedge \ldots \wedge \approx_{i_k}}(a,b,c,d) = F_{\approx_{i_1}}(a,b,c,d) \times \ldots \times F_{\approx_{i_k}}(a,b,c,d) \,.$$

Let us note that the GUHA procedure 4ft-Miner makes it possible to deal with such defined compound 4ft-quantifiers.

The well known 4ft-quantifier $\rightarrow_{p,s}$ with confidence p and support s used for analysis of market basket [1] (and called sometimes in this book as the MB-quantifier, see Sect. 1.1) has associated function $F_{\rightarrow_{p,s}}$ defined as

$$F_{\rightarrow_{p,s}} = \begin{cases} 1 & \text{if } \frac{a}{a+b} \ge p \wedge \frac{a}{a+b+c+d} \ge s \\ 0 & \text{otherwise.} \end{cases}$$

It can be also seen as conjunction $\Rightarrow_p \wedge \odot_s$ of quantifiers $\Rightarrow_p$ of p-implication and support $\odot_s$. In addition, the 4ft-quantifier $\rightarrow_{p,s}$ with confidence p and support s is an example of a supported 4ft-quantifier. Additional founded and supported 4ft-quantifiers are defined in Part II.

Chapter 5
Useful Results

The core of the book are results on classes of association rules. Classes of association rules are defined by classes of 4ft-quantifiers. Each class of 4ft-quantifiers is defined by a relatively simple condition concerning frequencies from 4ft-tables. However, the proof that a particular 4ft-quantifier belongs to a particular class of 4ft-quantifiers sometimes requires a relatively complex computation. This is especially true for statistically motivated 4ft-quantifiers. Useful results on statistically motivated 4ft-quantifiers are proved in this chapter. These results are later used when proving relations of particular statistically motivated 4ft-quantifiers to particular classes of 4ft-quantifiers as well as when proving additional theoretical results.

There are six 4ft-quantifiers defined using binomial distribution. Specifically, these are the 4ft-quantifiers $\Rightarrow^{!}_{p,\alpha}$ of lower critical implication, $\Rightarrow^{?}_{p,\alpha}$ of upper critical implication, $\Leftrightarrow^{!}_{p,\alpha}$ of lower critical double implication, $\Leftrightarrow^{?}_{p,\alpha}$ of upper critical double implication, $\equiv^{!}_{p,\alpha}$ of lower critical equivalence, and $\equiv^{?}_{p,\alpha}$ of upper critical equivalence, see rows 2, 3, 5, 6, 8, and 9 in Tables 4.2 and 4.4. We need to prove the properties of these quantifiers. The corresponding proofs are provided in theorem 4.5.4 in [18] for 4ft-quantifiers $\Rightarrow^{?}_{p,\alpha}$ and $\Rightarrow^{!}_{p,\alpha}$. The proofs use known facts from mathematical statistics.

The proofs presented here are based on properties of combinatorial numbers. Necessary properties of 4ft-quantifiers $\Rightarrow^{!}_{p,\alpha}$ of lower critical implication, $\Leftrightarrow^{!}_{p,\alpha}$ of lower critical double implication, and $\equiv^{!}_{p,\alpha}$ of lower critical equivalence are proved in Sect. 5.1. Corresponding properties of 4ft-quantifiers $\Rightarrow^{?}_{p,\alpha}$ of upper critical implication, $\Leftrightarrow^{?}_{p,\alpha}$ of upper critical double implication, and $\equiv^{?}_{p,\alpha}$ of upper critical equivalence are proved in Sect. 5.2.

In addition, we need to prove some properties of Fisher's quantifier, see row 11 in Tables 4.2 and 4.4 and of χ^2-quantifier, see row 12 in Tables 4.2 and 4.4. Properties of Fisher's quantifier and of χ^2-quantifier are proved in Sects. 5.3 and 5.4 respectively. The required properties of combinatorial numbers are proved in Sect. 5.5.

J. Rauch: *Observational Calculi and Association Rules*, SCI 469, pp. 39–61.
DOI: 10.1007/978-3-642-11737-4_5   © Springer-Verlag Berlin Heidelberg 2013

5.1 Properties of 4ft-quantifiers $\Rightarrow^!_{p,\alpha}$, $\Leftrightarrow^!_{p,\alpha}$, and $\equiv^!_{p,\alpha}$

We start with four lemmas.

Lemma 5.1. *Let $R > 0$, $0 \leq P \leq R$ be integer numbers and let $0 < p < 1$ be a real number. Then it holds*

$$\sum_{i=P+1}^{R+1} \binom{R+1}{i} p^i (1-p)^{R+1-i} \leq \sum_{i=P}^{R} \binom{R}{i} p^i (1-p)^{R-i} \ .$$

Proof. Let us denote $\sigma'_i = \binom{R+1}{i+1} p^{i+1} (1-p)^{R+1-(i+1)}$ and $\sigma_i = \binom{R}{i} p^i (1-p)^{R-i}$. Then

$$\sum_{i=P+1}^{R+1} \binom{R+1}{i} p^i (1-p)^{R+1-i} = \sum_{i=P}^{R} \binom{R+1}{i+1} p^{i+1} (1-p)^{R+1-(i+1)} = \sum_{i=P}^{R} \sigma'_i$$

and

$$\sum_{i=P}^{R} \binom{R}{i} p^i (1-p)^{R-i} = \sum_{i=P}^{R} \sigma_i \ .$$

We define $f(i)$ for $i = 0, \cdots, R$ such that $f(i) = \frac{\sigma'_i}{\sigma_i}$. It holds

$$f(i) = \frac{\frac{(R+1)!}{(i+1)!(R+1-(i+1))!} p^{i+1} (1-p)^{R+1-(i+1)}}{\frac{R!}{i!(R-i)!} p^i (1-p)^{R-i}} =$$

$$= p \frac{(R+1)! i! (R-i)!}{R! (i+1)! (R+1-(i+1))!} = p \frac{R+1}{i+1} \ .$$

The function $f(i)$ is decreasing, it holds $f(0) = p(R+1)$, $f(R) = p$ and we assume $p < 1$. There are two possibilities:

1. $f(0) = p(R+1) \leq 1$
2. $f(0) = p(R+1) > 1$

If $f(0) = p(R+1) \leq 1$ then $\frac{\sigma'_i}{\sigma_i} \leq 1$ and thus $\sigma'_i \leq \sigma_i$ for $i = 0, \cdots, R$. It holds

$$\sum_{i=P}^{R} \sigma'_i - \sum_{i=P}^{R} \sigma_i = \sum_{i=P}^{R} (\sigma'_i - \sigma_i) \leq 0$$

which means

$$\sum_{i=P+1}^{R+1} \binom{R+1}{i} p^i (1-p)^{R-i} \leq \sum_{i=P}^{R} \binom{R}{i} p^i (1-p)^{R-i} \ .$$

If $f(0) = p(R+1) > 1$ then we distinguish two cases:

2.a) $f(P) \leq 1$
2.b) $f(P) > 1$

If $f(P) \leq 1$ (i.e. in the case 2.a)) then $\sigma'_i \leq \sigma_i$ for $i = P, \cdots, R$ and the rest of the proof is the same as for the possibility 1 (i.e. for $f(0) = p(R+1) \leq 1$).

   If $f(P) > 1$ then we use the known fact that for each $K > 0$ and real p, $0 < p < 1$ it holds

$$\sum_{i=0}^{K} \binom{K}{i} p^i (1-p)^{K-i} = 1 \cdot$$

This means that

$$\sum_{i=P+1}^{R+1} \binom{R+1}{i} p^i (1-p)^{R+1-i} = 1 - \sum_{i=0}^{P} \binom{R+1}{i} p^i (1-p)^{R+1-i}$$

and

$$\sum_{i=P}^{R} \binom{R}{i} p^i (1-p)^{R-i} = 1 - \sum_{i=0}^{P-1} \binom{R}{i} p^i (1-p)^{R-i} \cdot$$

Thus the following is true:

$$\sum_{i=P+1}^{R+1} \binom{R+1}{i} p^i (1-p)^{R+1-i} - \sum_{i=P}^{R} \binom{R}{i} p^i (1-p)^{R-i} =$$

$$= \sum_{i=0}^{P-1} \binom{R}{i} p^i (1-p)^{R-i} - \sum_{i=0}^{P} \binom{R+1}{i} p^i (1-p)^{R+1-i} =$$

$$= \sum_{i=0}^{P-1} \binom{R}{i} p^i (1-p)^{R-i} - \sum_{i=1}^{P} \binom{R+1}{i} p^i (1-p)^{R+1-i} - \binom{R+1}{0} p^0 (1-p)^{R+1-0} =$$

$$= \sum_{i=0}^{P-1} \binom{R}{i} p^i (1-p)^{R-i} - \sum_{i=0}^{P-1} \binom{R+1}{i+1} p^{i+1} (1-p)^{R+1-(i+1)} - (1-p)^{R+1} =$$

$$= \sum_{i=0}^{P-1} \sigma_i - \sum_{i=0}^{P-1} \sigma'_i - (1-p)^{R+1} = \sum_{i=0}^{P-1} (\sigma_i - \sigma'_i) - (1-p)^{R+1}.$$

We assume $f(P) > 1$ and we know that $f(i) = \frac{\sigma'_i}{\sigma_i}$ is decreasing, thus also $f(i) > 1$ for $i = 0, \cdots, P-1$. This means that $\sigma'_i > \sigma_i$, and $(1-p)^{R+1} > 0$, and this way we obtain

$$\sum_{i=0}^{P-1} (\sigma_i - \sigma'_i) - (1-p)^{R+1} < 0 \cdot$$

This means that also in the case 2.b it holds

$$\sum_{i=P+1}^{R+1} \binom{R+1}{i} p^i (1-p)^{R+1-i} \le \sum_{i=P}^{R} \binom{R}{i} p^i (1-p)^{R-i} .$$

This finishes the proof.   □

Lemma 5.2. *Let $R > 0$, $0 \le P \le R$ and $K \ge 0$ be integer numbers and let $0 < p < 1$ be a real number. Then it holds*

$$\sum_{i=P+K}^{R+K} \binom{R+K}{i} p^i (1-p)^{R+K-i} \le \sum_{i=P}^{R} \binom{R}{i} p^i (1-p)^{R-i} .$$

Proof. There is nothing to prove for $K = 0$. If $K > 0$ then we have according to lemma 5.1:

$$\sum_{i=P+K}^{R+K} \binom{R+K}{i} p^i (1-p)^{R+K-i} \le \sum_{i=P+K-1}^{R+K-1} \binom{R+K-1}{i} p^i (1-p)^{R+K-1-i} \le \cdots$$

$$\cdots \le \sum_{i=P+1}^{R+1} \binom{R+1}{i} p^i (1-p)^{R+1-i} \le \sum_{i=P}^{R} \binom{R}{i} p^i (1-p)^{R-i} .$$

This finishes the proof.   □

Lemma 5.3. *Let $R > 0$, $0 \le P < R$ be integer numbers and let $0 < p < 1$ be a real number. Then it holds*

$$\sum_{i=P}^{R-1} \binom{R-1}{i} p^i (1-p)^{R-1-i} < \sum_{i=P}^{R} \binom{R}{i} p^i (1-p)^{R-i} .$$

Proof. Let us denote $\sigma_i' = \binom{R-1}{i} p^i (1-p)^{R-1-i}$ and $\sigma_i = \binom{R}{i} p^i (1-p)^{R-i}$. Then

$$\sum_{i=P}^{R-1} \binom{R-1}{i} p^i (1-p)^{R-1-i} - \sum_{i=P}^{R} \binom{R}{i} p^i (1-p)^{R-i} = \sum_{i=P}^{R-1} (\sigma_i' - \sigma_i) - \binom{R}{R} p^R (1-p)^{R-R} .$$

We define function $f(i)$ for $i = 0, \cdots, R-1$ such that $f(i) = \frac{\sigma_i'}{\sigma_i}$. It holds

$$f(i) = \frac{\frac{(R-1)!}{i!(R-1-i)!} p^i (1-p)^{R-1-i}}{\frac{R!}{i!(R-i)!} p^i (1-p)^{R-i}} = \frac{(R-1)! i! (R-i)!}{R! i! (R-1-i)!} \frac{1}{1-p} = \frac{R-i}{R} \frac{1}{1-p} .$$

The function $f(i)$ is decreasing, it holds $f(0) = \frac{1}{1-p} > 1$ and $f(R-1) = \frac{1}{R} \frac{1}{1-p}$. There are two possibilities:

1. $f(P) \le 1$
2. $f(P) > 1$

If $f(P) \le 1$ then for $i = P, \cdots, R$ we have $\frac{\sigma_i'}{\sigma_i} \le 1$ thus also $\sigma_i' \le \sigma_i$. It holds also $\binom{R}{R} p^R (1-p)^{R-R} = p^R > 0$, thus

$$\sum_{i=P}^{R-1} (\sigma'_i - \sigma_i) - \binom{R}{R} p^R (1-p)^{R-R} < 0,$$

which implies

$$\sum_{i=P}^{R-1} \binom{R-1}{i} p^i (1-p)^{R-1-i} \leq \sum_{i=P}^{R} \binom{R}{i} p^i (1-p)^{R-i}.$$

If $f(P) > 1$ we again use the fact that for each integer $K > 0$ and real p, $0 < p < 1$ it holds

$$\sum_{i=0}^{K} \binom{K}{i} p^i (1-p)^{K-i} = 1.$$

This means:

$$\sum_{i=P}^{R-1} \binom{R-1}{i} p^i (1-p)^{R-1-i} - \sum_{i=P}^{R} \binom{R}{i} p^i (1-p)^{R-i} =$$

$$\sum_{i=0}^{P-1} \binom{R}{i} p^i (1-p)^{R-i} - \sum_{i=0}^{P-1} \binom{R-1}{i} p^i (1-p)^{R-1-i} = \sum_{i=0}^{P-1} (\sigma_i - \sigma'_i).$$

We assume $f(P) > 1$ and we know that $f(i)$ is decreasing, thus it holds $f(i) > 1$ for $i = 0, \cdots, P-1$ which implies $\sigma'_i > \sigma_i$. Thus we have

$$\sum_{i=0}^{P-1} (\sigma_i - \sigma'_i) < 0$$

This means that also for $f(P) > 1$ it holds

$$\sum_{i=P}^{R-1} \binom{R-1}{i} p^i (1-p)^{R-1-i} \leq \sum_{i=P}^{R} \binom{R}{i} p^i (1-p)^{R-i}.$$

This finishes the proof. $\qquad\qquad\qquad\qquad\qquad\qquad\qquad\qquad\qquad\qquad$ $\square$

Lemma 5.4. *Let $R > 0$, $0 \leq P < R$, $0 \leq K \leq R - P$ be integer numbers and let $0 < p < 1$ be a real number. Then it holds:*

$$\sum_{i=P}^{R-K} \binom{R-K}{i} p^i (1-p)^{R-K-i} \leq \sum_{i=P}^{R} \binom{R}{i} p^i (1-p)^{R-i}.$$

Proof. There is nothing to prove for $K = 0$. If $K > 0$ then we have according to lemma 5.3:

$$\sum_{i=P}^{R-K} \binom{R-K}{i} p^i (1-p)^{R-K-i} < \sum_{i=P}^{R-(K-1)} \binom{R-(K-1)}{i} p^i (1-p)^{R-(K-1)-i} < \cdots$$

$$\cdots < \sum_{i=P}^{R-1} \binom{R-1}{i} p^i(1-p)^{R-1-i} < \sum_{i=P}^{R} \binom{R}{i} p^i(1-p)^{R-i}.$$

This finishes the proof. □

Now we can prove a theorem we will use to prove important properties of 4ft-quantifiers $\Rightarrow^!_{p,\alpha}$, $\Leftrightarrow^!_{p,\alpha}$, and $\equiv^!_{p,\alpha}$.

Theorem 5.1. *Let $u' \geq u \geq 0$ and $0 \leq v' \leq v$ be integer numbers and let $0 < p < 1$ be a real number. Then it holds*

$$\sum_{i=u'}^{u'+v'} \binom{u'+v'}{i} p^i(1-p)^{u'+v'-i} \leq \sum_{i=u}^{u+v} \binom{u+v}{i} p^i(1-p)^{u+v-i}.$$

Proof. It holds $u' \geq u$ and $v' \leq v$ and thus $u'+v' = u+v'+K$ and $u' = u+K$ where $K = u'-u \geq 0$. According to lemma 5.2 we get

$$\sum_{i=u+K}^{u+v'+K} \binom{u+v'+K}{i} p^i(1-p)^{u'+v'-i} \leq \sum_{i=u}^{u+v'} \binom{u+v'}{i} p^i(1-p)^{u+v'-i}$$

which means

$$\sum_{i=u'}^{u'+v'} \binom{u'+v'}{i} p^i(1-p)^{u'+v'-i} \leq \sum_{i=u}^{u+v'} \binom{u+v'}{i} p^i(1-p)^{u+v'-i}.$$

Inequality $v' \leq v$ implies $u+v' = u+v-K$ where $K = v-v' \geq 0$. According to lemma 5.4 we have

$$\sum_{i=u}^{u+v-K} \binom{u+v-K}{i} p^i(1-p)^{u+v-K-i} \leq \sum_{i=u}^{u+v} \binom{u+v}{i} p^i(1-p)^{u+v-i}$$

which means

$$\sum_{i=u}^{u+v'} \binom{u+v'}{i} p^i(1-p)^{u+v'-i} \leq \sum_{i=u}^{u+v} \binom{u+v}{i} p^i(1-p)^{u+v-i}$$

and this finishes the proof. □

5.2 Properties of 4ft-quantifiers $\Leftrightarrow^?_{p,\alpha}$, $\Leftrightarrow^?_{p,\alpha}$, and $\equiv^?_{p,\alpha}$

Again, we start with four lemmas.

Lemma 5.5. *Let $R > 0$, $0 \leq P \leq R$ be integer numbers and let $0 < p < 1$ be a real number. Then*

$$\sum_{i=0}^{P+1} \binom{R+1}{i} p^i(1-p)^{R+1-i} \geq \sum_{i=0}^{P} \binom{R}{i} p^i(1-p)^{R-i}.$$

Proof. We again use the fact that for each integer $K > 0$ and real p, $0 < p < 1$ it holds

$$\sum_{i=0}^{K} \binom{K}{i} p^i (1-p)^{K-i} = 1 \cdot \tag{5.1}$$

This means that it is enough to prove that

$$1 - \sum_{i=P+2}^{R+1} \binom{R+1}{i} p^i (1-p)^{R+1-i} \geq 1 - \sum_{i=P+1}^{R} \binom{R}{i} p^i (1-p)^{R-i},$$

which is equivalent to

$$\sum_{i=P+2}^{R+1} \binom{R+1}{i} p^i (1-p)^{R+1-i} \leq \sum_{i=P+1}^{R} \binom{R}{i} p^i (1-p)^{R-i} \cdot$$

The last assertion however directly follows from lemma 5.1 if we use $P+1$ instead of P. This finishes the proof. □

Lemma 5.6. *Let $R > 0$, $0 \leq P \leq R$, $0 \leq K \leq R - P$ be integer numbers and let $0 < p < 1$ be a real number. Then*

$$\sum_{i=0}^{P+K} \binom{R+K}{i} p^i (1-p)^{R+K-i} \geq \sum_{i=0}^{P} \binom{R}{i} p^i (1-p)^{R-i} \cdot$$

Proof. There it is nothing to prove for $K = 0$. If $K > 0$ then we have, according to lemma 5.5:

$$\sum_{i=0}^{P+K} \binom{R+K}{i} p^i (1-p)^{R+K-i} \geq \sum_{i=0}^{P+K-1} \binom{R+K-1}{i} p^i (1-p)^{R+K-1-i} \geq \cdots$$

$$\cdots \geq \sum_{i=0}^{P+1} \binom{R+1}{i} p^i (1-p)^{R+1-i} \geq \sum_{i=0}^{P} \binom{R}{i} p^i (1-p)^{R-i} \cdot$$

This finishes the proof. □

Lemma 5.7. *Let $R > 0$, $0 \leq P < R$ be integer numbers and let $0 < p < 1$ be a real number. Then*

$$\sum_{i=0}^{P} \binom{R-1}{i} p^i (1-p)^{R-1-i} > \sum_{i=0}^{P} \binom{R}{i} p^i (1-p)^{R-i} \cdot$$

Proof. There is nothing to prove for $P = R - 1$. If $P < R - 1$ then, similarly as in lemma 5.5, it is enough to prove the assertion

$$1 - \sum_{i=P+1}^{R-1} \binom{R-1}{i} p^i (1-p)^{R-1-i} > 1 - \sum_{i=P+1}^{R} \binom{R}{i} p^i (1-p)^{R-i},$$

which is equivalent to

$$\sum_{i=P+1}^{R-1} \binom{R-1}{i} p^i (1-p)^{R-1-i} < \sum_{i=P+1}^{R} \binom{R}{i} p^i (1-p)^{R-i}.$$

However, this directly follows from lemma 5.3 if we use $P+1$ instead of P. This finishes the proof. □

Lemma 5.8. *Let $R > 0$, $0 \le P \le R$, $0 \le K \le R - P$ be integer numbers and let $0 < p < 1$ be a real number. Then*

$$\sum_{i=0}^{P} \binom{R-K}{i} p^i (1-p)^{R-K-i} > \sum_{i=0}^{P} \binom{R}{i} p^i (1-p)^{R-i}.$$

Proof. There it is nothing to prove for $K = 0$. If $K > 0$ then we have according to lemma 5.7:

$$\sum_{i=0}^{P} \binom{R-K}{i} p^i (1-p)^{R-K-i} > \sum_{i=0}^{P} \binom{R-(K-1)}{i} p^i (1-p)^{R-(K-1)-i} > \cdots$$

$$\cdots > \sum_{i=0}^{P} \binom{R-1}{i} p^i (1-p)^{R-1-i} > \sum_{i=0}^{P} \binom{R}{i} p^i (1-p)^{R-i}.$$

This finishes the proof. □

Now we can prove a theorem we will use to prove important properties of 4ft-quantifiers $\Leftrightarrow_{p,\alpha}^?$, $\Leftrightarrow_{p,\alpha}^?$, and $\equiv_{p,\alpha}^?$.

Theorem 5.2. *Let $u' \ge u \ge 0$ and $0 \le v' \le v$ be integer numbers and let $0 < p < 1$ be a real number. Then it holds*

$$\sum_{i=0}^{u'} \binom{u'+v'}{i} p^i (1-p)^{u'+v'-i} \ge \sum_{i=0}^{u} \binom{u+v}{i} p^i (1-p)^{u+v-i}.$$

Proof. It holds $u' \ge u$ and thus $u' = u + K$ where $K = u' - u \ge 0$ and according to lemma 5.6 we get

$$\sum_{i=0}^{u+K} \binom{u+v'+K}{i} p^i (1-p)^{u+v'+K-i} \ge \sum_{i=0}^{u} \binom{u+v'}{i} p^i (1-p)^{u+v'-i}$$

which means

$$\sum_{i=0}^{u'} \binom{u'+v'}{i} p^i (1-p)^{u'+v'-i} \ge \sum_{i=0}^{u} \binom{u+v'}{i} p^i (1-p)^{u+v'-i}.$$

Inequality $v' \le v$ implies $v' = v - K$ and $u + v' = u + v - K$ where $K = v - v' \ge 0$. According to lemma 5.8 we have

$$\sum_{i=0}^{u} \binom{u+v-K}{i} p^i (1-p)^{u+v-K-i} > \sum_{i=0}^{u} \binom{u+v}{i} p^i (1-p)^{u+v-i}$$

which means

$$\sum_{i=0}^{u} \binom{u+v'}{i} p^i (1-p)^{u+v'-i} > \sum_{i=0}^{u} \binom{u+v}{i} p^i (1-p)^{u+v-i}$$

and this finishes the proof. □

5.3 Properties of Fisher's Quantifier

Let us remember the associated function $F_{\sim_\alpha^1}$ of Fisher's quantifier $\sim_\alpha^1$ – see rows 11 in Tables 4.2 and 4.4. By the definition, it is $F_{\sim_\alpha^1}(a,b,c,d) = 1$ if and only if

$$\sum_{i=a}^{\min(r,k)} \frac{\binom{k}{i}\binom{n-k}{r-i}}{\binom{n}{r}} \leq \alpha \wedge ad > bc$$

where $r = a+b$, $k = a+c$ and $n = a+b+c+d$. We define Fisher's measure and then we prove a theorem describing an important property of Fisher's quantifier. The theorem and its proof are based on theorem 4.5.1 in [18].

Definition 5.1. *Fisher's measure* is a function $\mathscr{F}_F(a,b,c,d)$ defined for all 4ft-tables $\langle a,b,c,d \rangle$ such that

$$\mathscr{F}_F(a,b,c,d) = \sum_{i=a}^{\min(r,k)} \frac{\binom{k}{i}\binom{n-k}{r-i}}{\binom{n}{r}}$$

where $r = a+b$, $k = a+c$ and $n = a+b+c+d$. We also denote

$$\sigma(i,r,k,n) = \frac{\binom{k}{i}\binom{n-k}{r-i}}{\binom{n}{r}}.$$

Before proving the theorem we prove two lemmas.

Lemma 5.9. *Fisher's measure* $\mathscr{F}_F(a,b,c,d)$ *satisfies*

1. $\mathscr{F}_F(a,b,c,d) = \mathscr{F}_F(a,c,b,d)$
2. $\mathscr{F}_F(a,b,c,d) = \mathscr{F}_F(d,b,c,a)$.

Proof

1. It holds

$$\mathscr{F}_F(a,b,c,d) = \sum_{i=a}^{\min(r,k)} \sigma(i,r,k,n) \quad and \quad \mathscr{F}_F(a,c,b,d) = \sum_{i=a}^{\min(k,r)} \sigma(i,k,r,n),$$

see 4ft-tables $\langle a,b,c,d \rangle$ and $\langle a,c,b,d \rangle$ in Fig. 5.1 where $r = a+b$, $s = c+d$, $k = a+c$, $l = b+d$ and $n = a+b+c+d$.
It is sufficient to prove $\sigma(i,r,k,n) = \sigma(i,k,r,n)$ for $i = a,\ldots,\min(r,k)$. We have

a	b	r
c	d	s
k	l	n

4ft-table $\langle a,b,c,d \rangle$

a	c	k
b	d	l
r	s	n

4ft-table $\langle a,c,b,d \rangle$

d	b	l
c	a	k
s	r	n

4ft-table $\langle d,b,c,a \rangle$

Fig. 5.1 4ft-tables $\langle a,b,c,d \rangle$, $\langle a,c,b,d \rangle$, and $\langle d,b,c,a \rangle$

$$\sigma(i,r,k,n) = \frac{\binom{k}{i}\binom{n-k}{r-i}}{\binom{n}{r}} = \frac{\frac{k!}{i!(k-i)!}\frac{(n-k)!}{(r-i)!(n-k-(r-i))!}}{\frac{n!}{r!(n-r)!}} = \frac{k!l!r!s!}{i!(k-i)!(r-i)!(n-k-r+i)!n!}$$

and

$$\sigma(i,k,r,n) = \frac{\binom{r}{i}\binom{n-r}{k-i}}{\binom{n}{k}} = \frac{\frac{r!}{i!(r-i)!}\frac{(n-r)!}{(k-i)!(n-r-(k-i))!}}{\frac{n!}{k!(n-k)!}} = \frac{r!s!k!l!}{i!(r-i)!(k-i)!(n-r-k+i)!n!} .$$

This finishes the proof of point 1.
2. *It holds*

$$\mathscr{F}_F(a,b,c,d) = \sum_{i=a}^{\min(r,k)} \sigma(i,r,k,n) \quad and \quad \mathscr{F}_F(d,b,c,a) = \sum_{i=d}^{\min(l,s)} \sigma(i,l,s,n) ,$$

see 4ft-tables $\langle a,b,c,d \rangle$ and $\langle d,b,c,a \rangle$ in Fig. 5.1. Let us assume $b \le c$. The case when $b > c$ is treated similarly. If $b \le c$ then

$$\min(r,k) = \min(a+b,a+c) = a+b \quad and \quad \min(l,s) = \min(d+b,d+c) = d+b ,$$

thus

$$\mathscr{F}_F(a,b,c,d) = \sum_{i=a}^{a+b} \sigma(i,r,k,n) \quad and \quad \mathscr{F}_F(a,b,c,d) = \sum_{i=d}^{d+b} \sigma(i,l,s,n) .$$

This means that it is sufficient to prove $\sigma(a+j,r,k,n) = \sigma(d+j,l,s,n)$ for $j = 1,\ldots,b$. It holds

$$\sigma(a+j,r,k,n) = \frac{\binom{k}{a+j}\binom{n-k}{r-(a+j)}}{\binom{n}{r}} = \frac{\frac{k!}{(a+j)!(k-(a+j))!}\frac{(n-k)!}{(r-(a+j))!(n-k-(r-(a+j)))!}}{\frac{n!}{r!(n-r)!}} =$$

$$= \frac{k!l!r!s!}{(a+j)!(c-j)!(b-j)!(d+j)!n!}$$

and

$$\sigma(d+j,l,s,n) = \frac{\binom{s}{d+j}\binom{n-s}{l-(d+j)}}{\binom{n}{l}} = \frac{\frac{s!}{(d+j)!(s-(d+j))!}\frac{(n-s)!}{(l-(d+j))!(n-s-(l-(d+j)))!}}{\frac{n!}{l!(n-l)!}} =$$

$$= \frac{s!\,r!\,l!\,k!}{(d+j)!(c-j)!(b-j)!(a+j)!n!}\;.$$

This finishes the proof. □

Lemma 5.10. *If* $ad > bc$ *then Fisher's measure* $\mathscr{F}_F(a,b,c,d)$ *satisfies*

1. $\mathscr{F}_F(a+1,b,c,d) \le \mathscr{F}_F(a,b,c,d)$
2. $\mathscr{F}_F(a,b-1,c,d) \le \mathscr{F}_F(a,b,c,d)$
3. $\mathscr{F}_F(a,b,c-1,d) \le \mathscr{F}_F(a,b,c,d)$
4. $\mathscr{F}_F(a,b,c,d+1) \le \mathscr{F}_F(a,b,c,d)$.

Proof

1. It holds

$$\mathscr{F}_F(a,b,c,d) = \sum_{i=a}^{\min(r,k)} \sigma(i,r,k,n) \tag{5.2}$$

and

$$\mathscr{F}_F(a+1,b,c,d) = \sum_{i=a+1}^{\min(r+1,k+1)} \sigma(i,r+1,k+1,n+1)$$

see also 4ft-tables $\langle a,b,c,d\rangle$ *and* $\langle a+1,b,c,d\rangle$ *in Fig. 5.2.*

a	b	r
c	d	s
k	l	n

a+1	b	r+1
c	d	s
k+1	l	n+1

a	b−1	r−1
c	d	s
k	l−1	n−1

4ft-table $\langle a,b,c,d\rangle$ 4ft-table $\langle a+1,b,c,d\rangle$ 4ft-table $\langle a,b-1,c,d\rangle$

Fig. 5.2 4ft-tables $\langle a,b,c,d\rangle$, $\langle a+1,b,c,d\rangle$, and $\langle a,b-1,c,d\rangle$

In addition

$$\sum_{i=a+1}^{\min(r+1,k+1)} \sigma(i,r+1,k+1,n+1) = \sum_{i=a}^{\min(r+1,k+1)-1} \sigma(i+1,r+1,k+1,n+1)$$

and thus we have

$$\mathscr{F}_F(a+1,b,c,d) = \sum_{i=a}^{\min(r+1,k+1)-1} \sigma(i+1,r+1,k+1,n+1). \tag{5.3}$$

The number of members in the sum 5.2 is the same to that in the sum 5.3. To prove that $\mathscr{F}_F(a+1,b,c,d) \le \mathscr{F}_F(a,b,c,d)$, *it is thus sufficient to prove that*

$$\sigma(i+1,r+1,k+1,n+1) \le \sigma(i,r,k,n)$$

for $i = a, \dots, \min(r,k)$, it holds of course $\min(r,k) = \min(r+1,k+1) - 1$. We have

$$\sigma(i+1,r+1,k+1,n+1) = \frac{\binom{k+1}{i+1}\binom{n+1-(k+1)}{r+1-(i+1)}}{\binom{n+1}{r+1}} =$$

$$= \frac{\frac{(k+1)!}{(i+1)!((k+1)-(i+1))!} \frac{(n+1-(k+1))!}{(r+1-(i+1))!(n+1-(k+1)-(r+1-(i+1)))!}}{\frac{(n+1)!}{(r+1)!((n+1)-(r+1))!}} =$$

$$= \frac{\frac{k+1}{i+1} \frac{k!}{i!(k-i)!} \frac{(n-k)!}{(r-i)!(n-k-(r-i))!}}{\frac{n+1}{r+1} \frac{n!}{r!(n-r)!}} = \frac{k+1}{i+1} \frac{r+1}{n+1} \frac{\binom{k}{i}\binom{n-k}{r-i}}{\binom{n+1}{r+1}} =$$

$$= \frac{(k+1)(r+1)}{(i+1)(n+1)}\sigma(i,r,k,n).$$

Thus it remains to prove $(i+1)(n+1) \ge (k+1)(r+1)$. We have

$$(i+1)(n+1) - (k+1)(r+1) = (i+1)(a+b+c+d+1) - (a+c+1)(a+b+1) =$$

$$= ia + ib + ic + id + i + a + b + c + d + 1 - (a^2 + ab + a + ca + cb + c + a + b + 1) =$$

$$= (ia - a^2) + (ib - ab) + (ic - ca) + (id - cb) + (i - a) + d.$$

It holds $i \ge a$ and it means also $ia \ge a^2$, $ib \ge ab$, $ic \ge ca$. We assume $ad > bc$ and it means also $id > cb$ and $d > 0$. This finishes the proof of point 1.

2. *We are going to prove $\mathscr{F}_F(a,b-1,c,d) \le \mathscr{F}_F(a,b,c,d)$. It holds*

$$\mathscr{F}_F(a,b-1,c,d) = \sum_{i=a}^{\min(r-1,k)} \sigma(i,r-1,k,n-1) \tag{5.4}$$

see also 4ft-table $\langle a,b-1,c,d \rangle$ in Fig. 5.2. We have to prove

$$\sum_{i=a}^{\min(r,k)} \sigma(i,r,k,n) - \sum_{i=a}^{\min(r-1,k)} \sigma(i,r-1,k,n-1) \ge 0.$$

There are only two possibilities: either $\min(r,k) = \min(r-1,k)$ or $\min(r,k) = \min(r-1,k) + 1$. This means that if $\min(r,k) = \min(r-1,k)$ then we have to prove

$$\sum_{i=a}^{\min(r,k)} (\sigma(i,r,k,n) - \sigma(i,r-1,k,n-1)) \ge 0$$

and if $\min(r,k) = \min(r-1,k) + 1$ then we have to prove

$$\sigma(\min(r,k),r,k,n) + \sum_{i=a}^{\min(r-1,k)} (\sigma(i,r,k,n) - \sigma(i,r-1,k,n-1)) \geq 0.$$

In both cases it is sufficient to prove $\sigma(i,r,k,n) \geq \sigma(i,r-1,k,n-1)$ *for* $i = a, \ldots, \min(r-1,k)$. *It holds*

$$\sigma(i,r-1,k,n-1) = \frac{\binom{k}{i}\binom{n-1-k}{r-1-i}}{\binom{n-1}{r-1}} = \frac{\binom{k}{i}\frac{(n-1-k)!}{(r-1-i)!(n-1-k-(r-1-i))!}}{\frac{(n-1)!}{(r-1)!((n-1)-(r-1))!}} =$$

$$= \frac{\binom{k}{i}\frac{(n-k)!}{(r-i)!(n-k-(r-i))!}\frac{r-i}{n-k}}{\frac{n!}{r!(n-r)!}\frac{r}{n}} = \frac{\binom{k}{i}\binom{n-k}{r-i}\frac{r-i}{n-k}}{\binom{n}{r}\frac{r}{n}} = \sigma(i,r,k,n)\frac{(r-i)n}{r(n-k)}$$

and it remains to prove that $r(n-k) \geq (r-i)n$. *We have*

$$r(n-k) - (r-i)n = (a+b)(b+d) - (a+b-i)(a+b+c+d) =$$

$$= ab + ad + b^2 + bd - (a^2 + ab + ac + ad + ab + b^2 + bc + bd) + ia + ib + ic + id =$$

$$= (ia - a^2) + (ic - ac) + (ib - ab) + (id - bc)$$

It holds $i \geq a$ *and it means also* $ia \geq a^2$, $ic \geq ca$, $ib \geq ab$. *We assume* $ad > bc$ *and it means also* $id > cb$. *This finishes the proof of point 2.*

3. *We have to prove* $\mathscr{F}_F(a,b,c-1,d) \leq \mathscr{F}_F(a,b,c,d)$. *According to point 1 of lemma 5.9 it holds* $\mathscr{F}_F(a,b,c-1,d) = \mathscr{F}_F(a,c-1,b,d)$. *From the just proved point 2 we have* $\mathscr{F}_F(a,c-1,b,d) \leq \mathscr{F}_F(a,c,b,d)$ *and point 1 of lemma 5.9 means* $\mathscr{F}_F(a,c,b,d) = \mathscr{F}_F(a,b,c,d)$. *This finishes the proof of point 3.*

4. *We have to prove* $\mathscr{F}_F(a,b,c,d+1) \leq \mathscr{F}_F(a,b,c,d)$. *According to point 2 of lemma 5.9 it holds* $\mathscr{F}_F(a,b,c,d+1) = \mathscr{F}_F(d+1,b,c,a)$. *From the point 1 of this lemma we have* $\mathscr{F}_F(d+1,b,c,a) \leq \mathscr{F}_F(d,b,c,a)$ *and point 2 of lemma 5.9 means* $\mathscr{F}_F(d,b,c,a) = \mathscr{F}_F(a,b,c,d)$.

This finishes the proof. □

Theorem 5.3. *If* $F_{\sim_\alpha^1}(a,b,c,d) = 1$ *and* $a' \geq a \wedge b' \leq b \wedge c' \leq c \wedge d' \geq d$, *where* $\sim_\alpha^1$ *is Fisher's quantifier (see line 11 in tables 4.2 and 4.4), then also* $F_{\sim_\alpha^1}(a',b',c',d') = 1$.

Proof. $F_{\sim_\alpha^1}(a,b,c,d) = 1$ *implies* $\mathscr{F}_F(a,b,c,d) \leq \alpha$ *and* $ad > bc$. *We have to prove*

$$\mathscr{F}_F(a',b',c',d') \leq \alpha \wedge a'd' > b'c'.$$

Assumption $a' \geq a \wedge b' \leq b \wedge c' \leq c \wedge d' \leq d$ *implies* $a'd' > b'c'$. *According to lemma 5.10 we then have:*

$$\mathscr{F}_F(a',b',c',d') \leq \mathscr{F}_F(a'-1,b',c',d') \leq \ldots \leq \mathscr{F}_F(a,b',c',d') \leq$$

$$\leq \mathscr{F}_F(a,b'+1,c',d') \leq \ldots \leq \mathscr{F}_F(a,b,c',d')$$

$$\leq \mathscr{F}_F(a,b,c'+1,d') \leq \ldots \leq \mathscr{F}_F(a,b,c,d') \leq$$
$$\leq \mathscr{F}_F(a,b,c,d'-1) \leq \ldots \leq \mathscr{F}_F(a,b,c,d) \leq \alpha \, .$$

This finishes the proof.   □

We prove an additional important property of Fisher's quantifier. The corresponding theorem and its proof are based on [60]. First we prove a lemma.

Lemma 5.11. *Let n, r, k be integer non-negative numbers satisfying $n \geq r$ and $n \geq k$. Then*

1. *Each 4ft-table $\langle a,b,c,d \rangle$ satisfying $a+b=r$, $a+c=k$, and $a+b+c+d=n$ satisfies also $\max\{0, r+k-n\} \leq a \leq k$, $b=r-a$, $c=k-a$, and $d=n-r-k+a$.*
2. *If $r \geq k$ then*

$$\sum_{i=a_{\min}}^{k} \frac{\binom{k}{i}\binom{n-k}{r-i}}{\binom{n}{r}} = 1$$

where $a_{\min} = \max\{0, r+k-n\}$.

Proof

1. *Possible values for frequencies a,b,c,d satisfying $a+b=r$, $a+c=k$, and $a+b+c+d=n$ for given n,r,k are illustrated in Fig. 5.3. The limitation for frequency c is $0 \leq c \leq k \wedge c \leq n-r$. After determining frequency c, frequencies a,b,d can be computed as follows: $a=k-c$, $b=r-a$, and $d=n-r-c$. Condition $c \leq n-r$ implies $a \geq k-(n-r)$ i.e. $a \geq r+k-n$ and of course it holds $a \geq 0$. We can conclude that $a \geq \max\{0, r+k-n\}$, equalities $b=r-a$, $c=k-a$, and $d=n-r-k+a$ are obvious. This finishes the proof of point 1.*
2. *Possible values of frequency a for given r, k, n where $r \geq k$ are $a_{\min}, \ldots, k$. Let $i \in \langle a_{\min}, k \rangle$ and let $Prb(i,r,k,n)$ denote the probability that 4ft-table $\langle i,r,k,n \rangle$ occurs among all possible 4ft-tables $\langle j,r,k,n \rangle$ for given r,k,n. Then we have*

$$\sum_{i=a_{\min}}^{k} Prb(i,r,k,n) = 1 \, .$$

However, it holds $Prb(i,r,k,n) = \frac{\binom{k}{i}\binom{n-k}{r-i}}{\binom{n}{r}}$ and thus $\sum_{i=a_{\min}}^{k} \frac{\binom{k}{i}\binom{n-k}{r-i}}{\binom{n}{r}} = 1$.

This finishes the proof.   □

Theorem 5.4. *If $b \geq c-1 \geq 0$ then the Fisher's measure $\mathscr{F}_F(a,b,c,d)$ satisfies $\mathscr{F}_F(a,b+1,c-1,d) \leq \mathscr{F}_F(a,b,c,d)$.*

	$a=k-c$	$b=r-a$	r
$c \leq k \wedge c \leq n-r$	$d=n-r-c$	$n-r$
	k	$n-k$	n

Fig. 5.3 Possible frequencies in 4ft-table $\langle a,b,c,d \rangle$ for given r,k,n

Proof. We distinguish two cases: $b = c - 1$ and $b \geq c$. If $b = c - 1$ then it holds
$\mathscr{F}_F(a,b,c,d) = \mathscr{F}_F(a,c-1,c,d)$ *and* $\mathscr{F}_F(a,b+1,c-1,d) = \mathscr{F}_F(a,c,c-1,d)$.
We have $\mathscr{F}_F(a,c-1,c,d) = \mathscr{F}_F(a,c,c-1,d)$ *according to point 1) of lemma 5.9*
and thus it is satisfied $\mathscr{F}_F(a,b+1,c-1,d) \leq \mathscr{F}_F(a,b,c,d)$.

Let us assume $b \geq c$. *This means that it holds* $\min(r,k) = k$ *and*
$\min(r+1,k-1) = k-1$, *see 4ft-tables* $\langle a,b,c,d \rangle$ *and* $\langle a,b+1,c-1,d \rangle$ *in Fig. 5.4.*

a	b	r
c	d	s
k	1	n

4ft-table $\langle a,b,c,d \rangle$

a	$b+1$	$r+1$
$c-1$	d	$s-1$
$k-1$	$l+1$	n

4ft-table $\langle a,b+1,c-1,d \rangle$

Fig. 5.4 4ft-tables $\langle a,b,c,d \rangle$ and $\langle a,b+1,c-1,d \rangle$

Thus, using definition 5.1 we have

$$\mathscr{F}_F(a,b,c,d) = \sum_{i=a}^{k} \sigma(i,r,k,n) \quad \text{where} \quad \sigma(i,r,k,n) = \frac{\binom{k}{i}\binom{n-k}{r-i}}{\binom{n}{r}}$$

and

$$\mathscr{F}_F(a,b+1,c-1,d) = \sum_{i=a}^{k-1} \sigma(i,r+1,k-1,n)$$

where

$$\sigma(i,r+1,k-1,n) = \frac{\binom{k-1}{i}\binom{n-(k-1)}{r+1-i}}{\binom{n}{r+1}}.$$

We finish the proof by proving the inequality

$$\sum_{i=a}^{k} \sigma(i,r,k,n) - \sum_{i=a}^{k-1} \sigma(i,r+1,k-1,n) \geq 0.$$

We use properties of ratio $\frac{\sigma(i,r,k,n)}{\sigma(i,r+1,k-1,n)}$. *It holds for* $i = 0,\ldots,k-1$:

$$\frac{\sigma(i,r,k,n)}{\sigma(i,r+1,k-1,n)} = \frac{\binom{k}{i}\binom{n-k}{r-i}}{\binom{n}{r}} \frac{\binom{n}{r+1}}{\binom{k-1}{i}\binom{n-(k-1)}{r+1-i}} =$$

$$= \frac{\frac{k!}{i!(k-i)!}\frac{(n-k)!}{(r-i)!(n-k-(r-i))!}}{\frac{n!}{r!(n-r)!}} \frac{\frac{n!}{(r+1)!(n-(r+1))!}}{\frac{(k-1)!}{i!(k-1-i)!}\frac{(n-(k-1))!}{(r+1-i)!(n-(k-1)-(r+1-i))!}} =$$

$$= \frac{k!(n-k)!r!(n-r)!}{i!(k-i)!(r-i)!(n-k-r+i)!n!} \frac{i!(k-1-i)!(r+1-i)!(n-k-r+i)!n!}{(k-1)!(n-(k-1))!(r+1)!(n-(r+1))!} =$$

$$= \frac{k(n-r)}{(n-(k-1))(r+1)} \cdot \frac{r+1-i}{k-i}.$$

Let us note that we assume $b \geq c-1 \geq 0$ *and* $b \geq c$. *This means that* $c \geq 1$ *and* $b \geq 1$, *thus* $n \geq k+1$ *and* $n-(k-1) > 0$ *which implies that ratio* $\frac{k(n-r)}{(n-(k-1))(r+1)}$ *is well defined.*

Let us denote $z = \frac{k(n-r)}{(n-(k-1))(r+1)}$ *and define function* $f(i)$ *for* $i = 0,\ldots,k-1$ *such that* $f(i) = \frac{r+1-i}{k-i}$. *This means that*

$$\frac{\sigma(i,r,k,n)}{\sigma(i,r+1,k-1,n)} = z \times f(i).$$

We assume $b \geq c$, *which implies* $r+1 > k$ *and for* $i = 0,\ldots,k-2$ *it holds*

$$f(i+1) - f(i) = \frac{r+1-(i+1)}{k-(i+1)} - \frac{r+1-i}{k-i} = \frac{r-i}{(k-i)-1} - \frac{(r-i)+1}{k-i} =$$

$$= \frac{(r-i)(k-i) - ((r-i)+1)((k-i)-1)}{((k-i)-1))(k-i)} =$$

$$= \frac{(r-i)(k-i) - ((r-i)(k-i) - (r-i) + (k-i) - 1)}{(k-(i+1))(k-i)} = \frac{r+1-k}{(k-(i+1))(k-i)} > 0$$

which in turn implies that $f(i)$ *is increasing.*

   There are two possibilities:

1. *For each* $i = 0,\ldots,k-1$ *it holds* $z \times f(i) < 1$
2. *There is* $j \in \{0,\ldots,k-1\}$ *such that* $z \times f(j) \geq 1$. *Function* $f(i)$ *is increasing and it means that there is* $i_0 \in \{0,\ldots,k-1\}$ *such that for* $i \geq i_0$ *it holds* $z \times f(i) \geq 1$ *and for* $i < i_0$ *it holds* $z \times f(i) < 1$.

To prove the inequality $\sum_{i=a}^{k} \sigma(i,r,k,n) - \sum_{i=a}^{k-1} \sigma(i,r+1,k-1,n) \geq 0$ *we distinguish cases a) and b):*

a)   *The second possibility occurs and it holds* $a \geq i_0$
b)   *The second possibility occurs and it holds* $a < i_0$ *or the first possibility occurs.*

a)   *If* $a \geq i_0$ *then we have* $z \times f(i) \geq 1$ *for* $i = a,\ldots,k-1$ *and thus it holds also* $\sigma(i,r,k,n) \geq \sigma(i,r+1,k-1,n)$ *which means*

$$\sum_{i=a}^{k} \sigma(i,r,k,n) - \sum_{i=a}^{k-1} \sigma(i,r+1,k-1,n) =$$

$$= \sum_{i=a}^{k-1} (\sigma(i,r,k,n) - \sigma(i,r+1,k-1,n)) + \sigma(i,r,k,n) > 0.$$

b)   *If* $a < i_0$ *or the first possibility occurs then* $\frac{\sigma(i,r,k,n)}{\sigma(i,r+1,k-1,n)} = z \times f(i) < 1$ *for* $i \leq a$ *and thus* $\sigma(i,r+1,k-1,n) > \sigma(i,r,k,n)$ *for* $i \leq a$. *We use the fact that*

$$\sum_{i=a_{min}}^{k} \frac{\binom{k}{i}\binom{n-k}{r-i}}{\binom{n}{r}} = \sum_{i=a_{min}}^{k-1} \frac{\binom{k-1}{i}\binom{n-(k-1)}{r+1-i}}{\binom{n}{r+1}} = 1$$

where $a_{min} = \max\{0, r+k-n\} = \max\{0, (r+1)+(k-1)-n\}$, see lemma 5.11. This fact means

$$\sum_{i=a_{min}}^{k} \sigma(i,r,k,n) = \sum_{i=a_{min}}^{k-1} \sigma(i,r+1,k-1,n) = 1.$$

If $a = a_{min}$ then $\sum_{i=a}^{k} \sigma(i,r,k,n) - \sum_{i=a}^{k-1} \sigma(i,r+1,k-1,n) = 0$.
If $a > a_{min}$ then

$$\sum_{i=a}^{k} \sigma(i,r,k,n) - \sum_{i=a}^{k-1} \sigma(i,r+1,k-1,n) =$$

$$= (1 - \sum_{i=a_{min}}^{a-1} \sigma(i,r,k,n)) - (1 - \sum_{i=a_{min}}^{a-1} \sigma(i,r+1,k-1,n)) =$$

$$= \sum_{i=a_{min}}^{a-1} (\sigma(i,r+1,k-1,n) - \sigma(i,r,k,n)) > 0.$$

This finishes the proof because $\sigma(i,r+1,k-1,n) > \sigma(i,r,k,n)$ for $i \leq a$.   □

5.4 Properties of χ^2-quantifier

The associated function $F_{\sim_\alpha^2}$ of χ^2-quantifier $\sim_\alpha^2$ (see rows 12 in Tables 4.2 and 4.4) is defined such that $F_{\sim_\alpha^2}(a,b,c,d) = 1$ if and only if

$$\frac{(ad-bc)^2}{rkls}n \geq \chi_\alpha^2 \wedge ad > bc$$

where χ_α^2 is the $1 - \alpha$ quantile of the χ^2-distribution function, see also 4ft-table $\langle a,b,c,d \rangle$ in Fig. 5.1. We define the χ^2-measure and then we prove a theorem describing an important property of the χ^2-quantifier. The theorem and its proof are based on theorem 4.5.1 in [18].

Definition 5.2. A χ^2-*measure* is a function $\mathscr{F}_{\chi^2}(a,b,c,d)$ defined for all 4ft-tables $\langle a,b,c,d \rangle$ satisfying $rskl > 0$ such that

$$\mathscr{F}_{\chi^2}(a,b,c,d) = \frac{(ad-bc)^2}{rkls}n,$$

see 4ft-table $\langle a,b,c,d \rangle$ in Fig. 5.1 for definition of r,k,l,s,n.

Before proving the theorem we prove two lemmas.

Lemma 5.12. *The χ^2-measure $\mathscr{F}_{\chi^2}(a,b,c,d)$ satisfies*

1. $\mathscr{F}_{\chi^2}(a,b,c,d) = \mathscr{F}_{\chi^2}(a,c,b,d)$
2. $\mathscr{F}_{\chi^2}(a,b,c,d) = \mathscr{F}_{\chi^2}(d,b,c,a)$.

Proof. The proof directly follows from definition 5.2 and definition of r,k,l,s,n in 4ft-table $\langle a,b,c,d\rangle$ in Fig. 5.1. □

Lemma 5.13. *If $ad > bc$ then the χ^2-measure $\mathscr{F}_{\chi^2}(a,b,c,d)$ satisfies*

1. $\mathscr{F}_{\chi^2}(a+1,b,c,d) \geq \mathscr{F}_{\chi^2}(a,b,c,d)$
2. $\mathscr{F}_{\chi^2}(a,b-1,c,d) \geq \mathscr{F}_{\chi^2}(a,b,c,d)$
3. $\mathscr{F}_{\chi^2}(a,b,c-1,d) \geq \mathscr{F}_{\chi^2}(a,b,c,d)$
4. $\mathscr{F}_{\chi^2}(a,b,c,d+1) \geq \mathscr{F}_{\chi^2}(a,b,c,d)$.

Proof

1. We have to prove

$$\frac{((a+1)d-bc)^2}{(r+1)(k+1)ls}(n+1) \geq \frac{(ad-bc)^2}{rkls}n,$$

see 4ft-tables $\langle a,b,c,d\rangle$ and $\langle a+1,b,c,d\rangle$ in Fig. 5.2. We prove a bit stronger inequality

$$\frac{((a+1)d-bc)^2}{(r+1)(k+1)} \geq \frac{(ad-bc)^2}{rk} \quad \text{which is the same as} \quad \frac{((ad-bc)+d)^2}{rk+r+k+1} \geq \frac{(ad-bc)^2}{rk}$$

and thus the same as

$$\frac{(ad-bc)^2+2d(ad-bc)+d^2}{rk+r+k+1} \geq \frac{(ad-bc)^2}{rk}.$$

We use a simple fact that

$$\frac{A+x}{B+y} \geq \frac{A}{B} \quad \text{is equivalent to} \quad \frac{x}{y} \geq \frac{A}{B}$$

for positive numbers A,B,x,y. Thus we have that

$$\frac{(ad-bc)^2+2d(ad-bc)+d^2}{rk+r+k+1} \geq \frac{(ad-bc)^2}{rk}$$

is equivalent to

$$\frac{2d(ad-bc)+d^2}{r+k+1} \geq \frac{(ad-bc)^2}{rk}$$

which is further equivalent to inequalities

$$2ad^2rk - 2bcdrk + d^2rk \geq a^2d^2(r+k+1) - 2abcd(r+k+1) + b^2c^2(r+k+1)$$

$$2ad^2rk + 2abcd(r+k+1) + d^2rk \geq a^2d^2(r+k+1) + 2bcdrk + b^2c^2(r+k+1).$$

We assume $ad > bc$ implying $abcd(r+k+1) \geq b^2c^2(r+k+1)$ and thus it is sufficient to prove

$$2ad^2rk + abcd(r+k+1) + d^2rk \geq a^2d^2(r+k+1) + 2bcdrk$$

which can be modified in several steps using $r = a+b$ and $k = a+c$ into several equivalent inequalities. The first one is

$$2ad^2(a^2 + ab + ac + bc) + abcd(2a+b+c+1) + d^2(a^2 + ab + ac + bc) \geq$$

$$\geq a^2d^2(2a+b+c+1) + 2bcd(a^2 + ab + ac + bc)$$

from which we get

$$2a^3d^2 + 2a^2bd^2 + 2a^2cd^2 + 2abcd^2 + 2a^2bcd + ab^2cd + abc^2d + abcd +$$

$$+ a^2d^2 + abd^2 + acd^2 + bcd^2 \geq$$

$$\geq 2a^3d^2 + a^2bd^2 + a^2cd^2 + a^2d^2 + 2a^2bcd + 2ab^2cd + 2abc^2d + 2b^2c^2d$$

and after omitting same members on both sides we have

$$a^2bd^2 + a^2cd^2 + 2abcd^2 + abcd + abd^2 + acd^2 + bcd^2 \geq ab^2cd + abc^2d + 2b^2c^2d.$$

We assume $ad > bc$ implying inequalities $a^2bd^2 \geq ab^2cd$, $a^2cd^2 \geq abc^2d$, and $2abcd^2 \geq 2b^2c^2d$, meaning that it remains to prove

$$abcd + abd^2 + acd^2 + bcd^2 \geq 0$$

which is obvious. This finishes the proof of point 1.

2. *We have to prove*

$$\frac{(ad - (b-1)c)^2}{(r-1)k(l-1)s}(n-1) \geq \frac{(ad-bc)^2}{rkls}n,$$

see 4ft-tables $\langle a,b,c,d \rangle$ and $\langle a,b-1,c,d \rangle$ in Fig. 5.2. This is equivalent to

$$\frac{(ad - bc + c)^2}{(ad - bc)^2} \geq \frac{r-1}{r}\frac{l-1}{l}\frac{n}{n-1}.$$

It holds $\frac{(ad-bc+c)^2}{(ad-bc)^2} \geq 1$, thus it is sufficient to prove $\frac{r-1}{r}\frac{l-1}{l}\frac{n}{n-1} \leq 1$. It holds

$$\frac{r-1}{r}\frac{n}{n-1} = \frac{rn-n}{rn-r} = \frac{1-\frac{1}{r}}{1-\frac{1}{n}} \leq 1$$

because $r \leq n$ and thus $1 - \frac{1}{r} \leq 1 - \frac{1}{n}$. In addition, $\frac{l-1}{l} < 1$ and so $\frac{r-1}{r} \frac{l-1}{l} \frac{n}{n-1} \leq 1$. This finishes the proof of point 2.

3. *We have to prove $\mathscr{F}_{\chi^2}(a,b,c-1,d) \geq \mathscr{F}_{\chi^2}(a,b,c,d)$. According to point 1 of lemma 5.12 it holds $\mathscr{F}_{\chi^2}(a,b,c-1,d) = \mathscr{F}_{\chi^2}(a,c-1,b,d)$. From the just proved point 2 we have $\mathscr{F}_{\chi^2}(a,c-1,b,d) \leq \mathscr{F}_{\chi^2}(a,c,b,d)$ and point 1 of lemma 5.12 means $\mathscr{F}_{\chi^2}(a,c,b,d) = \mathscr{F}_{\chi^2}(a,b,c,d)$. This finishes the proof of point 3.*

4. *We have to prove $\mathscr{F}_{\chi^2}(a,b,c,d+1) \geq \mathscr{F}_{\chi^2}(a,b,c,d)$. According to point 2 of lemma 5.12 it holds $\mathscr{F}_{\chi^2}(a,b,c,d+1) = \mathscr{F}_{\chi^2}(d+1,b,c,a)$. From point 1 of this lemma we have $\mathscr{F}_{\chi^2}(d+1,b,c,a) \leq \mathscr{F}_{\chi^2}(d,b,c,a)$ and point 2 of lemma 5.12 means $\mathscr{F}_{\chi^2}(d,b,c,a) = \mathscr{F}_{\chi^2}(a,b,c,d)$.*

This finishes the proof.   □

Theorem 5.5. *If $F_{\sim^2_\alpha}(a,b,c,d) = 1$ and $a' \geq a \wedge b' \leq b \wedge c' \leq c \wedge d' \geq d$ where $\sim^2_\alpha$ is the χ^2-quantifier, then also $F_{\sim^2_\alpha}(a',b',c',d') = 1$ (see line 12 in Tables 4.2 and 4.4).*

Proof. $F_{\sim^2_\alpha}(a,b,c,d) = 1$ implies $\mathscr{F}_{\chi^2}(a,b,c,d) \geq \chi^2_\alpha$ and $ad > bc$. We have to prove

$$\mathscr{F}_{\chi^2}(a',b',c',d') \geq \chi^2_\alpha \wedge a'd' > b'c' .$$

Assumption $a' \geq a \wedge b' \leq b \wedge c' \leq c \wedge d' \geq d$ implies $a'd' > b'c'$. According to lemma 5.13 we have:

$$\chi^2_\alpha \leq \mathscr{F}\chi^2(a,b,c,d) \leq \mathscr{F}\chi^2(a+1,b,c,d) \leq \ldots \leq \mathscr{F}\chi^2(a',b,c,d) \leq$$

$$\leq \mathscr{F}\chi^2(a',b-1,c,d) \leq \ldots \leq \mathscr{F}\chi^2(a',b',c,d) \leq$$

$$\leq \mathscr{F}\chi^2(a',b',c-1,d) \leq \ldots \leq \mathscr{F}\chi^2(a',b',c',d) \leq$$

$$\leq \mathscr{F}_{\chi^2}(a',b',c',d+1) \leq \ldots \leq \mathscr{F}\chi^2(a',b',c',d') .$$

This finishes the proof.   □

We prove an additional important property of χ^2-quantifier.

Theorem 5.6. *If $b \geq c - 1 \geq 0$, $rskl > 0$, and $(r+1)(k-1)(l+1)(s-1) > 0$ then the χ^2-measure $\mathscr{F}_{\chi^2}(a,b,c,d)$ satisfies $\mathscr{F}_{\chi^2}(a,b+1,c-1,d) \geq \mathscr{F}_{\chi^2}(a,b,c,d)$.*

Proof. It holds

$$\mathscr{F}_{\chi^2}(a,b,c,d) = \frac{(ad-bc)^2}{rkls}n$$

and

$$\mathscr{F}_{\chi^2}(a,b+1,c-1,d) = \frac{(ad-(b+1)(c-1))^2}{(r+1)(k-1)(l+1)(s-1)}n ,$$

see 4ft-tables $\langle a,b,c,d \rangle$ and $\langle a,b+1,c-1,d \rangle$ in Fig. 5.5 (which is the same as Fig. 5.4) and definition 5.2.

a	b	r
c	d	s
k	l	n

4ft-table $\langle a,b,c,d \rangle$

a	b+1	r+1
c-1	d	s-1
k-1	l+1	n

4ft-table $\langle a,b+1,c-1,d \rangle$

Fig. 5.5 4ft-tables $\langle a,b,c,d \rangle$ and $\langle a,b+1,c-1,d \rangle$

We modify $\mathscr{F}_{\chi^2}(a,b+1,c-1,d)$:

$$\frac{(ad-(b+1)(c-1))^2}{(r+1)(k-1)(l+1)(s-1)}n = \frac{(ad-bc+(b-c+1))^2}{(rk-(r-k+1))(ls-(l-s+1))}n =$$

$$= \frac{(ad-bc+X)^2}{(rk-X)(ls-X)}n$$

where $X = b-c+1$ and $X \geq 0$ because of $b \geq c-1 \geq 0$. Then we have
$(ad-bc+X)^2 \geq (ad-bc)^2$ and $(rk-X)(ls-X) \leq rkls$ which implies

$$\frac{(ad-(b+1)(c-1))^2}{(r+1)(k-1)(l+1)(s-1)}n \geq \frac{(ad-bc)^2}{rkls}n \,.$$

This finishes the proof.   □

5.5 Selected Results for Combinatorial Numbers

We prove several results on binomial coefficients in this section.

Lemma 5.14. Let us suppose that $0 < p < 1$ and that $i \geq 0$ is an integer number.
Then

$$\lim_{K \to \infty} \binom{K}{i} p^i (1-p)^{K-i} = 0 \,.$$

Proof. It holds:

$$\binom{K}{i} p^i (1-p)^{K-i} \leq K^i p^i (1-p)^K (1-p)^{-i} = K^i (1-p)^K \left(\frac{p}{1-p} \right)^i \,.$$

Thus it is enough to prove that for $r \in (0,1)$ and $i \geq 0$ it holds

$$\lim_{K \to \infty} K^i r^K = 0 \,.$$

To prove this it is enough to prove that for $r \in (0,1)$, a real x and an integer $i \geq 0$ it
holds

$$\lim_{x \to \infty} x^i r^x = 0 \,.$$

It holds $\lim_{x \to \infty} x^i = \infty$, $\lim_{x \to \infty} r^x = 0$ i.e. $\lim_{x \to \infty} r^{-x} = \infty$ and thus according to
l'Hospital's rule we have

$$\lim_{x \to \infty} x^i r^x = \lim_{x \to \infty} \frac{x^i}{r^{-x}} = \lim_{x \to \infty} \frac{(x^i)^{(i)}}{(r^{-x})^{(i)}} = \lim_{x \to \infty} \frac{i!}{(-\ln r)^i r^{-x}} = \lim_{x \to \infty} r^x = 0,$$

where $(x^i)^{(i)}$ is an i-th derivation of x^i and analogously for $(r^{-x})^{(i)}$. This finishes the proof. □

Lemma 5.15. *Let A be an integer non-negative number and let $0 < p < 1$ be real. Then the following holds for an integer N:*

1.

$$\lim_{N \to \infty} \sum_{i=0}^{A} \binom{A+N}{i} p^i (1-p)^{A+N-i} = 0 .$$

2.

$$\lim_{N \to \infty} \sum_{i=N-A}^{N} \binom{N}{i} p^i (1-p)^{N-i} = 0 .$$

3.

$$\lim_{N \to \infty} \sum_{i=N+1}^{N+A} \binom{N+A}{i} p^i (1-p)^{N+A-i} = 0 .$$

Proof. All sums have finite number of summands. Thus it is enough to prove that for integer K $\lim_{K \to \infty} \binom{K}{i} p^i (1-p)^{K-i} = 0$, which is proved in lemma 5.14. This finishes the proof. □

Lemma 5.16. *Let A be an integer non-negative number and let $0 < p < 1$ be real. Then the following holds for an integer N:*

1.

$$\lim_{N \to \infty} \sum_{i=A}^{A+N} \binom{A+N}{i} p^i (1-p)^{A+N-i} = 1 .$$

2.

$$\lim_{N \to \infty} \sum_{i=0}^{N-A} \binom{N}{i} p^i (1-p)^{N-i} = 1 .$$

3.

$$\lim_{N \to \infty} \sum_{i=0}^{N} \binom{N+A}{i} p^i (1-p)^{N+A-i} = 1 .$$

Proof

1. We use the fact that for integer $K > 0$ and real p, $0 < p < 1$, we have

$$\sum_{i=0}^{K} \binom{K}{i} p^i (1-p)^{K-i} = 1 .$$

Thus, according to point 1) of lemma 5.15 it holds

$$\lim_{N \to \infty} \sum_{i=A}^{A+N} \binom{A+N}{i} p^i (1-p)^{A+N-i} = \lim_{N \to \infty} (1 - \sum_{i=0}^{A-1} \binom{A+N}{i} p^i (1-p)^{A+N-i}) =$$

$$= 1 - \lim_{N \to \infty} \sum_{i=0}^{A-1} \binom{A+N}{i} p^i (1-p)^{A+N-i} = 1 .$$

2. *Analogously to point 1) we have (see also point 2) of lemma 5.15):*

$$\lim_{N \to \infty} \sum_{i=0}^{N-A} \binom{N}{i} p^i (1-p)^{N-i} = \lim_{N \to \infty} (1 - \sum_{i=N-A+1}^{N} \binom{N}{i} p^i (1-p)^{N-i}) =$$

$$= 1 - \lim_{N \to \infty} \sum_{i=N-A+1}^{N} \binom{N}{i} p^i (1-p)^{N-i} = 1 .$$

3. *Analogously to point 1) (see also point 3) of lemma 5.15) we have:*

$$\lim_{N \to \infty} \sum_{i=0}^{N} \binom{N+A}{i} p^i (1-p)^{N+A-i} = \lim_{N \to \infty} (1 - \sum_{i=N+1}^{N+A} \binom{N+A}{i} p^i (1-p)^{N+A-i}) =$$

$$= 1 - \lim_{N \to \infty} \sum_{i=N+1}^{N+A} \binom{N+A}{i} p^i (1-p)^{N+A-i} = 1 .$$

This finishes the proof. □

Lemma 5.17. *Let $a \geq 0$ and $b \geq 0$ be integer numbers. Then for each $k \in \langle 0, b \rangle$ and $0 < p < 1$ we have*

$$\lim_{a \to \infty} \binom{a+b}{a+k} p^{a+k} (1-p)^{b-k} = 0 .$$

Proof. It holds:

$$\binom{a+b}{a+k} p^{a+k} (1-p)^{b-k} = \binom{a+b}{a+b-(a+k)} p^{a+k} (1-p)^{b-k} =$$

$$= \binom{a+b}{b-k} p^{a+k} (1-p)^{b-k} \leq (a+b)^{b-k} p^{a+k} (1-p)^{b-k} .$$

Thus it is enough to prove that it holds

$$\lim_{a \to \infty} (a+b)^{b-k} p^a = 0 .$$

To prove this it is enough to prove that for $p \in (0,1)$, a real x and an integer $i \geq 0$ we have

$$\lim_{x \to \infty} (x+b)^i p^x = 0 .$$

The proof of this assertion is similar to the proof of the assertion $\lim_{x \to \infty} x^i r^x = 0$ in lemma 5.14. This finishes the proof. □

Part II
Classes of Association Rules

All important results presented in this book are related to classes of association rules. The classes of association rules are defined such that the properties of rules belonging to the same class of rules are identical with respect to several aspects. These aspects include namely deduction rules, dealing with missing information, and definability in classical predicate calculus. The way in which the classes of rules are defined and an overview of all classes is located in Chap. 6. The most examined classes of association rules are classes related to the GUHA procedure ASSOC, i.e. classes of implicational rules, weak implicational rules, double implicational rules, weak double implicational rules, equivalence rules, weak equivalence rules, and rules with the F-property. These are introduced in Chaps. 7 – 10.

Chapter 6
Overview of Classes

The properties of an association rule $\varphi \approx \psi$ are determined by the properties of the 4ft-quantifier $\approx$. This is the reason why classes of association rules are defined by classes of 4ft-quantifiers. The association rule $\varphi \approx \psi$ belongs to the *class of implicational association rules* if the 4ft-quantifier $\approx$ belongs to the *class of implicational quantifiers*. We also say that the association rule $\varphi \approx \psi$ is an *implicational rule* and that the 4ft-quantifier $\approx$ is an *implicational quantifier*. This is the same for additional classes of association rules.

One tool for defining classes of association rules is the so called TPC (*Truth Preservation Condition*), see Sect. 6.1. Suitable TPC's are used to define classes of implicational 4ft-quantifiers, double implicational 4ft-quantifiers, equivalence 4ft-quantifiers, symmetrical 4ft-quantifiers, and the class of 4ft-quantifiers with the F-property in Sects. 6.2 - 6.6. Examples of important 4ft-quantifiers belonging to particular classes are also given in corresponding sections. Important is dividing of classes of 4ft-quantifiers to $\mathcal{M}$-dependent classes and $\mathcal{M}$-independent classes, see Sect. 6.7. Relations among particular classes of 4ft-quantifiers are presented in Sect. 6.8.

Please note that the 4ft-quantifier is a symbol of language of association rules, see Sect. 3.1. Properties of 4ft-quantifier $\approx$ are given by properties of its association function $F_\approx$. Thus, when saying that 4ft-quantifier $\approx$ belongs to a class of quantifiers, we automatically assume that this quantifier will be used in the calculus of association rules see Sect. 3.4 and that there is a well-defined associated function of this quantifier.

Let us also note that the classes of implicational 4ft-quantifiers and symmetrical 4ft-quantifiers were defined in [18]. The class of equivalence 4ft-quantifiers was defined under the name of *association quantifiers* also in [18]. The reason why we use the name *equivalence 4ft-quantifiers* here is given in Sect. 6.4. The notion of truth preservation condition was in a bit less formal form introduced in [65]. The class of double implicational 4ft-quantifiers was defined in [67] on the basis of [24]. The class of 4ft-quantifiers with the F-property was introduced in [63].

In addition, let us note that we use the expressions *implicational 4ft-quantifier* and *implicational quantifier* as synonyms, similarly for other classes of quantifiers.

J. Rauch: *Observational Calculi and Association Rules*, SCI 469, pp. 65–79.
DOI: 10.1007/978-3-642-11737-4_6 © Springer-Verlag Berlin Heidelberg 2013

6.1 Truth Preservation Condition

Classes of 4ft-quantifiers are defined by TPC's – *Truth Preservation Conditions*. A TPC concerns two 4ft-tables $\langle a,b,c,d \rangle$ and $\langle a',b',c',d' \rangle$ and gives a condition ensuring that if $F_\approx(a,b,c,d) = 1$ then also $F_\approx(a',b',c',d') = 1$ for associated function $F_\approx$ of 4ft-quantifier $\approx$. This is formalized in definition 6.1.

Definition 6.1. A *truth preservation condition* (shortly TPC) is a $\{0,1\}$-valued function $C(a,b,c,d,a',b',c',d')$ defined for all 8-tuples $\langle a,b,c,d,a',b',c',d' \rangle$ of non-negative integer numbers.

The *class of 4ft-quantifiers* $4ft[C]$ *defined by truth preservation condition* C is a set of all 4ft-quantifiers $\approx$ satisfying for each pair of 4ft-tables $\langle a,b,c,d \rangle$ and $\langle a',b',c',d' \rangle$ the condition

$$\text{if } F_\approx(a,b,c,d) = 1 \;\wedge\; C(a,b,c,d,a',b',c',d') = 1 \text{ then } F_\approx(a',b',c',d') = 1 \,.$$

There is an important lemma concerning relation of two TPC's and corresponding classes of 4ft-quantifiers:

Lemma 6.1. *Let* C_1 *and* C_2 *be two TPC's such that*

$$C_1(a,b,c,d,a',b',c',d') = 1 \;\; implies \;\; C_2(a,b,c,d,a',b',c',d') = 1 \,.$$

Then if 4ft-quantifier $\approx$ *belongs to class* $4ft[C_2]$ *then it also belongs to class* $4ft[C_1]$.

Proof. Let us assume that $\approx$ *belongs to* $4ft[C_2]$. *We have to prove that* $\approx$ *belongs to* $4ft[C_1]$ *i.e. that it holds for each pair of 4ft-tables* $\langle a,b,c,d \rangle$ *and* $\langle a',b',c',d' \rangle$:

$$\text{if } F_\approx(a,b,c,d) = 1 \;\wedge\; C_1(a,b,c,d,a',b',c',d') = 1 \text{ then } F_\approx(a',b',c',d') = 1 \,.$$

Let $F_\approx(a,b,c,d) = 1 \;\wedge\; C_1(a,b,c,d,a',b',c',d') = 1$, *then according to the assumption that* $C_1(a,b,c,d,a',b',c',d') = 1$ *implies* $C_2(a,b,c,d,a',b',c',d') = 1$ *we have also* $F_\approx(a,b,c,d) = 1 \;\wedge\; C_2(a,b,c,d,a',b',c',d') = 1$. *We assume that* $\approx$ *belongs to* $4ft[C_2]$ *and thus* $F_\approx(a',b',c',d') = 1$. *This finishes the proof.* □

6.2 Implicational Quantifiers

The class of implicational quantifiers is defined in [18]. The definition can be formulated using a truth preservation condition for implicational quantifiers:

Definition 6.2. The *truth preservation condition* $TPC_\Rightarrow$ *for implicational quantifiers* is a $\{0,1\}$-valued function $TPC_\Rightarrow(a,b,c,d,a',b',c',d')$ defined for all 8-tuples $\langle a,b,c,d,a',b',c',d' \rangle$ of non-negative integer numbers such that

$$TPC_\Rightarrow(a,b,c,d,a',b',c',d') = \begin{cases} 1 & \text{if } a' \geq a \wedge b' \leq b \\ 0 & \text{otherwise.} \end{cases}$$

The *class of imlicational quantifiers* is the class $4ft[TPC_\Rightarrow]$ defined by truth preservation condition $TPC_\Rightarrow$. We say that 4ft-quantifier $\approx$ is *implicational* if it belongs to $4ft[TPC_\Rightarrow]$.

The condition $a' \geq a \wedge b' \leq b$ is called the *simple form of $TPC_\Rightarrow$* .

Definition 6.2 can also be formulated such that 4ft-quantifier $\approx$ is implicational if

$$F_\approx(a,b,c,d) = 1 \wedge a' \geq a \wedge b' \leq b \quad \text{implies} \quad F_\approx(a',b',c',d') = 1$$

for all 4ft-tables $\langle a,b,c,d \rangle$ and $\langle a',b',c',d' \rangle$.

The condition $a' \geq a$ and $b' \leq b$ means that the 4ft-table $\langle a',b',c',d' \rangle$ is "better with respect to implication" than the 4ft-table $\langle a,b,c,d \rangle$ (i-better, see [18]). If $\langle a,b,c,d \rangle$ is the 4ft-table of φ and ψ in data matrix $\mathscr{M}$ and if $\langle a',b',c',d' \rangle$ is the 4ft-table of φ and ψ in data matrix $\mathscr{M}'$, then the sentence *"better with respect to implication"* means: in data matrix $\mathscr{M}'$ there are more rows satisfying both φ and ψ than in data matrix $\mathscr{M}$ and in $\mathscr{M}'$ there are fewer rows satisfying φ and not satisfying ψ than in $\mathscr{M}$.

Thus, if $a' \geq a$ and $b' \leq b$, then it is reasonable to expect that if the implicational association rule $\varphi \approx \psi$ (i.e. the rule expressing implication by $\approx$) is true in the data matrix $\mathscr{M}$ then it is also true in data matrix $\mathscr{M}'$ which is better with respect to implication. This expectation is ensured by definition 6.2.

The 4ft-quantifier $\Rightarrow_p$ of p-implication (i.e. confidence or precision) is an example of an implicational quantifier, see the following lemma. Additional examples of implicational quantifiers are located in Chap. 7.

Lemma 6.2. *4ft-quantifier $\Rightarrow_p$ of p-implication, see rows 1 in Tables 4.2 and 4.4 and rows 2 in Tables 4.5 and 4.6, is implicational for $0 < p \leq 1$.*

Proof. It holds $F_{\Rightarrow_p}(a,b,c,d) = 1$ if and only if both $\frac{a}{a+b} \geq p$ and $a+b > 0$. If $F_{\Rightarrow_p}(a,b,c,d) = 1$ and $a' \geq a \wedge b' \leq b$ then also $a' + b' > 0$ and we have

$$\frac{a'}{a'+b'} \geq \frac{a}{a+b'} \geq \frac{a}{a+b} \geq p. \tag{6.1}$$

This finishes the proof. □

Note 6.1. It is easy to prove that the value $F_{\Rightarrow^*}(a,b,c,d)$ of associated function of implicational quantifier $\Rightarrow^*$ depends neither on c nor on d, see lemma 6.3. Thus we often write only $F_{\Rightarrow^*}(a,b)$ instead of $F_{\Rightarrow^*}(a,b,c,d)$ for the implicational quantifier $\Rightarrow^*$.

Lemma 6.3. *Let $\Rightarrow^*$ be an implicational 4ft-quantifier. Then the value $F_{\Rightarrow^*}(a,b,c,d)$ of its associated function depends neither on c nor on d.*

Proof. We have to prove that if $\langle a,b,c,d \rangle$ and $\langle a,b,c',d' \rangle$ are 4ft-tables then $F_{\Rightarrow^*}(a,b,c,d) = F_{\Rightarrow^*}(a,b,c',d')$.

Let $F_{\Rightarrow^*}(a,b,c,d) = 1$. Then we have $TPC_\Rightarrow(a,b,c,d,a,b,c',d') = 1$ because of $a \geq a \wedge b \leq b$ and thus $F_{\Rightarrow^*}(a,b,c',d') = 1$

Let $F_{\Rightarrow^*}(a,b,c',d') = 1$. Again, we have $TPC_\Rightarrow(a,b,c',d',a,b,c,d) = 1$ and this implies $F_{\Rightarrow^*}(a,b,c,d) = 1$. This finishes the proof. □

6.3   Double Implicational Quantifiers

The class of double implicational quantifiers is defined in [67]. The definition can be formulated using a truth preservation condition for double implicational quantifiers:

Definition 6.3. The *truth preservation condition* $TPC_\Leftrightarrow$ *for double implicational quantifiers* is a $\{0,1\}$-valued function $TPC_\Leftrightarrow(a,b,c,d,a',b',c',d')$ defined for all 8-tuples $\langle a,b,c,d,a',b',c',d'\rangle$ of non-negative integer numbers such that

$$TPC_\Leftrightarrow(a,b,c,d,a',b',c',d') = \begin{cases} 1 & \text{if } a' \geq a \wedge b' \leq b \wedge c' \leq c \\ 0 & \text{otherwise.} \end{cases}$$

The *class of double imlicational quantifiers* is the class $4ft[TPC_\Leftrightarrow]$ defined by truth preservation condition $TPC_\Leftrightarrow$. We say that 4ft-quantifier $\approx$ is *double implicational* if it belongs to $4ft[TPC_\Leftrightarrow]$.

The condition $a' \geq a \wedge b' \leq b \wedge c' \leq c$ is called the *simple form of* $TPC_\Leftrightarrow$.

Definition 6.3 can also be formulated such that 4ft-quantifier $\approx$ is double implicational if

$$F_\approx(a,b,c,d) = 1 \wedge a' \geq a \wedge b' \leq b \wedge c' \leq c \text{ implies } F_\approx(a',b',c',d') = 1$$

for all 4ft-tables $\langle a,b,c,d\rangle$ and $\langle a',b',c',d'\rangle$.

We can see the reason for such a definition in an analogy to propositional logic. If u and v are propositions and both $u \to v$ and $v \to u$ are true, then u is equivalent to v (the symbol "$\to$" is here a propositional connective of implication). Thus we can try to express the relation of equivalence of attributes φ and ψ using a "double implicational" 4ft-quantifier $\Leftrightarrow^*$ such that

$$\varphi \Leftrightarrow^* \psi \text{ if and only if both } \varphi \Rightarrow^* \psi \text{ and } \psi \Rightarrow^* \varphi , \qquad (6.2)$$

where $\Rightarrow^*$ is a suitable implicational quantifier.

If we apply the truth preservation condition for implicational quantifiers to rule $\varphi \Rightarrow^* \psi$, we obtain $a' \geq a \wedge b' \leq b$. If we apply it to rule $\psi \Rightarrow^* \varphi$, we obtain $a' \geq a \wedge c' \leq c$, ($c$ is used here instead of b, see Fig. 3.3). This leads to the truth preservation condition for double implicational quantifiers $a' \geq a \wedge b' \leq b \wedge c' \leq c$, see definition 6.3.

It is easy to prove the following lemma on the relation of implicational and double imlicational quantifiers.

Lemma 6.4. *Each implicational quantifier is also double implicational.*

Proof. Note that $a' \geq a \wedge b' \leq b \wedge c' \leq c$ implies $a' \geq a \wedge b' \leq b$. In other words $TPC_\Leftrightarrow(a,b,c,d,a',b',c',d') = 1$ implies $TPC_\Rightarrow(a,b,c,d,a',b',c',d') = 1$. Thus, according to lemma 6.1 each implicational quantifier is also a double imlicational quantifier. This finishes the proof.   □

The 4ft-quantifier $\Leftrightarrow_p$ of double p-implication (i.e. Jaccard, see Table 4.5) is an example of double implicational quantifiers, see the following lemma. It is also proved that the 4ft-quantifier $\Leftrightarrow_p$ is not implicational. Additional examples of double implicational quantifiers are located in Chap. 8.

Lemma 6.5. *The 4ft-quantifier of double p-implication $\Leftrightarrow_p$, see rows 4 in Tables 4.2 and 4.4 and rows 12 in Tables 4.5 and 4.6, is double implicational for $0 < p \le 1$. However, the 4ft-quantifier $\Leftrightarrow_p$ is not implicational for $0 < p < 1$.*

Proof. It holds $F_{\Leftrightarrow_p}(a,b,c,d) = 1$ if and only if $\frac{a}{a+b+c} \ge p$ and $a+b+c > 0$. If $F_{\Leftrightarrow_p}(a,b,c,d) = 1$ then it must be $a+b+c > 0$ and $a > 0$. In addition, $a' \ge a \wedge b' \le b \wedge c' \le c$ implies $a' + b' + c' > 0$ and we have

$$\frac{a'}{a'+b'+c'} \ge \frac{a}{a+b'+c'} \ge \frac{a}{a+b+c} \ge p, \tag{6.3}$$

thus $\Leftrightarrow_p$ is double implicational.

   Let be $0 < p < 1$, then there is an integer $u > 0$ such that $\frac{1}{1+u} < p$, which means both $F_{\Leftrightarrow_p}(1,0,u,0) = 0$ and $F_{\Leftrightarrow_p}(1,0,0,0) = 1$. We can conclude that $\Leftrightarrow_p$ is not implicational. This finishes the proof. □

Note 6.2. It is also easy to prove that the value $F_{\Leftrightarrow^*}(a,b,c,d)$ of associated function of double implicational quantifier $\Leftrightarrow^*$ does not depend on d, see lemma 6.6. Thus we often write only $F_{\Leftrightarrow^*}(a,b,c)$ instead of $F_{\Leftrightarrow^*}(a,b,c,d)$ for the double implicational quantifier $\Leftrightarrow^*$.

Lemma 6.6. *Let $\Leftrightarrow^*$ be a double implicational 4ft-quantifier. Then the value $F_{\Leftrightarrow^*}(a,b,c,d)$ of its associated function does not depend on d.*

Proof. We have to prove that if $\langle a,b,c,d \rangle$ and $\langle a,b,c,d' \rangle$ are 4ft-tables then $F_{\Leftrightarrow^*}(a,b,c,d) = F_{\Leftrightarrow^*}(a,b,c,d')$.

   Let $F_{\Leftrightarrow^*}(a,b,c,d) = 1$. Then we have $TPC_{\Leftrightarrow}(a,b,c,d,a,b,c,d') = 1$ because of $a \ge a \wedge b \le b \wedge c \le c$ and thus $F_{\Leftrightarrow^*}(a,b,c,d') = 1$

   Let $F_{\Leftrightarrow^*}(a,b,c,d') = 1$. Again, we have $TPC_{\Leftrightarrow}(a,b,c,d',a,b,c,d) = 1$ and it implies $F_{\Leftrightarrow^*}(a,b,c,d) = 1$. This finishes the proof. □

6.4 Equivalence Quantifiers

The class of equivalence quantifiers is defined in [18] under the name *associational quantifiers*. The definition can be formulated using a truth preservation condition for equivalence quantifiers:

Definition 6.4. The *truth preservation condition $TPC_{\equiv}$ for equivalence quantifiers* is a $\{0,1\}$-valued function $TPC_{\equiv}(a,b,c,d,a',b',c',d')$ defined for all 8-tuples $\langle a,b,c,d,a',b',c',d' \rangle$ of non-negative integer numbers such that

$$TPC_{\equiv}(a,b,c,d,a',b',c',d') = \begin{cases} 1 & \text{if } a' \ge a \wedge b' \le b \wedge c' \le c \wedge d' \ge d \\ 0 & \text{otherwise.} \end{cases}$$

The *class of equivalence quantifiers* is the class $4ft[TPC_{\equiv}]$ defined by truth preservation condition $TPC_{\equiv}$. We say that 4ft-quantifier $\approx$ is *equivalence* if it belongs to $4ft[TPC_{\equiv}]$.

The condition $a' \geq a \wedge b' \leq b \wedge c' \leq c \wedge d' \geq d$ is called a *simple form of* $TPC_{\equiv}$.

Definition 6.4 can also be formulated such that 4ft-quantifier $\approx$ is an equivalence 4ft-quantifier if

$$F_{\approx}(a,b,c,d) = 1 \wedge a' \geq a \wedge b' \leq b \wedge c' \leq c \wedge d' \geq d \text{ implies } F_{\approx}(a',b',c',d') = 1$$

for all 4ft-tables $\langle a,b,c,d \rangle$ and $\langle a',b',c',d' \rangle$.

The condition $a' \geq a \wedge b' \leq b \wedge c' \leq c \wedge d' \geq d$ means that the 4ft-table $\langle a',b',c',d' \rangle$ is "better with respect to association" than the 4ft-table $\langle a,b,c,d \rangle$ (a-better, see [18]). If $\langle a,b,c,d \rangle$ is the 4ft-table of φ and ψ in data matrix $\mathcal{M}$ and if $\langle a',b',c',d' \rangle$ is the 4ft-table of φ and ψ in data matrix $\mathcal{M}'$, then the sentence *"better with respect to association"* means that in data matrix $\mathcal{M}'$ there are more rows in which φ and ψ have the same value and less rows in which φ and ψ have different values, i.e. they are more associated than in data matrix $\mathcal{M}$. Having the same value means that both are true or both are false.

Thus if $a' \geq a \wedge b' \leq b \wedge c' \leq c \wedge d' \geq d$ then it is reasonable to expect that if the associational rule $\varphi \approx \psi$ (i.e. the rule expressing association by $\approx$) is true in the data matrix $\mathcal{M}$ then it is also true in data matrix $\mathcal{M}'$ which is better with respect to association. This expectation is ensured for associational quantifiers by definition 6.4. Please note, that the condition $a' \geq a \wedge b' \leq b \wedge c' \leq c \wedge d' \geq d$ can be understood also as expressing the assertion *"better with respect to equivalence"*.

The name of the class was changed from "associational" to "equivalence" to avoid confusion. The notion "association rule" is used in this book for the general relation of two Boolean attributes $\varphi \approx \psi$. There are 4ft-quantifiers expressing important associations between two Boolean attributes which are not equivalence 4ft-quantifiers (i.e. not associational 4ft-quantifiers), an example being the 4ft-quantifier of above average $\sim_q^+$, see theorem 6.3.

Using notion "associational" instead of "equivalence" thus leads to having important association rules which are not associational which is confusing. In addition, the well known 4ft-quantifier $\rightarrow_{p,s}$ with confidence p and support s used for analysis of market basket and introduced in [1] together with notion of association rule, also is not an equivalence 4ft-quantifier (i.e. not an associational 4ft-quantifier), see theorem 6.2.

The approach to express equivalence of Boolean attributes φ and ψ by an analogy to the equivalence of propositions u and v expressed by the implications $u \rightarrow v$ and $v \rightarrow u$ leads to the definition of double implicational quantifiers, see Sect. 6.3.

A similar analogy can be used for the equivalence quantifiers. If u and v are propositions and both $u \rightarrow v$ and $\neg u \rightarrow \neg v$ are true, then u is equivalent to v. Thus we can try to express the relation of equivalence of the attributes φ and ψ using an "equivalence" 4ft-quantifier $\equiv^*$ such that

$$\varphi \equiv^* \psi \text{ if and only if both } \varphi \Rightarrow^* \psi \text{ and } \neg\varphi \Rightarrow^* \neg\psi. \tag{6.4}$$

If we apply the truth preservation condition for implicational quantifiers to rule $\varphi \Rightarrow^* \psi$ we obtain $a' \geq a \wedge b' \leq b$. If we apply it to rule $\neg\varphi \Rightarrow^* \neg\psi$, we obtain $d' \geq d \wedge c' \leq c$, ($c$ is here instead of b and d is instead of a, see see Fig. 3.3). This leads to the definition of the equivalence quantifiers.

Again, it is easy to prove the following lemma on relation of double imlicational and equivalence quantifiers.

Lemma 6.7. *Each double implicational quantifier is also an equivalence quantifier.*

Proof. Note that $a' \geq a \wedge b' \leq b \wedge c' \leq c \wedge d' \geq d$ implies $a' \geq a \wedge b' \leq b \wedge c' \leq c$. Thus $TPC_{\equiv}(a,b,c,d,a',b',c',d') = 1$ implies $TPC_{\Leftrightarrow}(a,b,c,d,a',b',c',d') = 1$ and according to lemma 6.1 each double imlicational quantifier is also equivalence quantifier. This finishes the proof. □

The 4ft-quantifier $\equiv_p$ of p-equivalence (i.e. accuracy or success rate, see Table 4.5), is an example of equivalence quantifiers, see the following lemma. It is also proved that the 4ft-quantifier $\equiv_p$ is not double implicational. Additional examples of equivalence quantifiers are located in Chap. 9.

Lemma 6.8. *4ft-quantifier of p-equivalence $\equiv_p$, see rows 7 in Tables 4.2, 4.4, 4.5, and 4.6, is an equivalence 4ft-quantifier for $0 < p \leq 1$. However, the 4ft-quantifier $\equiv_p$ is not double implicational for $0 < p < 1$.*

Proof. It holds $F_{\equiv_p}(a,b,c,d) = 1$ if and only if $\frac{a+d}{a+b+c+d} \geq p$. If $F_{\Leftrightarrow_p}(a,b,c,d) = 1$ and $a' \geq a \wedge b' \leq b \wedge c' \leq c \wedge d' \geq d$, then we have

$$\frac{a'+d'}{a'+b'+c'+d'} \geq \frac{a+d}{a+b'+c'+d} \geq \frac{a+d}{a+b+c+d} \geq p, \tag{6.5}$$

thus $\equiv_p$ is an equivalence 4ft-quantifier.

Let be $0 < p < 1$, then there is an integer $u > 0$ such that $\frac{1}{1+u} < p$, which means both $F_{\equiv_p}(1,0,0,u) = 0$ and $F_{\equiv_p}(1,0,0,0) = 1$. We can conclude that $\equiv_p$ is not double implicational. This finishes the proof. □

6.5 Symmetrical and a-d Symmetrical Quantifiers

Symmetrical 4ft-quantifiers are defined in [18, 24]. The idea is that a 4ft-quantifier $\approx$ is symmetrical if it holds $Val(\varphi \approx \psi, \mathcal{M}) = Val(\psi \approx \varphi, \mathcal{M})$ for each data matrix $\mathcal{M}$ and all Boolean attributes φ, ψ. Definition of symmetrical quantifiers uses the fact that if $4ft(\varphi, \psi, \mathcal{M}) = \langle a,b,c,d \rangle$, then $4ft(\psi, \varphi, \mathcal{M}) = \langle a,c,b,d \rangle$, see Fig. 6.1.

It holds $Val(\varphi \approx \psi, \mathcal{M}) = F_{\approx}(a,b,c,d)$ and $Val(\psi \approx \varphi, \mathcal{M}) = F_{\approx}(a,c,b,d)$ where $F_{\approx}(a,b,c,d)$ is the associated function of 4ft-quantifier $\approx$. Thus a requirement $Val(\varphi \approx \psi, \mathcal{M}) = Val(\psi \approx \varphi, \mathcal{M})$ means $F_{\approx}(a,b,c,d) = F_{\approx}(a,c,b,d)$. This consideration leads to the following definition.

$\mathscr{M}$	ψ	$\neg\psi$
φ	a	b
$\neg\varphi$	c	d

$\mathscr{M}$	φ	$\neg\varphi$
ψ	a	c
$\neg\psi$	b	d

Fig. 6.1 4ft-tables 4ft($\varphi,\psi,\mathscr{M}$) and 4ft($\psi,\varphi,\mathscr{M}$)

Definition 6.5. A *4ft-quantifier* $\approx$ *is symmetrical if its associated function* $F_{\approx}(a,b,c,d)$ satisfies

$$F_{\approx}(a,b,c,d) = F_{\approx}(a,c,b,d).$$

The symmetrical quantifiers can be also defined using suitable truth preservation condition.

Definition 6.6. The *truth preservation condition* TPC_{SYM} *for symmetrical quantifiers* is a $\{0,1\}$-valued function $TPC_{SYM}(a,b,c,d,a',b',c',d')$ defined for all 8-tuples $\langle a,b,c,d,a',b',c',d'\rangle$ of non-negative integer numbers such that

$$TPC_{SYM}(a,b,c,d,a',b',c',d') = \begin{cases} 1 & \text{if } a'=a \wedge b'=c \wedge c'=b \wedge d'=d \\ 0 & \text{otherwise.} \end{cases}$$

The *class of symmetrical quantifiers* is the class $4ft[TPC_{SYM}]$ defined by truth preservation condition TPC_{SYM}. We say that a 4ft-quantifier $\approx$ is *symmetrical* if it belongs to $4ft[TPC_{SYM}]$.

We have to prove that definitions 6.5 and 6.6 are equivalent. This is done in the following lemma.

Lemma 6.9. *The 4ft-quantifier* $\approx$ *is symmetrical according to definition 6.5 if and only if* $\approx$ *belongs to the class* $4ft[TPC_{SYM}]$ *given by definition 6.6.*

Proof

1. *Let 4ft-quantifier* $\approx$ *be symmetrical according to definition 6.5. We have to show that* $\approx$ *belongs to* $4ft[TPC_{SYM}]$ *i.e. that*

if $F_{\approx}(a,b,c,d)=1 \wedge TPC_{SYM}(a,b,c,d,a',b',c',d')=1$ *then* $F_{\approx}(a',b',c',d')=1$.

According to definition 6.6 the assumption $TPC_{SYM}(a,b,c,d,a',b',c',d') = 1$ *means* $a'=a \wedge b'=c \wedge c'=b \wedge d'=d$. *Thus* $F_{\approx}(a',b',c',d') = F_{\approx}(a,c,b,d)$. *The 4ft-quantifier* $\approx$ *is symmetrical according to definition 6.5, thus* $F_{\approx}(a,c,b,d) = F_{\approx}(a,b,c,d)$. *We assume* $F_{\approx}(a,b,c,d) = 1$ *which means that* $F_{\approx}(a',b',c',d') = 1$.

2. *Let* $\approx$ *belongs to* $4ft[TPC_{SYM}]$, *i.e.*

if $F_{\approx}(a,b,c,d)=1 \wedge TPC_{SYM}(a,b,c,d,a',b',c',d')=1$ *then* $F_{\approx}(a',b',c',d')=1$.

We have to show that $\approx$ *is symmetrical according to definition 6.5 i.e. that* $F_{\approx}(a,b,c,d) = F_{\approx}(a,c,b,d)$.

Let $F_{\approx}(a,b,c,d) = 1$, we show $F_{\approx}(a,c,b,d) = 1$. Let $\langle a',b',c',d'\rangle = \langle a,c,b,d\rangle$, then

$$\langle a,b,c,d,a',b',c',d'\rangle = \langle a,b,c,d,a,c,b,d\rangle$$

and thus, according to definition 6.6 we have

$$TPC_{SYM}(a,b,c,d,a',b',c',d') = TPC_{SYM}(a,b,c,d,a,c,b,d) = 1 .$$

This means that $F_{\approx}(a,b,c,d) = 1$ implies $F_{\approx}(a,c,b,d) = 1$.
Let $F_{\approx}(a,c,b,d) = 1$, then we can show that $F_{\approx}(a,b,c,d) = 1$ in the same way using the fact that $TPC_{SYM}(a,c,b,d,a,b,c,d) = 1$. This finishes the proof. □

In some situations it is useful to know that the 4ft-quantifier $\approx$ in question satisfies $F_{\approx}(a,b,c,d) = F_{\approx}(d,b,c,a)$. We call such quantifiers a-d symmetrical. We introduce them similarly to symmetrical quantifiers.

Definition 6.7. A *4ft-quantifier $\approx$ is a-d symmetrical* if its associated function $F_{\approx}(a,b,c,d)$ satisfies

$$F_{\approx}(a,b,c,d) \;=\; F_{\approx}(d,b,c,a) .$$

Definition 6.8. The *truth preservation condition TPC_{adSYM} for a-d symmetrical quantifiers* is a $\{0,1\}$-valued function $TPC_{adSYM}(a,b,c,d,a',b',c',d')$ defined for all 8-tuples $\langle a,b,c,d,a',b',c',d'\rangle$ of non-negative integer numbers such that

$$TPC_{adSYM}(a,b,c,d,a',b',c',d') = \begin{cases} 1 & \text{if } a' = d \wedge b' = b \wedge c' = c \wedge d' = a \\ 0 & \text{otherwise.} \end{cases}$$

The *class of a-d symmetrical quantifiers* is the class $4ft[TPC_{adSYM}]$ defined by truth preservation condition TPC_{adSYM}. We say that a 4ft-quantifier $\approx$ is *a-d symmetrical* if it belongs to $4ft[TPC_{adSYM}]$.

Lemma 6.10. *The 4ft-quantifier $\approx$ is a-d symmetrical according to definition 6.7 if and only if $\approx$ belongs to the class $4ft[TPC_{SYM}]$ given by definition 6.8.*

Proof. The proof is similar to that of lemma 6.9. □

A simple example of a 4ft-quantifier which is both symmetrical and a-d symmetrical is 4ft-quantifier $\sim_q$ of simple deviation, see the following lemma.

Lemma 6.11. *4ft-quantifier $\sim_q$ of simple deviation, see rows 10 in Tables 4.2 and 4.4, is both symmetrical and a-d symmetrical.*

Proof. It holds $F_{\sim_q}(a,b,c,d) = 1$ if and only if $ad > e^q bc$ and the assertion of the lemma is obvious. □

6.6   4ft-quantifiers with F-property

The following definition of the class of 4ft-quantifiers with the F-property was inspired by considerations on dealing with missing information for Fisher's quantifier [60, 63].

Definition 6.9. The *4ft-quantifier* $\approx$ *has the F-property* if its associated function $F_{\approx}(a,b,c,d)$ satisfies

1. If $F_{\approx}(a,b,c,d) = 1$ and $b \geq c - 1 \geq 0$ then $F_{\approx}(a,b+1,c-1,d) = 1$.
2. If $F_{\approx}(a,b,c,d) = 1$ and $c \geq b - 1 \geq 0$ then $F_{\approx}(a,b-1,c+1,d) = 1$.

The quantifiers with the F-property can also be defined using a suitable truth preservation condition.

Definition 6.10. The *truth preservation condition* TPC_F *for quantifiers with the F-property* is a $\{0,1\}$-valued function $TPC_F(a,b,c,d,a',b',c',d')$ defined for all 8-tuples $\langle a,b,c,d,a',b',c',d' \rangle$ of non-negative integer numbers such that

$$TPC_F(a,b,c,d,a',b',c',d') = \begin{cases} 1 & \text{if } b \geq c - 1 \geq 0 \text{ and} \\ & a' = a \wedge b' = b+1 \wedge c' = c-1 \wedge d' = d \\ 1 & \text{if } c \geq b - 1 \geq 0 \text{ and} \\ & a' = a \wedge b' = b-1 \wedge c' = c+1 \wedge d' = d \\ 0 & \text{otherwise.} \end{cases}$$

The *class of quantifiers with the F-property* is the class $4ft[TPC_F]$ defined by truth preservation condition TPC_F. We say that *4ft-quantifier* $\approx$ *has the F-property* if it belongs to $4ft[TPC_F]$.

We have to prove that definitions 6.9 and 6.10 are equivalent. This is done in the following lemma.

Lemma 6.12. *The 4ft-quantifier $\approx$ has the F-property according to definition 6.9 if and only if $\approx$ belongs to the class $4ft[TPC_F]$ given by definition 6.10.*

Proof

1. *Let 4ft-quantifier $\approx$ have the F-property according to definition 6.9. We have to show that $\approx$ belongs to $4ft[TPC_F]$ i.e. that*

   if $F_{\approx}(a,b,c,d) = 1 \wedge TPC_F(a,b,c,d,a',b',c',d') = 1$ then $F_{\approx}(a',b',c',d') = 1$.

   The assumption $F_{\approx}(a,b,c,d) = 1 \wedge TPC_F(a,b,c,d,a',b',c',d') = 1$ means that I) or II) are satisfied:

   I) $b \geq c - 1 \geq 0$ and $a' = a \wedge b' = b+1 \wedge c' = c-1 \wedge d' = d$
   II) $c \geq b - 1 \geq 0$ and $a' = a \wedge b' = b-1 \wedge c' = c+1 \wedge d' = d$,

   see definition 6.10. If I) is satisfied then $F_{\approx}(a',b',c',d') = F_{\approx}(a,b+1,c-1,d)$ and if II) is satisfied then $F_{\approx}(a',b',c',d') = F_{\approx}(a,b-1,c+1,d)$. 4ft-quantifier $\approx$ has the F-property according to definition 6.9 and thus in both cases it holds $F_{\approx}(a',b',c',d') = 1$, see points 1. and 2. in definition 6.9.

2. *Let $\approx$ belong to $4ft[TPC_F]$, which means that*

   if $F_{\approx}(a,b,c,d) = 1 \wedge TPC_F(a,b,c,d,a',b',c',d') = 1$ then $F_{\approx}(a',b',c',d') = 1$.

We have to show that

a. if $F_{\approx}(a,b,c,d) = 1$ *and* $b \geq c - 1 \geq 0$ *then* $F_{\approx}(a,b+1,c-1,d) = 1$.
b. if $F_{\approx}(a,b,c,d) = 1$ *and* $c \geq b - 1 \geq 0$ *then* $F_{\approx}(a,b-1,c+1,d) = 1$,

see definition 6.9.
If $b \geq c - 1 \geq 0$ *then* $TPC_F(a,b,c,d,a,b+1,c-1,d) = 1$ *(see definition 6.10)*
and thus $F_{\approx}(a,b,c,d) = 1$ *and* $b \geq c - 1 \geq 0$ *means that* $F_{\approx}(a,b,c,d) = 1$ *and*
$TPC_F(a,b,c,d,a,b+1,c-1,d) = 1$ *and since* $\approx$ *belongs to* $4ft[TPC_F]$ *we have*
$F_{\approx}(a,b+1,c-1,d) = 1$.
It is the same in the case of $F_{\approx}(a,b,c,d) = 1$ *and* $c \geq b - 1 \geq 0$. *This finishes the*
proof. □

An example of a 4ft-quantifier with the F-property is the symmetrical 4ft-quantifier
$\sim_q$ of simple deviation, see lemma 6.13. There is also a 4ft-quantifier with the F-
property which is not symmetrical, see lemma 6.14.

Lemma 6.13. *The 4ft-quantifier* $\sim_q$ *of simple deviation, see row 10 in Tables 4.2*
and 4.4, has the the F-property.

Proof. It holds $F_{\sim_q}(a,b,c,d) = 1$ *if and only if* $ad > e^q bc$. *We have to prove*

a)   *If* $F_{\sim_q}(a,b,c,d) = 1$ *and* $b \geq c - 1 \geq 0$ *then* $F_{\sim_q}(a,b+1,c-1,d) = 1$.
b)   *If* $F_{\sim_q}(a,b,c,d) = 1$ *and* $c \geq b - 1 \geq 0$ *then* $F_{\sim_q}(a,b-1,c+1,d) = 1$.

It holds $F_{\sim_q}(a,b,c,d) = F_{\sim_q}(a,c,b,d)$ *and thus we prove only assertion a), assertion*
b) is a consequence of assertion a) because of the symmetry of $\sim_q$.
   It holds $(b+1)(c-1) = bc - (b+1-c)$. *We assume* $b \geq c - 1 \geq 0$ *which implies*
$b + 1 - c \geq 0$ *and thus* $bc \geq bc - (b+1-c) = (b+1)(c-1)$. *This means that*
if $F_{\sim_q}(a,b,c,d) = 1$ *and* $b \geq c - 1 \geq 0$ *then also* $F_{\sim_q}(a,b+1,c-1,d) = 1$. *This*
finishes the proof. □

Lemma 6.14. *There is a 4ft-quantifier with the F-property which is not symmetrical.*

Proof. Let us define 4ft-quantifier $\approx_{FNS}$ *with associated function* $F_{\approx_{FNS}}$ *given by*
Table 6.1. 4ft-quantifier $\approx_{FNS}$ *has the F-property, see a) and b):*

Table 6.1 4ft-quantifier $\approx_{FNS}$

| | definition of $F_{\approx_{FNS}}$ | | | | conditions satisfied | |
| | $\langle a,b,c,d \rangle$ | | | | yes or no | |
row	a	b	c	d	$F_{\approx_{FNS}}(a,b,c,d)$	$b \geq c - 1 \geq 0$	$c \geq b - 1 \geq 0$
1	any	0	0	any	1	no	no
2	any	1	0	any	1	no	yes
3	any	0	1	any	1	yes	no
4	any	2	0	any	1	no	yes
5	other				0	irrelevant to the proof	

a)   Condition $F_{\approx_{FNS}}(a,b,c,d) = 1$ and $b \geq c - 1 \geq 0$ (see point 1 in definition 6.9) is satisfied only for row 3 in Table 6.1 and $F_{\approx_{FNS}}(a, b+1, c-1, d) = 1$ is satisfied, see row 2.

b)   Condition $F_{\approx_{FNS}}(a,b,c,d) = 1$ and $c \geq b - 1 \geq 0$ (see point 2 in definition 6.9) is satisfied only for row 2 in Table 6.1 and $F_{\approx_{FNS}}(a, b-1, c+1, d) = 1$ is satisfied, see row 3.

4ft-quantifier $\approx_{FNS}$ is not symmetrical because $F_{\approx_{FNS}}(0,2,0,0) = 1$ and $F_{\approx_{FNS}}(0,0,2,0) = 0$, see rows 4 and 5. This finishes the proof.   □

6.7 $\mathscr{M}$-dependent and $\mathscr{M}$-independent Classes

If $TPC_{SYM}(a,b,c,d,a',b',c',d') = 1$ then $a' = a \wedge b' = c \wedge c' = b \wedge d' = d$ and thus $a+b+c+d = a'+b'+c'+d'$, see definition 6.6. According to definition 6.10, if $TPC_F(a,b,c,d,a',b',c',d') = 1$ then $a' = a \wedge b' = b+1 \wedge c' = c-1 \wedge d' = d$ or $a' = a \wedge b' = b-1 \wedge c' = c+1 \wedge d' = d$. In both cases it is again true that $a+b+c+d = a'+b'+c'+d'$.

However, this is not true for truth preservation conditions $TPC_{\Rightarrow}$, $TPC_{\Leftrightarrow}$, and $TPC_{\equiv}$. Specifically:

- $TPC_{\Rightarrow}(10,10,0,0,99,9,0,0) = 1$ and $10+10+0+0 \neq 99+9+0+0$
- $TPC_{\Leftrightarrow}(10,10,10,0,99,9,9,0) = 1$ and $10+10+10+0 \neq 99+9+9+0$
- $TPC_{\equiv}(10,10,10,10,99,9,9,99) = 1$ and $10+10+10+10 \neq 99+9+9+99$.

$a+b+c+d = a'+b'+c'+d'$ could be interpreted such that the 4ft-tables $\langle a,b,c,d \rangle$ and $\langle a',b',c',d' \rangle$ concern one data matrix $\mathscr{M}$ (or at least that $\langle a,b,c,d \rangle$ and $\langle a',b',c',d' \rangle$ concern two data matrices $\mathscr{M}$ and $\mathscr{M}'$ with the same number of rows). The first interpretation can be seen such that the TPC depends on one data matrix, this is important when dealing with deduction rules, see Chap. 11. This leads to the definition of $\mathscr{M}$-dependent and $\mathscr{M}$-independent TPC's.

Definition 6.11. We define:

1. A *truth preservation condition* $C(a,b,c,d,a',b',c',d')$ *is $\mathscr{M}$-dependent* if it satisfies

$$C(a,b,c,d,a',b',c',d') = 1 \text{ implies } a+b+c+d = a'+b'+c'+d' .$$

2. A *class of 4ft-quantifiers* $4ft[C]$ is $\mathscr{M}$-dependent if truth preservation condition $C(a,b,c,d,a',b',c',d')$ is $\mathscr{M}$-dependent.
3. A *4ft-quantifier* $\approx$ is $\mathscr{M}$-dependent if it belongs to an $\mathscr{M}$-dependent class of 4ft-quantifiers.
4. A *truth preservation condition* C is $\mathscr{M}$-independent if there are 4ft-tables $\langle a,b,c,d \rangle$ and $\langle a',b',c',d' \rangle$ such that

$$C(a,b,c,d,a',b',c',d') = 1 \text{ and } a+b+c+d \neq a'+b'+c'+d' .$$

5. A *class of 4ft-quantifiers* $4ft[C]$ *is $\mathcal{M}$-independent* if truth preservation condition
   $C(a,b,c,d,a',b',c',d')$ is $\mathcal{M}$-independent.
6. A *4ft-quantifier $\approx$ is $\mathcal{M}$-independent* if it belongs to an $\mathcal{M}$-independent class of
   4ft-quantifiers.

There are old results on $\mathcal{M}$-independent classes derived from $\mathcal{M}$-independent truth
preservation conditions $TPC_{\Rightarrow}$, $TPC_{\Leftrightarrow}$, and $TPC_{\equiv}$ [63, 67, 68] and new emerging
results for classes derived from $TPC_{\Rightarrow}^{\mathcal{M}}$, $TPC_{\Leftrightarrow}^{\mathcal{M}}$, and $TPC_{\equiv}^{\mathcal{M}}$. Here $TPC_{\Rightarrow}^{\mathcal{M}}$ is a truth
preservation condition defined by adding $a+b+c+d = a'+b'+c'+d'$ to $TPC_{\Rightarrow}$
and analogously for $TPC_{\Leftrightarrow}^{\mathcal{M}}$ and $TPC_{\equiv}^{\mathcal{M}}$, see [74]. That's why we define an $\mathcal{M}$-
dependent class of quantifiers derived from an $\mathcal{M}$-independent class of quantifiers.

Definition 6.12. Let $C(a,b,c,d,a',b',c',d')$ be an $\mathcal{M}$-independent truth preservation condition. Then $\{0,1\}$-valued function $C^{\mathcal{M}}(a,b,c,d,a',b',c',d')$ defined such
that

$$C^{\mathcal{M}}(a,b,c,d,a',b',c',d') = \begin{cases} 1 & \text{if } C(a,b,c,d,a',b',c',d') = 1 \text{ and} \\ & a+b+c+d = a'+b'+c'+d' \\ 0 & \text{otherwise.} \end{cases}$$

is called an *$\mathcal{M}$-dependent truth preservation condition derived from $\mathcal{M}$ - independent truth preservation condition* $C(a,b,c,d,a',b',c',d')$.

The class $4ft[C^{\mathcal{M}}]$ of 4ft-quantifiers is an *$\mathcal{M}$-dependent class of 4ft-quantifiers
derived from $\mathcal{M}$-independent class* $4ft[C]$ *of 4ft-quantifiers*.

There is an important lemma:

Lemma 6.15. *Let $4ft[C^{\mathcal{M}}]$ be $\mathcal{M}$-dependent class of 4ft-quantifiers derived from
$\mathcal{M}$-independent class $4ft[C]$. Then, if 4ft-quantifier $\approx$ belongs to class $4ft[C]$, it
also belongs to class $4ft[C^{\mathcal{M}}]$.*

*Proof. It follows from definition 6.12 that if $C^{\mathcal{M}}(a,b,c,d,a',b',c',d') = 1$ then
$C(a,b,c,d,a',b',c',d') = 1$. Thus, according to lemma 6.1 if $\approx$ belongs to $4ft[C]$
then $\approx$ belongs to $4ft[C^{\mathcal{M}}]$. This finishes the proof.* □

We will define and study $\mathcal{M}$-dependent classes $TPC_{\Rightarrow}^{\mathcal{M}}$, $TPC_{\Leftrightarrow}^{\mathcal{M}}$, and $TPC_{\equiv}^{\mathcal{M}}$ of 4ft-
quantifiers in Chaps. 7 – 9.

6.8 Relations among Classes

We have introduced various properties of classes of 4ft-quantifiers. These are summarized in the following theorem.

Theorem 6.1

1. There are implicational 4ft-quantifiers.
2. Each implicational 4ft-quantifier is also double implicational.
3. There are double implicational 4ft-quantifiers which are not implicatinal.

4. Each double implicational 4ft-quantifier is an equivalence 4ft-quantifier.
5. There are equivalence 4ft-quantifiers which are not double implicational.
6. There are symmetrical and a-d symmetrical 4ft-quantifiers.
7. There are symmetrical 4ft-quantifiers with the F-property.
8. There is a 4ft-quantifier with the F-property which is not symmetrical.

Proof

1. See lemma 6.2.
2. See lemma 6.4.
3. See lemma 6.5.
4. See lemma 6.7.
5. See lemma 6.8.
6. See lemma 6.11.
7. See lemma 6.11 and lemma 6.13.
8. See lemma 6.14.

This finishes the proof.   □

Additional examples of 4ft-quantifiers satisfying particular relations listed in theorem 6.1 are introduced in Chaps. 7, 8, and 9.

Various subclasses of the class of double implicational rules and of the class of equivalence rules are defined in Chaps. 8 and 9. Relations among these subclasses are summarized in Sects. 8.9 and 9.9.

There are important 4ft-quantifiers which belong to $4ft[TPC_{\Rightarrow}^{\mathscr{M}}]$ but do not belong to $4ft[TPC_{\Rightarrow}]$, important 4ft-quantifiers which belong to $4ft[TPC_{\Leftrightarrow}^{\mathscr{M}}]$ but do not belong to $4ft[TPC_{\Leftrightarrow}]$, and important 4ft-quantifiers which belong to $4ft[TPC_{\equiv}^{\mathscr{M}}]$ but do not belong to $4ft[TPC_{\equiv}]$, see Sects. 7.3, 8.3, and 9.3 respectively.

We prove that the MB-quantifier $\rightarrow_{p,s}$ with confidence p and support s used for analysis of market basket and introduced in [1] together with the notion of association rules is not an equivalence 4ft-quantifier [67].

Theorem 6.2. *The MB-quantifier $\rightarrow_{p,s}$ defined for $0 < p \leq 1$ and $0 < s \leq 1$ such that*

$$F_{\rightarrow_{p,s}}(a,b,c,d) = \begin{cases} 1 & \text{if } \frac{a}{a+b} \geq p \wedge \frac{a}{a+b+c+d} \geq s \\ 0 & \text{otherwise.} \end{cases}$$

is not an equivalence 4ft-quantifier.

Proof. We have to show that for each $0 < p \leq 1$ and $0 < s \leq 1$ there are 4ft-tables $\langle a,b,c,d \rangle$, $\langle a',b',c',d' \rangle$ such that $a' \geq a \wedge b' \leq b \wedge c' \leq c \wedge d' \geq d$, $F_{\rightarrow_{p,s}}(a,b,c,d) = 1$ and $F_{\rightarrow_{p,s}}(a',b',c',d') = 0$.

It holds $F_{\rightarrow_{p,s}}(1,0,0,0) = 1$. For sure there is an integer u such that $\frac{1}{1+u} < s$ and thus $F_{\rightarrow_{p,s}}(1,0,0,u) = 0$. This finishes the proof.   □

Note 6.3. Please note that the MB-quantifier $\rightarrow_{p,s}$ can be also seen as a supported p-implication, see theorem 7.4.

Finally we prove that the important 4ft-quantifier $\sim_q^+$ of above average is not an equivalence 4ft-quantifier, see also [5]. Let us note that it is proved in theorem 10.2 that this quantifier is symmetrical and has the F-property.

Theorem 6.3. *The above average quantifier $\sim_q^+$, see row 13 in Tables 4.2 and 4.4, is not an equivalence 4ft-quantifier.*

Proof. The associated function $F_{\sim_q^+}$ of above average quantifier $\sim_q^+$ is for $q > 0$ defined such that

$$F_{\sim_q^+}(a,b,c,d) = \begin{cases} 1 & \text{if } \frac{a}{a+b} \geq (1+q)\frac{a+c}{a+b+c+d} \\ 0 & \text{otherwise.} \end{cases}$$

see Table 4.4.

We have to show that there are 4ft-tables $\langle a,b,c,d \rangle$, $\langle a',b',c',d' \rangle$ such that $a' \geq a \wedge b' \leq b \wedge c' \leq c \wedge d' \geq d$, $F_{\sim_q^+}(a,b,c,d) = 1$ and $F_{\sim_q^+}(a',b',c',d') = 0$.

Let $q > 0$ be a real number and $u > 0$ and $v > 0$ be integer numbers, then there is an integer $w > 0$ such that $w \geq q(u+v)$. Let us define $\langle a,b,c,d \rangle = \langle u,v,0,w \rangle$ and then the following inequalities are equivalent

$$\frac{a}{a+b} \geq (1+q)\frac{a}{a+b+c+d}$$

$$\frac{u}{u+v} \geq (1+q)\frac{u}{u+v+w}$$

$$u+v+w \geq (1+q)(u+v)$$

$$w \geq q(u+v)$$

which means $F_{\sim_q^+}(a,b,c,d) = 1$.

For sure there is an integer u' such that $u' \geq u$ and $w < q(u'+v)$. Let us define $\langle a',b',c',d' \rangle = \langle u',v,0,w \rangle$ and then the following inequalities are equivalent

$$\frac{a'}{a'+b} < (1+q)\frac{a'}{a'+b+c+d}$$

$$\frac{u'}{u'+v} < (1+q)\frac{u'}{u'+v+w}$$

$$u'+v+w < (1+q)(u'+v)$$

$$w < q(u'+v)$$

which means $F_{\sim_q^+}(a',b,c,d) = 0$. We have also $a' \geq a \wedge b' \leq b \wedge c' \leq c \wedge d' \geq d$ and this finishes the proof. $\square$

Chapter 7
Implicational Rules

The 4ft-quantifier $\Rightarrow_p$ of p-implication is implicational and it is easy to prove that it is $\mathscr{M}$-independent. However, it can be also easily proved that MB-quantifier $\rightarrow_{p,s}$ with confidence p and support s which is used for analysis of market basket [1] and shortly mentioned in Sect. 1.1 is neither $\mathscr{M}$-independent nor implicational. This leads to notions of weakly implicational quantifiers and rules, see Sect. 7.1. Several practically important implicational and weakly implicational quantifiers and rules are then introduced in Sects. 7.2 and 7.3 respectively.

Some of these 4ft-quantifiers are called interesting implicational quantifiers because they have important properties with respect to deduction rules. Interesting implicational quantifiers are introduced in Sect. 7.4. Tables of critical frequencies are useful tools for the study of 4ft-quantifiers. These are defined for most classes of 4ft-quantifiers and there are various types of such tables. There is, however, only one type of tables of critical frequencies defined for implicational quantifiers. We introduce it in Sect. 7.5.

7.1 Implicational and Weakly Implicational Quantifiers

Implicational quantifiers are defined in [18], see also definition 6.2 where the truth preservation condition $TPC_{\Rightarrow}$ is used. Truth preservation condition $TPC_{\Rightarrow}$ and the implicational quantifier are $\mathscr{M}$-independent, see following lemma.

Lemma 7.1

1. Truth preservation condition $TPC_{\Rightarrow}$ is $\mathscr{M}$-independent.
2. Implicational quantifier is $\mathscr{M}$-independent.

Proof

1. According to point 4) of definition 6.11 we have to prove that there are 4ft-tables $\langle a,b,c,d \rangle$ and $\langle a',b',c',d' \rangle$ such that

$$TPC_{\Rightarrow}(a,b,c,d,a',b',c',d') = 1 \text{ and } a+b+c+d \neq a'+b'+c'+d'.$$

J. Rauch: *Observational Calculi and Association Rules*, SCI 469, pp. 81–97.
DOI: 10.1007/978-3-642-11737-4_7 © Springer-Verlag Berlin Heidelberg 2013

4ft-tables $\langle 10,10,0,0 \rangle$ and $\langle 99,9,0,0 \rangle$ satisfy this condition, it holds

$$TPC_\Rightarrow(10,10,0,0,99,9,0,0) = 1 \text{ and } 10+10+0+0 \neq 99+9+0+0.$$

This finishes the proof of point 1.
2. It follows from just proved point 1) and point 6) of definition 6.11.

This finishes the proof.   □

Definition 7.1. Let $\mathscr{LC}$ be a logical calculus of association rules and let $\varphi \approx \psi$ be an association rule of $\mathscr{LC}$. If 4ft-quantifier $\approx$ is implicational then the rule $\varphi \approx \psi$ is an *implicational rule*.

We derive a new $\mathscr{M}$-dependent class of 4ft-quantifiers from the class of implicational 4ft-quantifiers in the way given by definition 6.12. We will call 4ft-quantifiers belonging to this class as weakly implicational 4ft-quantifiers. Consequently, we define weakly implicational rules.

Definition 7.2. The *truth preservation condition* $TPC_\Rightarrow^{\mathscr{M}}$ for weakly implicational quantifiers is a $\{0,1\}$-valued function $TPC_\Rightarrow^{\mathscr{M}}(a,b,c,d,a',b',c',d')$ defined such that

$$TPC_\Rightarrow^{\mathscr{M}}(a,b,c,d,a',b',c',d') = \begin{cases} 1 & \text{if } a' \geq a \wedge b' \leq b \text{ and} \\ & a+b+c+d = a'+b'+c'+d' \\ 0 & \text{otherwise.} \end{cases}$$

The class $4ft[TPC_\Rightarrow^{\mathscr{M}}]$ is a *class of weakly implicational quantifiers*. If 4ft-quantifier $\approx$ belongs to $4ft[TPC_\Rightarrow^{\mathscr{M}}]$ then it is a *weakly implicational quantifier*.

Definition 7.3. Let $\mathscr{LC}_{\mathscr{T}}$ be a logical calculus of association rules and let $\varphi \approx \psi$ be an association rule of $\mathscr{LC}_{\mathscr{T}}$. If 4ft-quantifier $\approx$ is weakly implicational then the rule $\varphi \approx \psi$ is a *weakly implicational rule*.

The following lemma says that each implicational quantifier is also a weakly implicational quantifier.

Lemma 7.2. *If $\Rightarrow^*$ is an implicational quantifier then $\Rightarrow^*$ is also a weakly implicational quantifier.*

Proof. The lemma is a direct consequence of lemma 6.15.   □

Note 7.1. The class of weakly implicational quantifiers is introduced in [74] under the name of a class of $\mathscr{M}$-dependent implicational quantifiers. We will use the name *weakly implicational quantifiers* instead of *$\mathscr{M}$-dependent implicational quantifiers* because the name *$\mathscr{M}$-dependent implicational quantifiers* can be cumbersome. An example follows from the fact that the implicational quantifier is $\mathscr{M}$-independent (see lemma 7.1) and that also each implicational quantifier is weak implicational (see lemma 7.2). Thus, if we use *$\mathscr{M}$-dependent implicational quantifiers* instead of *weakly implicational quantifiers*, we get:

- the 4ft-quantifier $\Rightarrow_p$ is implicational and $\mathscr{M}$-independent, see lemmas 6.2 and 7.1
- the 4ft-quantifier $\Rightarrow_p$ is $\mathscr{M}$-dependent implicational, see lemma 7.2.

7.2   Selected Implicational Quantifiers

Eight of the 4ft-quantifiers introduced in Tables 4.4 and 4.6 are implicational, see the following theorem.

Theorem 7.1. *The following 4ft-quantifiers are implicational:*

1. *p-implication $\Rightarrow_p$ (i.e. confidence or precision), see rows 1 in Tables 4.2 and 4.4 and rows 2 in Tables 4.5 and 4.6*
2. *Likely p-implication (i.e. lower critical implication) $\Rightarrow^!_{p,\alpha}$, see rows 2 in Tables 4.2 and 4.4*
3. *Suspicious critical implication (i.e. upper critical implication) $\Rightarrow^?_{p,\alpha}$, see rows 3 in Tables 4.2 and 4.4*
4. *Base $\oplus_{Base}$, see rows 18 in Tables 4.2 and 4.4*
5. *Laplace correction $\Rightarrow^L_p$, see rows 18 in Tables 4.5 and 4.6*
6. *Sebag - Schoenauer $\Rightarrow^S_q$, see rows 25 in Tables 4.5 and 4.6*
7. *Least contradiction $\Rightarrow^C_u$, see rows 26 in Tables 4.5 and 4.6*
8. *Example and counterexample rate $\Rightarrow^E_v$, see rows 28 in Tables 4.5 and 4.6.*

Proof. We have to prove for each quantifier $\approx$ from the above eight 4ft-quantifiers:

$$\text{if } F_\approx(a,b,c,d) = 1 \text{ and } a' \geq a \wedge b' \leq b \text{ then } F_\approx(a',b',c',d') = 1,$$

see definition 6.2. Here $F_\approx(a,b,c,d)$ is the associated function of the 4ft-quantifier $\approx$.

1. *See lemma 6.2.*
2. *It holds that $F_{\Rightarrow^!_{p,\alpha}}(a,b,c,d) = 1$ if and only if $\sum_{i=a}^{a+b} \binom{a+b}{i} p^i(1-p)^{a+b-i} \leq \alpha$ and $a+b > 0$. We have to prove that if $F_{\Rightarrow^!_{p,\alpha}}(a,b,c,d) = 1$ and $a' \geq a \wedge b' \leq b$ then*

$$\sum_{i=a'}^{d'+b'} \binom{a'+b'}{i} p^i(1-p)^{a'+b'-i} \leq \alpha.$$

This however directly follows from theorem 5.1 if we use $u = a$, $v = b$, $u' = a'$, and $v' = b'$.

3. *It holds that $F_{\Rightarrow^?_{p,\alpha}}(a,b,c,d) = 1$ if and only if $\sum_{i=0}^{a} \binom{a+b}{i} p^i(1-p)^{a+b-i} > \alpha$. We have to prove that if $F_{\Rightarrow^?_{p,\alpha}}(a,b,c,d) = 1$ and $a' \geq a \wedge b' \leq b$ then*

$$\sum_{i=0}^{d'} \binom{a'+b'}{i} p^i(1-p)^{a'+b'-i} > \alpha.$$

This however directly follows from theorem 5.2 if we use $u = a$, $v = b$, $u' = a'$, and $v' = b'$.

4. *It holds that $F_{\oplus_{Base}}(a,b,c,d) = 1$ if and only if $a \geq Base$. Thus, it is obvious that $F_{\oplus_{Base}}(a,b,c,d) = 1$ and $a' \geq a \wedge b' \leq b$ implies $F_{\oplus_{Base}}(a',b',c',d') = 1$.*

5. *It holds that $F_{\Rightarrow^L_p}(a,b,c,d) = 1$ if and only if $\frac{a+1}{a+b+2} \geq p$. If $F_{\Rightarrow^L_p}(a,b,c,d) = 1$ and $a' \geq a \wedge b' \leq b$ then we have*

$$\frac{a'+1}{a'+b'+2} \geq \frac{a+1}{a+b'+2} \geq \frac{a+1}{a+b+2} \geq p, \tag{7.1}$$

and this finishes the proof for $\Rightarrow_p^L$.

6. It holds that $F_{\Rightarrow_q^S}(a,b,c,d) = 1$ if and only if $a > 0 \wedge b = 0$ or $\frac{a}{b} \geq q$. If $F_{\Rightarrow_q^S}(a,b,c,d) = 1$ and $a' \geq a \wedge b' \leq b$ then it holds that $a > 0 \wedge b = 0$ or $\frac{a}{b} \geq q$ and we have:

- If $a > 0 \wedge b = 0$ then $a' \geq a \wedge b' \leq b$ implies $a' > 0 \wedge b' = 0$, thus $F_{\Rightarrow_q^S}(a',b',c,d) = 1$.
- If $\frac{a}{b} \geq q$ and $a' \geq a \wedge b' \leq b$ then $\frac{a}{b} \geq q$ means $a > 0$ (because of $q > 0$) and $b > 0$ and there are two possibilities:
  - If $b' = 0$ then $F_{\Rightarrow_q^S}(a',b',c,d) = 1$ because of $a' > a > 0 \wedge b' = 0$.
  - If $b' > 0$ then $F_{\Rightarrow_q^S}(a',b',c,d) = 1$ because of

$$\frac{a'}{b'} \geq \frac{a}{b'} \geq \frac{a}{b} \geq q. \tag{7.2}$$

This finishes the proof for $\Rightarrow_q^S$.

7. It holds that $F_{\Rightarrow_u^C}(a,b,c,d) = 1$ if and only if $\frac{a-b}{a+b} \geq u$ and $a + b > 0$. If $F_{\Rightarrow_u^C}(a,b,c,d) = 1$ and $a' \geq a \wedge b' \leq b$ then we have $a'b \geq ab'$, $a + b > 0$ and thus $a' + b' > 0$ and:

$$2a'b \geq 2ab'$$
$$a'b - ab' \geq ab' - a'b$$
$$a'a + a'b - b'a - b'b \geq aa' + ab' - ba' - bb'$$
$$a'(a+b) - b'(a+b) \geq a(a'+b') - b(a'+b')$$
$$(a'-b')(a+b) \geq (a-b)(a'+b')$$
$$\frac{a'-b'}{a'+b'} \geq \frac{a-b}{a+b} \geq u \tag{7.3}$$

and this finishes the proof for $\Rightarrow_u^C$.

8. It holds that $F_{\Rightarrow_v^E}(a,b,c,d) = 1$ if and only if $\frac{a-b}{a} \geq v$ and $a > 0$. If $F_{\Rightarrow_v^E}(a,b,c,d) = 1$ and $a' \geq a \wedge b' \leq b$ then we have $a > 0$ and $a'b \geq ab'$ which implies $a' > 0$ and $-ab' \geq -a'b$ and thus:

$$a'a - ab' \geq a'a - a'b$$
$$a(a'-b') \geq a'(a-b)$$
$$\frac{a'-b'}{a'} \geq \frac{a-b}{a} \geq u \tag{7.4}$$

and this finishes the proof for $\Rightarrow_v^E$.

This finishes the proof.   □

The practically important quantifier $\Rightarrow_{p,Base}$ of *founded p-implication* is mentioned in Sects. 1.1 and 4.4. Its associated function $F_{\Rightarrow_{p,Base}}$ is defined such that

$$F_{\Rightarrow_{p,Base}}(a,b,c,d) = \begin{cases} 1 & \text{if } \frac{a}{a+b} \geq p \wedge a \geq Base \\ 0 & \text{otherwise.} \end{cases}$$

This quantifier is a composition $\Rightarrow_p \wedge \oplus_{Base}$ of 4ft-quantifiers $\Rightarrow_p$ and $\oplus_{Base}$ according to definition 4.3. We have just proved that both $\Rightarrow_p$ and $\oplus_{Base}$ are implicational. It is a natural question whether the compound 4ft-quantifier $\Rightarrow_p \wedge \oplus_{Base}$ is also implicational. The answer is yes and it is a consequence of the following simple theorem.

Theorem 7.2. *Let $\mathscr{LC}$ be a logical calculus of association rules with 4ft-quantifiers $\approx_1, \ldots \approx_Q$ with associated functions $F_{\approx_1}, \ldots, F_{\approx_Q}$ respectively. Then the compound 4ft-quantifier $\approx_{i_1} \wedge \ldots \wedge \approx_{i_k}$ is implicational if and only if all 4ft-quantifiers $\approx_{i_1}, \ldots, \approx_{i_k}$ are implicational.*

Proof. The associated function $F_{\approx_{i_1} \wedge \ldots \wedge \approx_{i_k}}$ of $\approx_{i_1} \wedge \ldots \wedge \approx_{i_k}$ is defined as a product:

$$F_{\approx_{i_1} \wedge \ldots \wedge \approx_{i_k}}(a,b,c,d) = F_{\approx_{i_1}}(a,b,c,d) \times \ldots \times F_{\approx_{i_k}}(a,b,c,d),$$

see definition 4.3. We have to prove:

1. If all $\approx_{i_1}, \ldots \approx_{i_k}$ are implicational then $\approx_{i_1} \wedge \ldots \wedge \approx_{i_k}$ is implicational
2. If $\approx_{i_1} \wedge \ldots \wedge \approx_{i_k}$ is implicational then all $\approx_{i_1}, \ldots \approx_{i_k}$ are implicational.

ad 1.: We have to prove that if $F_{\approx_{i_1} \wedge \ldots \wedge \approx_{i_k}}(a,b,c,d) = 1$ and $a' \geq a \wedge b' \leq b$ then $F_{\approx_{i_1} \wedge \ldots \wedge \approx_{i_k}}(a',b',c',d') = 1$. The assumption $F_{\approx_{i_1} \wedge \ldots \wedge \approx_{i_k}}(a,b,c,d) = 1$ means $F_{\approx_{i_1}}(a,b,c,d) \times \ldots \times F_{\approx_{i_k}}(a,b,c,d) = 1$ which implies $F_{\approx_{i_1}}(a,b,c,d) = 1$, \ldots, $F_{\approx_{i_k}}(a,b,c,d) = 1$. We assume that all $\approx_{i_1}, \ldots, \approx_{i_k}$ are implicational and this means that also $F_{\approx_{i_1}}(a',b',c',d') = 1$, \ldots, $F_{\approx_{i_k}}(a',b',c',d') = 1$ which implies

$$F_{\approx_{i_1} \wedge \ldots \wedge \approx_{i_k}}(a',b',c',d') = F_{\approx_{i_1}}(a',b',c',d') \times \ldots \times F_{\approx_{i_k}}(a',b',c',d') = 1.$$

This finishes the proof of point 1.

ad 2.: We have to prove that if $F_{\approx_{i_1}}(a,b,c,d) = 1$, \ldots, $F_{\approx_{i_k}}(a,b,c,d) = 1$ and $a' \geq a \wedge b' \leq b$ then $F_{\approx_{i_1}}(a',b',c',d') = 1$, \ldots, $F_{\approx_{i_k}}(a',b',c',d') = 1$. The assumption $F_{\approx_{i_1}}(a,b,c,d) = 1$, \ldots, $F_{\approx_{i_k}}(a,b,c,d) = 1$ implies

$$F_{\approx_{i_1}}(a,b,c,d) \times \ldots \times F_{\approx_{i_k}}(a,b,c,d) = F_{\approx_{i_1} \wedge \ldots \wedge \approx_{i_k}}(a,b,c,d) = 1.$$

We assume that $\approx_{i_1} \wedge \ldots \wedge \approx_{i_k}$ is implicational and $a' \geq a \wedge b' \leq b$, thus $F_{\approx_{i_1} \wedge \ldots \wedge \approx_{i_k}}(a',b',c',d') = 1$. This means that

$$F_{\approx_{i_1}}(a',b',c',d') \times \ldots \times F_{\approx_{i_k}}(a',b',c',d') = F_{\approx_{i_1} \wedge \ldots \wedge \approx_{i_k}}(a',b',c',d') = 1$$

and this implies that $F_{\approx_{i_1}}(a',b',c',d') = 1, \ldots, F_{\approx_{i_k}}(a',b',c',d') = 1$. *This finishes the proof.* □

Theorem 7.2 can be used to prove the following theorem on implicational compound quantifiers.

Theorem 7.3. *The following 4ft-quantifiers are implicational:*

1. *Founded p-implication* $\Rightarrow_{p,Base}$ *defined as* $\Rightarrow_p \land \oplus_{Base}$
2. *Founded lower critical implication* $\Rightarrow^!_{p,\alpha,Base}$ *defined as* $\Rightarrow^!_{p,\alpha} \land \oplus_{Base}$
3. *Founded upper critical implication* $\Rightarrow^?_{p,\alpha,Base}$ *defined as* $\Rightarrow^?_{p,\alpha} \land \oplus_{Base}$
4. *Founded Laplace correction* $\Rightarrow^L_{p,Base}$ *defined as* $\Rightarrow^L_{p,Base} \land \oplus_{Base}$
5. *Founded Sebag-Schoenauer quantifier* $\Rightarrow^S_{q,Base}$ *defined as* $\Rightarrow^S_q \land \oplus_{Base}$
6. *Founded least contradiction* $\Rightarrow^C_{u,Base}$ *defined as* $\Rightarrow^C_u \land \oplus_{Base}$
7. *Founded example and counterexample rate* $\Rightarrow^E_{v,Base}$ *defined as* $\Rightarrow^E_v \land \oplus_{Base}$.

Proof. Directly follows from theorems 7.1 and 7.2. □

7.3 Selected Weakly Implicational Quantifiers

The well known MB-quantifier $\rightarrow_{p,s}$ with confidence p and support s [1] with associated function $F_{\rightarrow_{p,s}}$ defined as

$$F_{\rightarrow_{p,s}} = \begin{cases} 1 & \text{if } \frac{a}{a+b} \geq p \land \frac{a}{a+b+c+d} \geq s \\ 0 & \text{otherwise} \end{cases}$$

introduced in Sect. 1.1 is not implicational but only weakly implicational. It can be also seen as conjunction $\Rightarrow_p \land \odot_s$ of quantifiers $\Rightarrow_p$ of p-implication and support $\odot_s$ (see row 1 in Table 4.6).

The quantifier $\Rightarrow_p$ of p-implication is implicational according to theorem 7.1, however the same cannot be said for the 4ft-quantifier $\odot_s$ of support, see the following lemma.

Lemma 7.3. *4ft-quantifier* $\odot_s$ *of support with* $0 < s \leq 1$ *is not implicational but it is weakly implicational.*

Proof. It holds $\odot_s(a,b,c,d) = 1$ *if and only if* $\frac{a}{a+b+c+d} \geq s$. *Let be* $0 < s \leq 1$ *a real number and* $F_{\odot_s}$ *be the associated function of* $\odot_s$. *Put* $\langle a,b,c,d \rangle = \langle 1,0,0,0 \rangle$ *and* $\langle a',b',c',d' \rangle = \langle 1,0,0,D \rangle$ *where D is an integer number such that* $D > \frac{1-s}{s}$. *Then*

- $a' \geq a \land b' \leq b$
- $F_{\odot_s}(a,b,c,d) = 1$ *because of* $\frac{a}{a+b+c+d} = \frac{1}{1+0+0+0} \geq s$
- $F_{\odot_s}(a',b',c',d') = 0$ *because of* $\frac{a'}{a'+b'+c'+d'} = \frac{1}{1+0+0+D} < \frac{1}{1+0+0+\frac{1-s}{s}} = \frac{1}{\frac{1}{s}} = s$

which means that $\odot_s$ *is not implicational.*

Let be $0 < s \leq 1$ *a real number,* $F_{\odot_s}(a,b,c,d) = 1$, $a' \geq a \land b' \leq b$ *and let be* $a+b+c+d = a'+b'+c'+d'$. *Then we have* $\frac{a}{a+b+c+d} \geq s$ *and*

$$\frac{a'}{a'+b'+c'+d'} = \frac{a'}{a+b+c+d} \geq \frac{a}{a+b+c+d} \geq s$$

which means that $F_{\odot_s}(a',b',c',d') = 1$ and thus $\odot_s$ is weakly implicational. This finishes the proof. □

We have introduced seven compound 4ft-quantifiers as a conjunction of an implicational quantifier and additional implicational quantifier $\oplus_{Base}$ in theorem 7.3. All of them are called *founded quantifiers*. Similarly we are going to introduce seven compound 4ft-quantifiers as a conjunction of an implicational quantifier and weakly implicational quantifier $\odot_s$ in theorem 7.4. All resulting quantifiers will be weakly implicational quantifiers but not implicational quantifiers. We will call these quantifiers *supported* quantifiers. First we prove a useful lemma.

Lemma 7.4. *Let $\approx_1$ be an implicational 4ft-quantifier with associated function $F_{\approx_1}$ and let $\approx_2$ be a weakly implicational quantifier with associated function $F_{\approx_2}$ which is not implicational. Then*

1. *compound 4ft-quantifier $\approx_1 \wedge \approx_2$ is a weakly implicational quantifier*
2. *if there are 4ft-tables $\langle a_0,b_0,c_0,d_0 \rangle$, $\langle a'_0,b'_0,c'_0,d'_0 \rangle$ such that $a'_0 \geq a_0 \wedge b'_0 \leq b_0$, $F_{\approx_1}(a_0,b_0,c_0,d_0) = 1$, $F_{\approx_2}(a_0,b_0,c_0,d_0) = 1$, and $F_{\approx_2}(a'_0,b'_0,c'_0,d'_0) = 0$ then compound 4ft-quantifier $\approx_1 \wedge \approx_2$ is not an implicational quantifier.*

Proof

1. We have to prove: if $F_{\approx_1}(a,b,c,d) \times F_{\approx_2}(a,b,c,d) = 1$, $a' \geq a \wedge b' \leq b$, and $a+b+c+d = a'+b'+c'+d'$ then $F_{\approx_1}(a',b',c',d') \times F_{\approx_2}(a',b',c',d') = 1$. Assumption $F_{\approx_1}(a,b,c,d) \times F_{\approx_2}(a,b,c,d) = 1$ implies both $F_{\approx_1}(a,b,c,d)$ and $F_{\approx_2}(a,b,c,d) = 1$. Quantifier $\approx_1$ is implicational and thus $a' \geq a \wedge b' \leq b$ means that $F_{\approx_1}(a',b',c',d') = 1$. Quantifier $\approx_2$ is weakly implicational and thus $a' \geq a \wedge b' \leq b$ and $a+b+c+d = a'+b'+c'+d'$ means $F_{\approx_2}(a',b',c',d') = 1$. We can conclude that $F_{\approx_1}(a',b',c',d') \times F_{\approx_2}(a',b',c',d') = 1$. This finishes the proof of point 1.
2. It holds that $F_{\approx_1}(a_0,b_0,c_0,d_0) = 1$ and $F_{\approx_2}(a_0,b_0,c_0,d_0) = 1$, which implies $F_{\approx_1}(a_0,b_0,c_0,d_0) \times F_{\approx_2}(a_0,b_0,c_0,d_0) = 1$ and thus $F_{\approx_1 \wedge \approx_2}(a_0,b_0,c_0,d_0) = 1$. In addition it holds $F_{\approx_2}(a'_0,b'_0,c'_0,d'_0) = 0$, which implies

$$F_{\approx_1}(a'_0,b'_0,c'_0,d'_0) \times F_{\approx_2}(a'_0,b'_0,c'_0,d'_0) = F_{\approx_1}(a'_0,b'_0,c'_0,d'_0) \times 0 = 0$$

and thus $F_{\approx_1 \wedge \approx_2}(a'_0,b'_0,c'_0,d'_0) = 0$. However, $a'_0 \geq a_0 \wedge b'_0 \leq b_0$ and we can conclude that compound 4ft-quantifier $\approx_1 \wedge \approx_2$ is not an implicational quantifier. This finishes the proof of point 2. □

Theorem 7.4. *The following 4ft-quantifiers are weakly implicational but not implicational:*

1. *Supported p-implication $\rightarrow_{p,s}$ defined as $\Rightarrow_p \wedge \odot_s$*
2. *Supported lower critical implication $\rightarrow^!_{p,\alpha,s}$ defined as $\Rightarrow^!_{p,\alpha} \wedge \odot_s$*
3. *Supported upper critical implication $\rightarrow^?_{p,\alpha,s}$ defined as $\Rightarrow^?_{p,\alpha} \wedge \odot_s$*

4. Supported Laplace correction $\to_{p,s}^{L}$ defined as $\Rightarrow_{p}^{L} \land \odot_{s}$

5. Supported Sebag-Schoenauer quantifier $\to_{q,s}^{S}$ defined as $\Rightarrow_{q}^{S} \land \odot_{s}$

6. Supported least contradiction $\to_{u,s}^{C}$ defined as $\Rightarrow_{u}^{C} \land \odot_{s}$

7. Supported example and counterexample rate $\to_{v,s}^{E}$ defined as $\Rightarrow_{v}^{E} \land \odot_{s}$.

Proof. It directly follows from theorem 7.1, lemma 7.3 and point 1 of lemma 7.4 that all the supported quantifiers defined above are weakly implicational.

We use point 2 of lemma 7.4 to prove that all the above defined supported quantifiers are not implicational. Let $\langle a_0, b_0, c_0, d_0 \rangle = \langle 1, 0, 0, 0 \rangle$ and $\langle a_0', b_0', c_0', d_0' \rangle = \langle 1, 0, 0, D \rangle$ where D is an integer number such that $D > \frac{1-s}{s}$. Then for $0 < s \leq 1$:

- $a_0' \geq a_0 \land b_0' \leq b_0$
- $F_{\odot_s}(a_0, b_0, c_0, d_0) = 1$ since $\frac{1}{1+0+0+0} \geq s$
- $F_{\odot_s}(a_0', b_0', c_0', d_0') = 0$ since $\frac{1}{1+0+0+D} < \frac{1}{1+0+0+\frac{1-s}{s}} = \frac{1}{\frac{1}{s}} = s$.

In addition we have:

- It holds that $F_{\Rightarrow_p}(a, b, c, d) = 1$ if and only if both $\frac{a}{a+b} \geq p$ and $a+b > 0$. $F_{\Rightarrow_p}(a_0, b_0, c_0, d_0) = 1$ for $0 < p \leq 1$ since $\frac{1}{1+0} \geq p$ and thus supported p-implication $\to_{p,s}$ is not implicational, see point 1 above.
- It holds that $F_{\Rightarrow_{p,\alpha}^{?}}(a, b, c, d) = 1$ if and only if $\sum_{i=0}^{a} \binom{a+b}{i} p^i (1-p)^{a+b-i} > \alpha$. $F_{\Rightarrow_{p,\alpha}^{?}}(a_0, b_0, c_0, d_0) = 1$ for $0 < p < 1$ and $0 < \alpha \leq 0.5$ since $\sum_{i=0}^{1} \binom{1+0}{i} p^i (1-p)^{1+0-i} = \binom{1}{0} p^0 (1-p)^{1-0} + \binom{1}{1} p^1 (1-p)^{1-1} = 1 - p + p > \alpha$ and thus supported upper critical implication $\to_{p,\alpha,s}^{?}$ is not implicational, see point 3 above.
- It holds that $F_{\Rightarrow_u^C}(a, b, c, d) = 1$ if and only if $\frac{a-b}{a+b} \geq u$ and $a+b > 0$. $F_{\Rightarrow_u^C}(a_0, b_0, c_0, d_0) = 1$ for $-1 \leq u \leq 1$ since $\frac{1-0}{1+0} \geq u$ and thus supported least contradiction quantifier $\to_{u,s}^{C}$ is not implicational, see point 6 above.
- It holds that $F_{\Rightarrow_v^E}(a, b, c, d) = 1$ if and only if $\frac{a-b}{a} \geq v$ and $a > 0$. $F_{\Rightarrow_v^E}(a_0, b_0, c_0, d_0) = 1$ for $v \leq 1$ since $\frac{1-0}{1} \geq v$ and thus supported example and counterexample rate $\to_{v,s}^{E}$ is not implicational, see point 7 above.

It remains to prove that the supported 4ft-quantifiers $\to_{p,\alpha,s}^{!}$ (see point 2), and $\to_{p,s}^{L}$ (see point 4), and $\to_{q,s}^{S}$ (see point 5) are not implicational.

Proof for $\to_{p,\alpha,s}^{!}$, point 2.: It holds that $F_{\Rightarrow_{p,\alpha}^{!}}(a, b, c, d) = 1$ if and only if $\sum_{i=a}^{a+b} \binom{a+b}{i} p^i (1-p)^{a+b-i} \leq \alpha$ and $a+b > 0$. Let there be $\langle a_0, b_0, c_0, d_0 \rangle = \langle A, 0, 0, 0 \rangle$ such that $p^A \leq \alpha$ and let $\langle a_0', b_0', c_0', d_0' \rangle = \langle A, 0, 0, D \rangle$ where D is an integer number such that $D > \frac{A-sA}{s}$. Then it holds for $0 < s \leq 1$:

- $a_0' \geq a_0 \land b_0' \leq b_0$
- $F_{\odot_s}(a_0, b_0, c_0, d_0) = 1$ because of $\frac{A}{A+0+0+0} \geq s$
- $F_{\odot_s}(a_0', b_0', c_0', d_0') = 0$ because of $\frac{A}{A+0+0+D} < \frac{A}{A+0+0+\frac{A-sA}{s}} = \frac{A}{\frac{A}{s}} = s$.

In addition, it holds that $F_{\Rightarrow^!_{p,\alpha}}(a_0,b_0,c_0,d_0) = 1$ for $0 < p < 1$ and $0 < \alpha \leq 0.5$ because $\sum_{i=A}^{A+0} \binom{A+0}{i} p^i (1-p)^{A+0-i} = \binom{A}{A} p^A (1-p)^{A-A} = p^A \leq \alpha$. This finishes the proof for $\Rightarrow^!_{p,\alpha,s}$.

*Proof for $\Rightarrow^L_{p,s}$, point 4.: It holds that $F_{\Rightarrow^L_p}(a,b,c,d) = 1$ if and only if $\frac{a+1}{a+b+2} \geq p$.
Let be $\langle a_0,b_0,c_0,d_0 \rangle = \langle A,0,0,0 \rangle$ such that $\frac{A+1}{A+2} \geq p$ (remember that $0 < p < 1$ for 4ft-quantifier $\Rightarrow^L_{p,s}$, see definition 4.2). In addition, let $\langle a'_0,b'_0,c'_0,d'_0 \rangle = \langle A,0,0,D \rangle$ where D is an integer number satisfying $D > \frac{A-sA}{s}$. Then it holds for $0 < s \leq 1$:*

- $a'_0 \geq a_0 \wedge b'_0 \leq b_0$
- $F_{\ominus_s}(a_0,b_0,c_0,d_0) = 1$ because of $\frac{A}{A+0+0+0} \geq s$
- $F_{\ominus_s}(a'_0,b'_0,c'_0,d'_0) = 0$ because of $\frac{A}{A+0+0+D} < \frac{A}{A+0+0+\frac{A-sA}{s}} = \frac{A}{\frac{A}{s}} = s$.

In addition, it holds that $F_{\Rightarrow^L_p}(a_0,b_0,c_0,d_0) = 1$ for $0 < p < 1$ because of $\frac{A+0+1}{A+0+2} \geq p$. This finishes the proof for $\Rightarrow^L_{p,s}$.

Proof for $\Rightarrow^S_{q,s}$ point 5.: It holds that $F_{\Rightarrow^S_q}(a,b,c,d) = 1$ if and only if $a > 0 \wedge b = 0$ or $\frac{a}{b} \geq q$. We distinguish two cases $0 < s < 1$ and $s = 1$.

$0 < s < 1$: Let there be $\langle a_0,b_0,c_0,d_0 \rangle = \langle A,1,0,0 \rangle$ such that $A \geq q \wedge \frac{A}{A+1} \geq s$ and let $\langle a'_0,b'_0,c'_0,d'_0 \rangle = \langle A,1,0,D \rangle$ where D is an integer number satisfying $D > \frac{A-sA}{s}$. Then it holds for $0 < s < 1$:

- $a'_0 \geq a_0 \wedge b'_0 \leq b_0$
- $F_{\ominus_{Base}}(a_0,b_0,c_0,d_0) = 1$ because of $\frac{A}{A+1+0+0} \geq s$
- $F_{\ominus_{Base}}(a'_0,b'_0,c'_0,d'_0) = 0$ because of $\frac{A}{A+1+0+D} < \frac{A}{A+0+0+\frac{A-sA}{s}} = \frac{A}{\frac{A}{s}} = s$.

In addition, it holds that $F_{\Rightarrow^S_q}(a_0,b_0,c_0,d_0) = 1$ for $0 < q$ because of $\frac{A}{1} = A \geq q$. This finishes the proof for $\Rightarrow^S_{q,s}$ and $0 < s < 1$.

$s = 1$: Let there be $\langle a_0,b_0,c_0,d_0 \rangle = \langle A,0,0,0 \rangle$ such that $A \geq q$ and let $\langle a'_0,b'_0,c'_0,d'_0 \rangle = \langle A,0,0,D \rangle$ where D is an integer number such that $D > \frac{A-sA}{s}$. Then the following holds even for $0 < s \leq 1$:

- $a'_0 \geq a_0 \wedge b'_0 \leq b_0$
- $F_{\ominus_{Base}}(a_0,b_0,c_0,d_0) = 1$ because $\frac{A}{A+0+0+0} \geq s$
- $F_{\ominus_{Base}}(a'_0,b'_0,c'_0,d'_0) = 0$ because $\frac{A}{A+0+0+D} < \frac{A}{A+0+0+\frac{A-sA}{s}} = \frac{A}{\frac{A}{s}} = s$.

In addition, it holds that $F_{\Rightarrow^S_q}(a_0,b_0,c_0,d_0) = 1$ because of point M_{18}) in definition 4.2. Please note that the proof for $s = 1$ works for $0 < s \leq 1$ but the version above for $0 < s < 1$ better illustrates the properties of the 4ft-quantifier $\Rightarrow^S_{q,s}$ This finishes the proof for $\Rightarrow^S_{q,s}$ and the proof of the theorem. □

7.4 Interesting Implicational Quantifiers

Important theoretical results concern only specific implicational quantifiers which we call interesting implicational quantifiers. To define them we need a notion of 4ftquantifiers dependent on particular frequencies from 4ft-table. Please remember that for an implicational quantifier $\Rightarrow^*$ we can write $F_{\Rightarrow^*}(a,b)$ instead of $F_{\Rightarrow^*}(a,b,c,d)$, see note 6.1.

Definition 7.4. Let $\approx$ be a 4ft-quantifier. Then

1. $\approx$ is *a-dependent* if there are non-negative integers a, a', b, c, d such that

$$a+b+c+d>0 \ \wedge \ a'+b+c+d>0 \ \wedge \ F_{\approx}(a,b,c,d) \ \neq \ F_{\approx}(a',b,c,d) \ .$$

2. $\approx$ is *b-dependent* if there are non-negative integers a, b, b', c, d such that

$$a+b+c+d>0 \ \wedge \ a+b'+c+d>0 \ \wedge \ F_{\approx}(a,b,c,d) \ \neq \ F_{\approx}(a,b',c,d) \ .$$

Definition 7.5. *Implicational quantifier* $\Rightarrow^*$ *is interesting if the following conditions are satisfied:*

- $\Rightarrow^*$ is a-dependent
- $\Rightarrow^*$ is b-dependent
- $F_{\Rightarrow^*}(0,0) = 0$.

We prove that most of the implicational quantifiers from theorem 7.1 are interesting implicational quantifiers for most of or for all possible values of their parameters, see theorem 7.5. The only exceptions are the quantifiers $\oplus_{Base}$ and upper critical implication $\Rightarrow^?_{p,\alpha}$. It holds that $F_{\oplus_{Base}}(a,b,c,d) = 1$ if and only if $a \geq Base$ and thus the quantifier $\oplus_{Base}$ is not b-dependent. 4ft-quantifier $\Rightarrow^?_{p,\alpha}$ of upper critical implication is both a-dependent and b-dependent but it does satisfy $F_{\Rightarrow^?_{p,\alpha}}(0,0) = 0$, see theorem 7.6.

Theorem 7.5. *The following 4ft-quantifiers are interesting implicational quantifiers:*

1. *p-implication $\Rightarrow_p$ (i.e. confidence or precision), see rows 1 in Tables 4.2 and 4.4 and rows 2 in Tables 4.5 and 4.6*
2. *Likely p-implication (i.e. lower critical implication) $\Rightarrow^!_{p,\alpha}$ for $0 < p < 1$ and $0 < \alpha \leq 0.5$, see rows 2 in Tables 4.2 and 4.4*
3. *Laplace correction $\Rightarrow^L_p$ for $0.5 < p < 1$, see rows 18 in Tables 4.5 and 4.6*
4. *Sebag - Schoenauer $\Rightarrow^S_q$ for $q > 0$, see rows 25 in Tables 4.5 and 4.6*
5. *Least contradiction $\Rightarrow^C_u$ for $-1 < u < 1$, see rows 26 in Tables 4.5 and 4.6*
6. *Example and counterexample rate $\Rightarrow^E_v$ for $v < 1$, see rows 28 in Tables 4.5 and 4.6.*

Proof. All the above six 4ft-quantifiers are implicational according to theorem 7.1. We have to prove for each 4ft-quantifier $\Rightarrow^*$ from them the following facts D_a), D_b), and $D_{0,0}$):

D_a) *there are non-negative integers a, a', b, c, d such that*

$$a+b+c+d>0 \ \wedge \ a'+b+c+d>0 \ \wedge \ F_{\Rightarrow^*}(a,b,c,d) \ \neq \ F_{\Rightarrow^*}(a',b,c,d)$$

D_b)   *there are non-negative integers a, b, b', c, d such that*

$$a+b+c+d > 0 \ \wedge \ a+b'+c+d > 0 \ \wedge \ F_{\Rightarrow^*}(a,b,c,d) \neq F_{\Rightarrow^*}(a,b',c,d)$$

$D_{0,0}$) $F_{\Rightarrow^*}(0,0) = 0$ *(remember that associated function of implicational quantifier $\Rightarrow^*$ depends neither on c nor on d, see note 6.1 and lemma 6.3).*

Proofs for particular 4ft-quantifiers follow.

1. *It holds that* $F_{\Rightarrow_p}(a,b,c,d) = 1$ *if and only if* $\frac{a}{a+b} \geq p \wedge a+b > 0$ *where* $0 < p \leq 1$, *which means:*

    D_a) $F_{\Rightarrow_p}(1,0,0,1) = 1$ *and* $F_{\Rightarrow_p}(0,0,0,1) = 0$ *(see point G_1) in definition 4.1); please note that for $0 < p < 1$ it also holds that $F_{\Rightarrow_p}(A,1,0,0) = 1$ for $A \geq \frac{p}{p+1}$ and $F_{\Rightarrow_p}(0,1,0,0) = 0$*

    D_b) $F_{\Rightarrow_p}(1,0,0,0) = 1$ *and for* $b > \frac{1-p}{p}$ *also* $F_{\Rightarrow_p}(1,b,0,0) = 0$ *since*
    $$\frac{1}{1+b} < \frac{1}{1+\frac{1-p}{p}} = \frac{p}{p+1-p} = p$$

    $D_{0,0}$) $F_{\Rightarrow_p}(0,0) = 0$, *see point G_1 in definition 4.1.*

2. *It holds that* $F_{\Rightarrow_{p,\alpha}^!}(a,b,c,d) = 1$ *if and only if* $\sum_{i=a}^{a+b} \binom{a+b}{i} p^i (1-p)^{a+b-i} \leq \alpha$, *we assume $0 < p < 1$ and $0 < \alpha \leq 0.5$ which means:*

    D_a)   *for $a > 0$ satisfying $p^a < \alpha$ it holds that* $F_{\Rightarrow_{p,\alpha}^!}(a,0,0,1) = 1$ *because*
    $$\sum_{i=a}^{a+0} \binom{a+0}{i} p^i (1-p)^{a+0-i} = \binom{a+0}{a} p^a (1-p)^{a+0-a} = p^a < \alpha$$
    and $F_{\Rightarrow_{p,\alpha}^!}(0,0,0,1) = 0$ (see point G_2) in definition 4.1)

    D_b)   *for $a > 0$ satisfying $p^a < \alpha$ it holds that* $F_{\Rightarrow_{p,\alpha}^!}(a,0,0,0) = 1$, *see above; according to point 1 in lemma 5.16 it holds that* $\lim_{b \to \infty} \sum_{i=a}^{a+b} \binom{a+b}{i} p^i (1-p)^{a+b-i} = 1$, *thus there is b such that* $\sum_{i=a}^{a+b} \binom{a+b}{i} p^i (1-p)^{a+b-i} > \alpha$ *which means that there is $b > 0$ such that* $F_{\Rightarrow_{p,\alpha}^!}(a,b,0,0) = 0$

    $D_{0,0}$) $F_{\Rightarrow_{p,\alpha}^!}(0,0) = 0$, *see point G_2) in definition 4.1.*

3. *It holds that* $F_{\Rightarrow_p^L}(a,b,c,d) = 1$ *if and only if* $\frac{a+1}{a+b+2} \geq p$, *we assume $0.5 < p < 1$ which means:*

    D_a)   *there are b_1 such that $\frac{1}{b_1+2} < p$ and a_1 such that $\frac{a_1+1}{a_1+b_1+2} \geq p$ and thus* $F_{\Rightarrow_p^L}(a_1,b_1,0,0) = 1$ *and* $F_{\Rightarrow_p^L}(0,b_1,0,0) = 0$

    D_b)   *there are a_2 such that $\frac{a_2+1}{a_2+2} \geq p$ and b_2 such that $\frac{a_2+1}{a_2+b_2+2} < p$ and thus* $F_{\Rightarrow_p^L}(a_2,0,0,0) = 1$ *and* $F_{\Rightarrow_p^L}(a_2,b_2,0,0) = 0$

    $D_{0,0}$) $F_{\Rightarrow_p^L}(0,0) = 0$ *because* $\frac{0+1}{0+0+2} = 0.5 < p$.

4. *It holds that* $F_{\Rightarrow_q^S}(a,b,c,d) = 1$ *if and only if $a > 0 \wedge b = 0$ or $\frac{a}{b} \geq q$, we assume $q > 0$ which means:*

    D_a)   *there are b_1 such that $\frac{1}{b_1} < q$ and a_1 such that $\frac{a_1}{b_1} \geq q$ and thus* $F_{\Rightarrow_q^S}(a_1,b_1,0,0) = 1$ *and* $F_{\Rightarrow_q^S}(1,b_1,0,0) = 0$

$D_b)$ there are a_2 such that $\frac{a_2}{1} \geq q$ and b_2 such that $\frac{a_2}{b_2} < q$ and thus $F_{\Rightarrow_q^S}(a_2,1,0,0) = 1$ and $F_{\Rightarrow_q^S}(a_2,b_2,0,0) = 0$.

$D_{0,0})$ $F_{\Rightarrow_q^S}(0,0) = 0$, see point M_{18} in definition 4.2.

5. It holds that $F_{\Rightarrow_u^C}(a,b,c,d) = 1$ if and only if $\frac{a-b}{a+b} \geq u$ and $a+b > 0$. We assume $-1 < u < 1$ which means:

$D_a)$ if $a' > \frac{u+1}{1-u}$ then it holds that $\frac{a'-1}{a'+1} > \frac{\frac{u+1}{1-u}-1}{\frac{u+1}{1-u}+1} = u$ (please note that if $x > y > 0$

    then $\frac{x-1}{x+1} > \frac{y-1}{y+1}$) and thus $F_{\Rightarrow_u^C}(a',1,0,0) = 1$ and $F_{\Rightarrow_u^C}(0,1,0,0) = 0$

$D_b)$ if $b' > \frac{1-u}{1+u}$ then it holds that $\frac{1-b'}{1+b'} < \frac{1-\frac{1-u}{1+u}}{1+\frac{1-u}{1+u}} = u$ (please note that if $x > y > 0$

    then $\frac{1-x}{1+x} < \frac{1-y}{1+y}$) and thus $F_{\Rightarrow_u^C}(1,b',0,0) = 0$ and $F_{\Rightarrow_u^C}(1,0,0,0) = 1$.

$D_{0,0})$ $F_{\Rightarrow_u^C}(0,0) = 0$, see point M_{19} in definition 4.2.

6. It holds that $F_{\Rightarrow_v^E}(a,b,c,d) = 1$ if and only if $\frac{a-b}{a} \geq v$ and $a > 0$. We assume $v < 1$ which means:

$D_a)$ it holds that $F_{\Rightarrow_v^E}(0,1,0,1) = 0$ and there is $a > 0$ such that $\frac{a-1}{a} \geq v$ which

    means $F_{\Rightarrow_v^E}(a,1,0,1) = 0$

$D_b)$ it holds that $F_{\Rightarrow_v^E}(1,0,0,0) = 1$, in addition $F_{\Rightarrow_v^E}(1,1,0,0) = 0$ for $v > 0$

    and $F_{\Rightarrow_v^E}(1,b,0,0) = 0$ where $b > |v| + 2$ for $v \leq 0$.

$D_{0,0})$ $F_{\Rightarrow_v^E}(0,0) = 0$, see point M_{21} in definition 4.2.

This finishes the proof.   □

Theorem 7.6. *The 4ft-quantifier $\Rightarrow_{p,\alpha}^?$ of suspicious critical implication (i.e. upper critical implication), see rows 3 in Tables 4.2 and 4.4, is both a-dependent and b-dependent for $0 < p < 1$ and $0 < \alpha \leq 0.5$, but it does not satisfy the condition $F_{\Rightarrow_{p,\alpha}^?}(0,0) = 0$.*

Proof. It holds that $F_{\Rightarrow_{p,\alpha}^?}(a,b,c,d) = 1$ if and only if $\sum_{i=0}^{a}\binom{a+b}{i}p^i(1-p)^{a+b-i} > \alpha$, we assume $0 < \alpha \leq 0.5$ and this means (we use the same notation $D_a)$, $D_b)$, and $D_{0,0})$ as in theorem 7.5):

$D_a)$ it holds that $\sum_{i=0}^{0}\binom{0+b}{i}p^i(1-p)^{0+b-i} = (1-p)^b$ and thus there is b

    such that $(1-p)^b < \alpha$ which means $F_{\Rightarrow_{p,\alpha}^?}(0,b,0,1) = 0$; according to point 3

    of lemma 5.16 there is a' such that $\sum_{i=0}^{a'}\binom{a'+b}{i}p^i(1-p)^{a'+b-i} > \alpha$ which means

    $F_{\Rightarrow_{p,\alpha}^?}(a',b,0,1) = 1$

$D_b)$ it holds that $F_{\Rightarrow_{p,\alpha}^?}(0,0,0,1) = 1$ because $\sum_{i=0}^{0}\binom{0+0}{i}p^i(1-p)^{0+0-i} = 1$; and

    there is also b' such that $F_{\Rightarrow_{p,\alpha}^?}(0,b',0,1) = 0$ (see $D_a)$)

$D_{0,0})$ it holds that also $\sum_{i=0}^{0}\binom{0+0}{i}p^i(1-p)^{0+0-i} = 1 > \alpha$ for $0 < \alpha \leq 0.5$ and

    thus we have $F_{\Rightarrow_{p,\alpha}^?}(0,0,c,d) = 1$ for all $c+d > 0$ which means that condition

    $F_{\Rightarrow^*}(0,0) = 0$ is not satisfied.

This finishes the proof.   □

In addition we prove that founded implicational quantifiers from theorem 7.3 are interesting implicational quantifiers for most of their parameters.

Theorem 7.7. *The following founded 4ft-quantifiers are interesting implicational quantifiers.*

1. *Founded p-implication* $\Rightarrow_{p,Base}$ *defined as* $\Rightarrow_p \wedge \oplus_{Base}$
2. *Founded lower critical implication* $\Rightarrow^!_{p,\alpha,Base}$ *defined as* $\Rightarrow^!_{p,\alpha} \wedge \oplus_{Base}$, *for* $0 < p < 1$ *and* $0 < \alpha \leq 0.5$
3. *Founded upper critical implication* $\Rightarrow^?_{p,\alpha,Base}$ *defined as* $\Rightarrow^?_{p,\alpha} \wedge \oplus_{Base}$, *for* $0 < p < 1$ *and* $0 < \alpha \leq 0.5$
4. *Founded Laplace correction* $\Rightarrow^L_{p,Base}$ *defined as* $\Rightarrow^L_{p,Base} \wedge \oplus_{Base}$, *for* $0.5 < p < 1$
5. *Founded Sebag - Schoenauer quantifier* $\Rightarrow^S_{q,Base}$ *defined as* $\Rightarrow^S_q \wedge \oplus_{Base}$, *for* $q > 0$
6. *Founded least contradiction* $\Rightarrow^C_{u,Base}$ *defined as* $\Rightarrow^C_p \wedge \oplus_{Base}$, *, for* $-1 < u < 1$
7. *Founded example and counterexample rate* $\Rightarrow^E_{v,Base}$ *defined as* $\Rightarrow^E_v \wedge \oplus_{Base}$, *for* $v < 1$.

Proof. All the above seven 4ft-quantifiers are implicational according to theorem 7.3. We have to prove for each of these 4ft-quantifiers $\Rightarrow^*$ *the following facts* D_a), D_b), *and* $D_{0,0}$):

D_a) *there are non-negative integers a, a', b, c, d such that*

$$a+b+c+d > 0 \wedge a'+b+c+d > 0 \wedge F_{\Rightarrow^*}(a,b,c,d) \neq F_{\Rightarrow^*}(a',b,c,d)$$

D_b) *there are non-negative integers a, b, b', c, d such that*

$$a+b+c+d > 0 \wedge a+b'+c+d > 0 \wedge F_{\Rightarrow^*}(a,b,c,d) \neq F_{\Rightarrow^*}(a,b',c,d)$$

$D_{0,0}$) $F_{\Rightarrow^*}(0,0) = 0$ *(remember that associated function of implicational quantifier* $\Rightarrow^*$ *depends neither on c nor on d, see note 6.1 and lemma 6.3.*

All these quantifiers are founded, this means that $F_{\Rightarrow^*}(0,0) = 0$ *for all of them because Base* > 0 *and thus* $F_{\oplus_{Base}}(0,0,c,d) = 0$ *which means* $F_{\Rightarrow^*}(0,0) = 0$. *Therefore it is sufficient to prove only* D_a) *and* D_b) *for each 4ft-quantifier* $\Rightarrow^*$ *from the above seven 4ft-quantifiers. In most cases we can use proofs of theorems 7.5 and 7.6. Proofs for particular 4ft-quantifiers follow.*

1. *It holds that* $F_{\Rightarrow_{p,Base}}(a,b,c,d) = 1$ *if and only if* $\frac{a}{a+b} \geq p \wedge a \geq Base$ *where* $0 < p \leq 1$ *and this means:*

   D_a) $F_{\Rightarrow_{p,Base}}(Base,0,0,1) = 1$ *and* $F_{\Rightarrow_{p,Base}}(0,0,0,1) = 0$.
   D_b) $F_{\Rightarrow_{p,Base}}(Base,0,0,0) = 1$ *and for* $b > Base\frac{1-p}{p}$ *also*
   $F_{\Rightarrow_{p,Base}}(Base,b,0,0) = 0$ *because* $\frac{Base}{Base+b} < \frac{Base}{Base+Base\frac{1-p}{p}} = p$.

2. $F_{\Rightarrow^!_{p,\alpha,Base}}(a,b,c,d) = 1$ *if and only if* $\sum_{i=a}^{a+b} \binom{a+b}{i} p^i (1-p)^{a+b-i} \leq \alpha \wedge a \geq Base$, *we assume* $0 < p < 1$ *and* $0 < \alpha \leq 0.5$ *and this means:*

D_a)   *for a > Base satisfying $p^a < \alpha$ it holds that $F_{\Rightarrow^!_{p,\alpha}}(a,0,0,1) = 1$ because*

$\sum_{i=a}^{a+0} \binom{a+0}{i} p^i (1-p)^{a+0-i} = \binom{a+0}{a} p^a (1-p)^{a+0-a} = p^a < \alpha$

and $F_{\Rightarrow^!_{p,\alpha,Base}}(0,0,0,1) = 0$

D_b)   *for a > Base satisfying $p^a < \alpha$ it holds $F_{\Rightarrow^!_{p,\alpha,Base}}(a,0,0,0) = 1$, see above;*

point 1 in lemma 5.16 implies $\lim_{b\to\infty} \sum_{i=a}^{a+b} \binom{a+b}{i} p^i (1-p)^{a+b-i} = 1$, thus

there is b such that $\sum_{i=a}^{a+b} \binom{a+b}{i} p^i (1-p)^{a+b-i} > \alpha$ which means

$F_{\Rightarrow^!_{p,\alpha}}(a,b,0,0) = 0$.

3. $F_{\Rightarrow^?_{p,\alpha,Base}}(a,b,c,d) = 1$ *if and only if $\sum_{i=0}^{a} \binom{a+b}{i} p^i (1-p)^{a+b-i} > \alpha \wedge a \geq Base$,*
   we assume $0 < p < 1$ and $0 < \alpha \leq 0.5$ and this means:

D_a) $F_{\Rightarrow^?_{p,\alpha,Base}}(Base,0,0,1) = 1$ *because $\sum_{i=0}^{Base} \binom{Base+0}{i} p^i (1-p)^{Base+0-i} = 1$*

and $F_{\Rightarrow^?_{p,\alpha,Base}}(0,0,0,1) = 0$.

D_b) $F_{\Rightarrow^?_{p,\alpha,Base}}(Base,0,0,1) = 1$, *see D_a); according to point 1 in lemma 5.15 it*

holds that $\lim_{b\to\infty} \sum_{i=0}^{Base} \binom{Base+b}{i} p^i (1-p)^{Base+b-i} = 0$ and thus there is b such

that $\sum_{i=0}^{Base} \binom{Base+b}{i} p^i (1-p)^{Base+b-i} < \alpha$ and $F_{\Rightarrow^?_{p,\alpha,Base}}(Base,b,0,0) = 0$.

4. *It holds that $F_{\Rightarrow^L_{p,Base}}(a,b,c,d) = 1$ if and only if $\frac{a+1}{a+b+2} \geq p \wedge a \geq Base$, we*
   assume $0.5 < p < 1$ and this means:

D_a)   *there is $a \geq Base$ such that $\frac{a+1}{a+2} \geq p$ and thus $F_{\Rightarrow^L_{p,Base}}(a,0,0,1) = 1$ and*

$F_{\Rightarrow^L_{p,Base}}(0,0,0,1) = 0$

D_b)   *there are $a_2 \geq Base$ such that $\frac{a_2+1}{a_2+2} \geq p$ and b_2 such that $\frac{a_2+1}{a_2+b_2+2} < p$ and*

thus $F_{\Rightarrow^L_{p,Base}}(a_2,0,0,0) = 1$ and $F_{\Rightarrow^L_{p,Base}}(a_2,b_2,0,0) = 0$.

5. *It holds that $F_{\Rightarrow^S_{q,Base}}(a,b,c,d) = 1$ if and only if $a \geq Base \wedge b = 0$ or*
   $\frac{a}{b} \geq q \wedge a \geq Base$, we assume $q > 0$ which means:

D_a)   *there are b_1 such that $\frac{1}{b_1} < q$ and $a_1 \geq Base$ such that $\frac{a_1}{b_1} \geq q$ and*

thus $F_{\Rightarrow^S_{q,Base}}(a_1,b_1,0,0) = 1$ and $F_{\Rightarrow^S_q}(1,b_1,0,0) = 0$

D_b)   *there are $a_2 \geq Base$ such that $\frac{a_2}{1} \geq q$ and b_2 such that $\frac{a_2}{b_2} < q$ and*

thus $F_{\Rightarrow^S_{q,Base}}(a_2,1,0,0) = 1$ and $F_{\Rightarrow^S_{q,Base}}(a_2,b_2,0,0) = 0$.

6. *It holds that $F_{\Rightarrow^C_{u,Base}}(a,b,c,d) = 1$ if and only if $\frac{a-b}{a+b} \geq u \wedge a \geq Base$, we*
   assume $-1 < u < 1$ which means:

D_a)   *if $a' > \max\{Base, \frac{u+1}{1-u}\}$ then it holds $\frac{a'-1}{a'+1} > \frac{\frac{u+1}{1-u}-1}{\frac{u+1}{1-u}+1} = u$ (please note that*

if $x > y > 0$ then $\frac{x-1}{x+1} > \frac{y-1}{y+1}$) and thus also $F_{\Rightarrow^C_{u,Base}}(a',1,0,0) = 1$ and

$F_{\Rightarrow^C_{u,Base}}(0,1,0,0) = 0$

$D_b)$ if $b' > Base\frac{1-u}{1+u}$ then it holds $\frac{Base-b'}{Base+b'} < \frac{Base-Base\frac{1-u}{1+u}}{Base+Base\frac{1-u}{1+u}} = u$ (please note

that if $x > y > 0$ then $\frac{1-x}{1+x} < \frac{1-y}{1+y}$) and thus $F_{\Rightarrow^C_{u,Base}}(Base,b',0,0) = 0$ and
$F_{\Rightarrow^C_{u,Base}}(Base,0,0,0) = 1$.

7. It holds that $F_{\Rightarrow^E_{v,Base}}(a,b,c,d) = 1$ if and only if $\frac{a-b}{a} \geq v \wedge a \geq Base$, we assume
$v < 1$ which means:

$D_a)$ it holds that $F_{\Rightarrow^E_{v,Base}}(0,Base,0,0) = 0$ and there is $a > Base$ such that
$\frac{a-Base}{a} \geq v$ which means $F_{\Rightarrow^E_v}(a,Base,0,0) = 1$
$D_b)$ it holds that $F_{\Rightarrow^E_{v,Base}}(Base,0,0,0) = 1$, in addition
$F_{\Rightarrow^E_v}(Base,Base,0,0) = 0$ for $v > 0$ and $F_{\Rightarrow^E_v}(Base,b,0,0) = 0$ where
$b > Base(|v|+2)$ for $v \leq 0$.

This finishes the proof. □

7.5 Tables of Critical Frequencies for Implicational Quantifiers

Evaluation of an associated function requires complex computation for some 4ft-quantifiers corresponding to statistical hypothesis. Such computation can be avoided by suitable tables of critical frequencies which have also additional interesting properties. Among the above studied implicational ones, there are two quantifiers the evaluation of which requires complex computation:

- Associated function $F_{\Rightarrow^!_{p,\alpha}}(a,b,c,d)$ of 4ft-quantifier $\Rightarrow^!_{p,\alpha}$ of lower critical implication is defined such that $F_{\Rightarrow^!_{p,\alpha}}(a,b) = 1$ if the condition $\sum_{i=a}^{a+b}\binom{a+b}{i}p^i(1-p)^{a+b-i} \leq \alpha$ is satisfied, see row 2 in Table 4.4.
- Associated function $F_{\Rightarrow^?_{p,\alpha}}(a,b,c,d)$ of 4ft-quantifier $\Rightarrow^?_{p,\alpha}$ of upper critical implication is defined such that $F_{\Rightarrow^?_{p,\alpha}}(a,b,c,d) = 1$ if the condition $\sum_{i=0}^{a}\binom{a+b}{i}p^i(1-p)^{a+b-i} > \alpha$ is satisfied, see row 3 in Table 4.4

This computation can be avoided by tables of maximal b for implicational quantifiers. The table of maximal b is a table of critical frequencies for implicational quantifiers. It was originally introduced in [18, 24], and introduced in the form defined below in [63]. It was also shown in [63] that there are additional important results related to various tables of critical frequencies. Results concern namely the definability of association rules in classical predicate calculus, see Chap. 13.

Before we define the table of maximal b we introduce the set $\mathcal{N}^+$ i.e. the set of non-negative integer numbers enhanced by ∞.

Definition 7.6. Set $\mathcal{N}^+$ is defined as a union $\mathcal{N}^+ = \mathcal{N} \cup \{\infty\}$ where $\mathcal{N}$ is the set of all non-negative integer numbers, $\mathcal{N} = \{0,1,2,,\ldots\}$. It holds

- $n < \infty, n+\infty = \infty, \infty - n = \infty$ for each $n \in \mathcal{N}$
- $\max(\mathcal{N}) = \infty, \min(\mathcal{N}^+) = \infty$.

Below, we use the fact that the value $F_{\Rightarrow^*}(a,b,c,d)$ of associated function of implicational quantifier $\Rightarrow^*$ depends neither on c nor on d, see note 6.1 and lemma 6.3. Thus we write only $F_{\Rightarrow^*}(a,b)$ instead of $F_{\Rightarrow^*}(a,b,c,d)$ for the implicational quantifier $\Rightarrow^*$.

Definition 7.7. Let $\Rightarrow^*$ be an implicational quantifier. Then a *table* $Tb_{\Rightarrow^*}$ *of maximal b for* $\Rightarrow^*$ is a function $Tb_{\Rightarrow^*}$ assigning to each integer non-negative number $a \in \mathcal{N}$ a value $Tb_{\Rightarrow^*}(a)$ from $\mathcal{N}^+$ such that

$$Tb_{\Rightarrow^*}(a) = \min\{e|F_{\Rightarrow^*}(a,e) = 0\} \ .$$

The following theorem shows that we can use the table $Tb_{\Rightarrow^*}$ of maximal b of implicational quantifier $\Rightarrow^*$ instead of evaluating possibly complex conditions related to the associated function $F_{\Rightarrow^*}$ of $\Rightarrow^*$.

Theorem 7.8. *Let $Tb_{\Rightarrow^*}$ be a table of maximal b for implicational quantifier $\Rightarrow^*$ with associated function $F_{\Rightarrow^*}$. Then*

1. *$F_{\Rightarrow^*}(a,b) = 1$ if and only if $b < Tb_{\Rightarrow^*}(a)$.*
2. *$Tb_{\Rightarrow^*}(a)$ is a nondecreasing function of a.*

Proof. We use the fact that $F_{\Rightarrow^*}(a, Tb_{\Rightarrow^*}(a)) = 0$. It follows from definition $Tb_{\Rightarrow^*}(a) = \min\{e|F_{\Rightarrow^*}(a,e) = 0\}$.

1. • *Let $F_{\Rightarrow^*}(a,b) = 1$, we have to show $b < Tb_{\Rightarrow^*}(a)$. It cannot be $Tb_{\Rightarrow^*}(a) \leq b$ because of $F_{\Rightarrow^*}(a,b) = 1$ and $Tb_{\Rightarrow^*}(a) \leq b$ implies $F_{\Rightarrow^*}(a, Tb_{\Rightarrow^*}(a)) = 1$ and it contradicts to $F_{\Rightarrow^*}(a, Tb_{\Rightarrow^*}(a)) = 0$. Thus it must hold that $b < Tb_{\Rightarrow^*}(a)$.*
   • *Let $F_{\Rightarrow^*}(a,b) = 0$, we have to show that not $b < Tb_{\Rightarrow^*}(a)$. However, the definition $Tb_{\Rightarrow^*}(a) = \min\{e|F_{\Rightarrow^*}(a,e) = 0\}$ implies $Tb_{\Rightarrow^*}(a) \leq b$. Thus it is not true that $b < Tb_{\Rightarrow^*}(a)$.*
   This finishes the proof of point 1.
2. *Let be $a' > a$. We have to prove $Tb_{\Rightarrow^*}(a') \geq Tb_{\Rightarrow^*}(a)$. Let us assume $Tb_{\Rightarrow^*}(a') < Tb_{\Rightarrow^*}(a)$. This means $F_{\Rightarrow^*}(a, Tb_{\Rightarrow^*}(a')) = 1$ because of point 1. The 4ft-quantifier $\Rightarrow^*$ is implicational and $a' > a$, thus $F_{\Rightarrow^*}(a', Tb_{\Rightarrow^*}(a')) = 1$ which contradicts to $F_{\Rightarrow^*}(a', Tb_{\Rightarrow^*}(a')) = 0$. Thus it must be $Tb_{\Rightarrow^*}(a') \geq Tb_{\Rightarrow^*}(a)$.*

This finishes the proof.   □

We prove an additional useful simple theorem.

Theorem 7.9. *Let $T_\approx$ be a non-negative nondecreasing function assigning to each integer non-negative number u a value $T_\approx(u)$ from $\mathcal{N}^+$. Let us define a 4ft-quantifier $\approx$ with associated function $F_\approx$ such that*

$$F_\approx(a,b,c,d) = 1 \ if \ and \ only \ if \ b < T_\approx(a) \ .$$

Then

1. $\approx$ *is an implicational quantifier*
2. *function $T_\approx$ is a table of maximal b of quantifier $\approx$.*

Proof

1. *We have to prove that if $F_\approx(a,b,c,d) = 1$ and $a' \geq a \;\wedge\; b' \leq b$ then also $F_\approx(a',b',c',d') = 1$. Let $F_\approx(a,b,c,d) = 1$ and $a' \geq a \;\wedge\; b' \leq b$. Then $b < T_\approx(a)$, $T_\approx$ is a non-negative nondecreasing function, thus $T_\approx(a) \leq T_\approx(a')$ and all of this implies $b' \leq b < T_\approx(a) \leq T_\approx(a')$ i.e. $b' < T_\approx(a')$. This however means $F_\approx(a',b',c',d') = 1$ which finishes the proof of point 1).*
2. *We have to prove that $T_\approx(a) = \min\{e | F_{\approx^*}(a,e) = 0\}$ see definition 7.7. This however follows from the definition of $F_\approx$ and from the fact proved above that quantifier $\approx$ is implicational.*

This finishes the proof. $\qquad\qquad\qquad\qquad\qquad\qquad\qquad\qquad\qquad\qquad\qquad$ $\square$

Chapter 8
Double Implicational Rules

We have introduced implicational quantifiers and weakly implicational quantifiers and rules in Chap. 7. Double implicational quantifiers and rules are introduced in a similar way. Again, it is easy to see that double implicational quantifiers are $\mathscr{M}$-independent and it is also reasonable to introduce weakly double implicational quantifiers.

The double implicational quantifiers dealing with sum $b + c$ of frequencies of 4ft-table $\langle a, b, c, d \rangle$ are of particular interest. This leads to the definition of Σ-double implicational quantifiers and weakly Σ-double implicational quantifiers. We define and study two important subclasses of Σ-double implicational quantifiers – interesting Σ-double implicational quantifiers and typical Σ-double implicational quantifiers. Tables of critical frequencies are one important tool for the study of double implicational quantifiers. One such table is used to define the class of typical Σ-double implicational quantifiers.

The definition of double implicational quantifiers can be seen as a combination of two definitions of implicational quantifiers. This leads to definitions of additional important or at least interesting subclasses of double implicational quantifiers – pure double implicational quantifiers and strong double implicational quantifiers.

Weakly double implicational quantifiers, Σ-double implicational quantifiers and weakly Σ-double implicational quantifiers are introduced in Sect. 8.1. There are several practically important double implicational quantifiers defined in Tables 4.4 and 4.6. All of these are Σ-double implicational quantifiers, see Sect. 8.2. Weakly double implicational quantifiers defined in Sect. 8.3 are derived from the double implicational quantifiers introduced in Sect. 8.2. The class of interesting Σ-double implicational quantifiers is defined in Sect. 8.4. It is also shown in Sect. 8.4 that practically important double implicational quantifiers are interesting Σ-double implicational quantifiers.

Tables of critical frequencies for double implicational quantifiers are introduced in Sect. 8.5. These are used to define typical Σ-double implicational quantifiers in Sect. 8.6. Pure and strong double implicational quantifiers are defined and studied in Sects. 8.7 and 8.8. The relationships between subclasses of double implicational quantifiers are summarized in Sect. 8.9.

J. Rauch: *Observational Calculi and Association Rules*, SCI 469, pp. 99–126.
DOI: 10.1007/978-3-642-11737-4_8 © Springer-Verlag Berlin Heidelberg 2013

8.1 Double Implicational and Weakly Double Implicational Quantifiers

Double implicational quantifiers are defined in definition 6.3, the truth preservation condition $TPC_\Leftrightarrow$ is used. Remember that definition 6.3 can also be formulated such that 4ft-quantifier $\approx$ is double implicational if

$$F_\approx(a,b,c,d) = 1 \;\wedge\; a' \geq a \;\wedge\; b' \leq b \;\wedge\; c' \leq c \text{ implies } F_\approx(a',b',c',d') = 1$$

for all 4ft-tables $\langle a,b,c,d \rangle$ and $\langle a',b',c',d' \rangle$, see Sect. 6.3. Truth preservation condition $TPC_\Leftrightarrow$ is $\mathscr{M}$-independent as well as the double implicational quantifiers, see a following lemma.

Lemma 8.1

1. Truth preservation condition $TPC_\Leftrightarrow$ is $\mathscr{M}$-independent.
2. Double implicational quantifier is $\mathscr{M}$-independent.

Proof

1. According to point 4) of definition 6.11 we have to prove that there are 4ft-tables $\langle a,b,c,d \rangle$ and $\langle a',b',c',d' \rangle$ such that

$$TPC_\Leftrightarrow(a,b,c,d,a',b',c',d') = 1 \text{ and } a+b+c+d \neq a'+b'+c'+d'.$$

4ft-tables $\langle 10,10,10,0 \rangle$ and $\langle 99,9,9,0 \rangle$ satisfy this condition:

$$TPC_\Leftrightarrow(10,10,10,0,99,9,9,0) = 1 \text{ and } 10+10+10+0 \neq 99+9+9+0.$$

This finishes the proof of point 1.
2. This follows from the proof of point 1) shown above and point 6) of definition 6.11.

This finishes the proof. $\square$

We use the notion of double implicational quantifier to define the notion of a *double implicational rule*.

Definition 8.1. Let $\mathscr{LC}$ be a logical calculus of association rules and let $\varphi \approx \psi$ be an association rule of $\mathscr{LC}$. If quantifier $\approx$ is double implicational then the rule $\varphi \approx \psi$ is a *double implicational rule* .

We derive a new $\mathscr{M}$-dependent class of 4ft-quantifiers from the class of double implicational 4ft-quantifiers in the way given by definition 6.12. We will call 4ft-quantifiers belonging to this class as weakly double implicational 4ft-quantifiers. Consequently, we define weakly double implicational rules. Please note that the class of weakly double implicational quantifiers is introduced in [74] as a class of $\mathscr{M}$-dependent double implicational quantifiers. We will use the name *weakly double implicational quantifiers* instead of $\mathscr{M}$-*dependent implicational quantifiers* for the same reason which we use the name *weakly implicational quantifiers* instead of $\mathscr{M}$-*dependent implicational quantifiers*, see note 7.1.

Definition 8.2. The *truth preservation condition* $TPC_{\Leftrightarrow}^{\mathscr{M}}$ *for weakly double implicational quantifiers* is a $\{0,1\}$-valued function $TPC_{\Leftrightarrow}^{\mathscr{M}}(a,b,c,d,a',b',c',d')$ defined such that

$$TPC_{\Leftrightarrow}^{\mathscr{M}}(a,b,c,d,a',b',c',d') = \begin{cases} 1 & \text{if } a' \geq a \wedge b' \leq b \wedge c' \leq c \text{ and} \\ & a+b+c+d = a'+b'+c'+d' \\ 0 & \text{otherwise.} \end{cases}$$

The class $4ft[TPC_{\Leftrightarrow}^{\mathscr{M}}]$ is a *class of weakly double implicational quantifiers*. If quantifier $\approx$ belongs to $4ft[TPC_{\Leftrightarrow}^{\mathscr{M}}]$ then it is a *weakly double implicational quantifier*.

Definition 8.3. Let $\mathscr{LC}$ be a logical calculus of association rules and let $\varphi \approx \psi$ be an association rule of $\mathscr{LC}$. If quantifier $\approx$ is weakly double implicational then the rule $\varphi \approx \psi$ is a *weakly double implicational rule* .

The associated function $F_{\Leftrightarrow_p}(a,b,c,d)$ of 4ft-quantifier of double p-implication $\Leftrightarrow_p$ is defined such that $F_{\Leftrightarrow_p}(a,b,c,d) = 1$ if and only if $\frac{a}{a+b+c} \geq p$, see Table 4.4. Please note that the sum $b+c$ plays the same role in this definition as the frequency b plays in the definition of associated function $F_{\Rightarrow_p}(a,b,c,d)$ of 4ft-quantifier of p-implication $\Rightarrow_p$ which is defined such that $F_{\Rightarrow_p}(a,b,c,d) = 1$ if and only if $\frac{a}{a+b} \geq p$. The sum $b+c$ plays a similarly important role in the definition of associated functions of 4ft-quantifiers $\Leftrightarrow_{p,\alpha}^!$ and $\Leftrightarrow_{p,\alpha}^?$, see Table 4.4. This leads to a definition of Σ-double implicational quantifiers and weakly Σ-double implicational quantifiers and rules.

Definition 8.4. The *truth preservation condition* $TPC_{\Sigma,\Leftrightarrow}$ *for Σ-double implicational quantifiers* is a $\{0,1\}$-valued function $TPC_{\Sigma,\Leftrightarrow}(a,b,c,d,a',b',c',d')$ defined for all 8-tuples $\langle a,b,c,d,a',b',c',d' \rangle$ of non-negative integer numbers such that

$$TPC_{\Sigma,\Leftrightarrow}(a,b,c,d,a',b',c',d') = \begin{cases} 1 & \text{if } a' \geq a \wedge b'+c' \leq b+c \\ 0 & \text{otherwise.} \end{cases}$$

The *class of Σ-double imlicational quantifiers* is the class $4ft[TPC_{\Sigma,\Leftrightarrow}]$ defined by truth preservation condition $TPC_{\Leftrightarrow}^{\Sigma}$. We say that 4ft-quantifier $\approx$ is *Σ-double implicational* if it belongs to $4ft[TPC_{\Sigma,\Leftrightarrow}]$.

The condition $a' \geq a \wedge b'+c' \leq b+c$ is called the *simple form of $TPC_{\Sigma,\Leftrightarrow}$*.

Definition 8.5. Let $\mathscr{LC}$ be a logical calculus of association rules and let $\varphi \approx \psi$ be an association rule of $\mathscr{LC}$. If quantifier $\approx$ is Σ-double implicational then the rule $\varphi \approx \psi$ is a *Σ-double implicational rule* .

Definition 8.6. The *truth preservation condition* $TPC_{\Sigma,\Leftrightarrow}^{\mathscr{M}}$ *for weakly Σ-double implicational quantifiers* is a $\{0,1\}$-valued function $TPC_{\Sigma,\Leftrightarrow}^{\mathscr{M}}(a,b,c,d,a',b',c',d')$ defined such that

$$TPC_{\Sigma,\Leftrightarrow}^{\mathscr{M}}(a,b,c,d,a',b',c',d') = \begin{cases} 1 & \text{if } a' \geq a \wedge b'+c' \leq b+c \text{ and} \\ & a+b+c+d = a'+b'+c'+d' \\ 0 & \text{otherwise.} \end{cases}$$

The class $4ft[TPC^{\mathscr{M}}_{\Sigma,\Leftrightarrow}]$ is a *class of weakly Σ-double implicational quantifiers*. If quantifier $\approx$ belongs to $4ft[TPC^{\mathscr{M}}_{\Sigma,\Leftrightarrow}]$ then it is a *weakly Σ-double implicational quantifier*.

Definition 8.7. Let $\mathscr{L}\mathscr{C}$ be a logical calculus of association rules and let $\varphi \approx \psi$ be an association rule of $\mathscr{L}\mathscr{C}$. If quantifier $\approx$ is weakly Σ-double implicational then the rule $\varphi \approx \psi$ is a *weakly Σ-double implicational rule*.

It is easy to prove that each Σ-double implicational quantifier is double implicational quantifier:

Theorem 8.1. *Each Σ-double implicational quantifier is a double implicational quantifier.*

Proof. We have to prove that if quantifier $\approx$ belongs to class $4ft[TPC_{\Sigma,\Leftrightarrow}]$ then it belongs to class $4ft[TPC_{\Leftrightarrow}]$. According to lemma 6.1 it is enough to prove that

$$TPC_{\Leftrightarrow}(a,b,c,d,a',b',c',d') = 1 \text{ implies } TPC_{\Sigma,\Leftrightarrow}(a,b,c,d,a',b',c',d') = 1 \,.$$

This means that we have to prove that

$$a' \geq a \wedge b' \leq b \wedge c' \leq c \text{ implies } a' \geq a \wedge b' + c' \leq b + c$$

and this is obvious. This finishes the proof.   □

The following lemma says that each double implicational quantifier is also a weakly double implicational quantifier and that each Σ-double implicational quantifier is also a weakly Σ-double implicational quantifier.

Lemma 8.2

1. *If $\Leftrightarrow^*$ is a double implicational quantifier then $\Leftrightarrow^*$ is also a weakly double implicational quantifier.*
2. *If $\Leftrightarrow^*$ is a Σ-double implicational quantifier then $\Leftrightarrow^*$ is also a weakly Σ-double implicational quantifier.*

Proof. The lemma is a direct consequence of lemma 6.15.   □

We also prove two simple but important lemmas.

Lemma 8.3. *Let $\Leftrightarrow^*$ be a Σ-double implicational quantifier and let $F_{\Leftrightarrow^*}$ be its associated function. In addition, let $\langle a,b,c,d \rangle$ and $\langle a,b',c',d' \rangle$ be 4ft-tables such that $b' + c' = b + c$. Then*

$$F_{\Leftrightarrow^*}(a,b,c,d) = F_{\Leftrightarrow^*}(a,b',c',d') \,.$$

Proof. Let $F_{\Leftrightarrow^*}(a,b,c,d) = 1$. We have $b' + c' = b + c$ which means $b' + c' \leq b + c$. 4ft-quantifier $\Leftrightarrow^*$ is Σ-double implicational and thus $F_{\Leftrightarrow^*}(a,b,c,d) = 1$ implies $F_{\Leftrightarrow^*}(a,b',c',d') = 1$. In addition, $F_{\Leftrightarrow^*}(a,b',c',d') = 1$ implies in a similar way $F_{\Leftrightarrow^*}(a,b,c,d) = 1$. This finishes the proof.   □

Lemma 8.4. *Let $\Leftrightarrow^*$ be a weakly Σ-double implicational quantifier and let $F_{\Leftrightarrow^*}$ be its associated function. In addition, let $\langle a,b,c,d \rangle$ and $\langle a,b',c',d \rangle$ be 4ft-tables such that $b' + c' = b + c$. Then*

$$F_{\Leftrightarrow^*}(a,b,c,d) = F_{\Leftrightarrow^*}(a,b',c',d).$$

Proof. Let $F_{\Leftrightarrow^*}(a,b,c,d) = 1$. We have $b' + c' = b + c$ which means $b' + c' \leq b + c$ and $a + b' + c' + d = a + b + c + d$. 4ft-quantifier $\Leftrightarrow^*$ is weakly Σ-double implicational and thus $F_{\Leftrightarrow^*}(a,b,c,d) = 1$ implies $F_{\Leftrightarrow^*}(a,b',c',d) = 1$. In addition, $F_{\Leftrightarrow^*}(a,b',c',d) = 1$ implies in a similar way $F_{\Leftrightarrow^*}(a,b,c,d) = 1$. This finishes the proof. □

8.2 Selected Double Implicational Quantifiers

First we show that four of the 4ft-quantifiers introduced in Tables 4.4 and 4.6 are Σ-double implicational.

Theorem 8.2. *The following 4ft-quantifiers are Σ-double implicational:*

1. *Double p-implication $\Leftrightarrow_p$ (i.e. Jaccard), see rows 4 in Tables 4.2 and 4.4 and rows 12 in Tables 4.5 and 4.6*
2. *Lower critical double implication $\Leftrightarrow^!_{p,\alpha}$, see rows 5 in Tables 4.2 and 4.4*
3. *Upper critical double implication $\Leftrightarrow^?_{p,\alpha}$, see rows 6 in tables 4.2 and 4.4*
4. *Base $\oplus_{Base}$, see rows 18 in tables 4.2 and 4.4*

Proof. We have to prove for each quantifier $\approx$ from the above four 4ft-quantifiers:

$$\text{if } F_{\approx}(a,b,c,d) = 1 \text{ and } a' \geq a \wedge b' + c' \leq b + c \text{ then } F_{\approx}(a',b',c',d') = 1$$

where $F_{\approx}(a,b,c,d)$ is the associated function of the 4ft-quantifier $\approx$, see definition 8.6.

1. $F_{\Leftrightarrow_p}(a,b,c,d) = 1$ if and only if it holds both $\frac{a}{a+b+c} \geq p$ and $a+b+c > 0$. If $F_{\Leftrightarrow_p}(a,b,c,d) = 1$ and $a' \geq a \wedge b' + c' \leq b + c$ then it holds $a+b+c > 0$ and thus also $a' + b' + c' > 0$ and we have

$$\frac{a'}{a'+b'+c'} \geq \frac{a}{a+b'+c'} \geq \frac{a}{a+b+c} \geq p, \tag{8.1}$$

which finishes the proof for $\Leftrightarrow_p$.

2. $F_{\Leftrightarrow^!_{p,\alpha}}(a,b,c,d) = 1$ if and only if $\sum_{i=a}^{a+b+c} \binom{a+b+c}{i} p^i (1-p)^{a+b+c-i} \leq \alpha$. We have to prove that if $F_{\Leftrightarrow^!_{p,\alpha}}(a,b,c,d) = 1$ and $a' \geq a \wedge b' + c' \leq b + c$ then

$$\sum_{i=a'}^{a'+b'+c'} \binom{a'+b'+c'}{i} p^i (1-p)^{a'+b'+c'-i} \leq \alpha.$$

This however directly follows from theorem 5.1 if we use $u = a$, $v = b + c$, $u' = a'$, and $v' = b' + c'$.

3. $F_{\Leftrightarrow_{p,\alpha}^?}(a,b,c,d) = 1$ *if and only if* $\sum_{i=0}^{a} \binom{a+b+c}{i} p^i (1-p)^{a+b+c-i} > \alpha$. *We have to prove that if* $F_{\Leftrightarrow_{p,\alpha}^?}(a,b,c,d) = 1$ *and* $a' \geq a \wedge b' + c' \leq b + c$ *then*

$$\sum_{i=0}^{a'} \binom{a' + b' + c'}{i} p^i (1-p)^{a'+b'+c'-i} > \alpha .$$

This however directly follows from theorem 5.2 if we use $u = a$, $v = b + c$, $u' = a'$, and $v' = b' + c'$.

4. $F_{\oplus_{Base}}(a,b,c,d) = 1$ *if and only if* $a \geq Base$. *Thus, it is obvious that* $F_{\oplus_{Base}}(a,b,c,d) = 1$ *and* $a' \geq a \wedge b' + c' \leq b + c$ *implies* $F_{\oplus_{Base}}(a',b',c',d') = 1$.

This finishes the proof.   □

The above theorem has a direct consequence in the following theorem.

Theorem 8.3. *The following 4ft-quantifiers are double implicational:*

1. *Double p-implication $\Leftrightarrow_p$ (i.e. Jaccard), see rows 4 in Tables 4.2 and 4.4 and rows 12 in Tables 4.5 and 4.6*
2. *Lower critical double implication $\Leftrightarrow_{p,\alpha}^!$, see rows 5 in Tables 4.2 and 4.4*
3. *Upper critical double implication $\Leftrightarrow_{p,\alpha}^?$, see rows 6 in tables 4.2 and 4.4*
4. *Base $\oplus_{Base}$, see rows 18 in tables 4.2 and 4.4*

Proof. The theorem follows from theorems 8.1 and 8.2.   □

We have derived several founded implicational 4ft-quantifiers from important implicational 4ft-quantifiers in Sect. 7.2, see theorems 7.2 and 7.3. We apply a similar approach to double implicational quantifiers.

Theorem 8.4. *Let $\mathscr{LC}$ be a logical calculus of association rules with 4ft-quantifiers $\approx_1, \ldots \approx_Q$ with associated functions $F_{\approx_1}, \ldots, F_{\approx_Q}$ respectively. Then*

1. *the compound 4ft-quantifier $\approx_{i_1} \wedge \ldots \wedge \approx_{i_k}$ is double implicational if and only if all 4ft-quantifiers $\approx_{i_1}, \ldots, \approx_{i_k}$ are double implicational*
2. *the compound 4ft-quantifier $\approx_{i_1} \wedge \ldots \wedge \approx_{i_k}$ is Σ-double implicational if and only if all 4ft-quantifiers $\approx_{i_1}, \ldots, \approx_{i_k}$ are Σ-double implicational.*

Proof

1. *The associated function $F_{\approx_{i_1} \wedge \ldots \wedge \approx_{i_k}}$ of $\approx_{i_1} \wedge \ldots \wedge \approx_{i_k}$ is defined as a product:*

$$F_{\approx_{i_1} \wedge \ldots \wedge \approx_{i_k}}(a,b,c,d) = F_{\approx_{i_1}}(a,b,c,d) \times \ldots \times F_{\approx_{i_k}}(a,b,c,d) ,$$

see definition 4.3. We have to prove:

a) *If all $\approx_{i_1}, \ldots, \approx_{i_k}$ are double implicational then $\approx_{i_1} \wedge \ldots \wedge \approx_{i_k}$ is double implicational*

b) If $\approx_{i_1} \wedge \ldots \wedge \approx_{i_k}$ is double implicational then all $\approx_{i_1}, \ldots, \approx_{i_k}$ are double implicational.

ad a): We have to prove that if

$$F_{\approx_{i_1} \wedge \ldots \wedge \approx_{i_k}}(a,b,c,d) = 1 \quad and \, a' \geq a \wedge b' \leq b \wedge c' \leq c$$

then $F_{\approx_{i_1} \wedge \ldots \wedge \approx_{i_k}}(a',b',c',d') = 1$. *The assumption* $F_{\approx_{i_1} \wedge \ldots \wedge \approx_{i_k}}(a,b,c,d) = 1$ *means* $F_{\approx_{i_1}}(a,b,c,d) \times \ldots \times F_{\approx_{i_k}}(a,b,c,d) = 1$ *which implies* $F_{\approx_{i_1}}(a,b,c,d) = 1$, ..., $F_{\approx_{i_k}}(a,b,c,d) = 1$. *We assume that all* $\approx_{i_1}, \ldots \approx_{i_k}$ *are double implicational and it together with* $a' \geq a \wedge b' \leq b \wedge c' \leq c$ *means that also* $F_{\approx_{i_1}}(a',b',c',d') = 1$, ..., $F_{\approx_{i_k}}(a',b',c',d') = 1$ *which implies*

$$F_{\approx_{i_1} \wedge \ldots \wedge \approx_{i_k}}(a',b',c',d') = F_{\approx_{i_1}}(a',b',c',d') \times \ldots \times F_{\approx_{i_k}}(a',b',c',d') = 1 \ .$$

This finishes the proof of point a.

ad b): We have to prove that if $F_{\approx_{i_1}}(a,b,c,d) = 1$, ..., $F_{\approx_{i_k}}(a,b,c,d) = 1$ *and* $a' \geq a \wedge b' \leq b \wedge c' \leq c$ *then* $F_{\approx_{i_1}}(a',b',c',d') = 1$, ..., $F_{\approx_{i_k}}(a',b',c',d') = 1$. *The assumption* $F_{\approx_{i_1}}(a,b,c,d) = 1$, ..., $F_{\approx_{i_k}}(a,b,c,d) = 1$ *implies*

$$F_{\approx_{i_1}}(a,b,c,d) \times \ldots \times F_{\approx_{i_k}}(a,b,c,d) = F_{\approx_{i_1} \wedge \ldots \wedge \approx_{i_k}}(a,b,c,d) = 1 \ .$$

We assume that $\approx_{i_1} \wedge \ldots \wedge \approx_{i_k}$ *is double implicational and in addition that* $a' \geq a \wedge b' \leq b \wedge c' \leq c$, *thus* $F_{\approx_{i_1} \wedge \ldots \wedge \approx_{i_k}}(a',b',c',d') = 1$. *This means that*

$$F_{\approx_{i_1} \wedge \ldots \wedge \approx_{i_k}}(a',b',c',d') = F_{\approx_{i_1}}(a',b',c',d') \times \ldots \times F_{\approx_{i_k}}(a',b',c',d') = 1 \ ,$$

which implies that $F_{\approx_{i_1}}(a',b',c',d') = 1$, ..., $F_{\approx_{i_k}}(a',b',c',d') = 1$.
This finishes the proof of point 1.
2. *The proof is analogous to the proof of point 1.*

This finishes the proof. □

Theorem 8.4 can be used to prove the following theorem on double implicational compound quantifiers.

Theorem 8.5. *The following 4ft-quantifiers are both Σ-double implicational and double implicational:*

1. *Founded double p-implication* $\Leftrightarrow_{p,Base}$ *defined as* $\Leftrightarrow_p \wedge \oplus_{Base}$
2. *Founded lower critical double implication* $\Leftrightarrow^{!}_{p,\alpha,Base}$ *defined as* $\Leftrightarrow^{!}_{p,\alpha} \wedge \oplus_{Base}$
3. *Founded upper critical double implication* $\Leftrightarrow^{?}_{p,\alpha,Base}$ *defined as* $\Leftrightarrow^{?}_{p,\alpha} \wedge \oplus_{Base}$

Proof. Directly follows from theorems 8.2, 8.3, and 8.4. □

4ft-quantifier $\approx$ is symmetrical if its associated function $F_{\approx}$ satisfies

$$F_{\approx}(a,b,c,d) = F_{\approx}(a,c,b,d),$$

see definitions 6.5 and 6.6 which are equivalent according to lemma 6.9. The following theorem is their direct consequence.

Theorem 8.6. *The following 4ft-quantifiers are symmetrical:*

1. *Double p-implication* $\Leftrightarrow_p$
2. *Lower critical double implication* $\Leftrightarrow^!_{p,\alpha}$
3. *Upper critical double implication* $\Leftrightarrow^?_{p,\alpha}$
4. *Base* $\oplus_{Base}$
5. *Founded double p-implication* $\Leftrightarrow_{p,Base}$
6. *Founded lower critical double implication* $\Leftrightarrow^!_{p,\alpha,Base}$
7. *Founded upper critical double implication* $\Leftrightarrow^?_{p,\alpha,Base}$

Proof. Directly follows from the definitions of these quantifiers and from definition 6.5. □

8.3   Selected Weakly Double Implicational Quantifiers

We have derived several supported weakly implicational 4ft-quantifiers from important implicational 4ft-quantifiers and the quantifier $\odot_s$ of support, see Sect. 7.3. The same approach can be applied to double implicational quantifiers.

Lemma 8.5. *The following is true for 4ft-quantifier $\odot_s$ of support with $0 < s \le 1$:*

1. *$\odot_s$ is not double implicational but it is weakly double implicational.*
2. *$\odot_s$ is not Σ-double implicational but it is weakly Σ-double implicational.*

Proof. Remember that $\odot_s(a,b,c,d) = 1$ if and only if $\frac{a}{a+b+c+d} \ge s$ for a real number $0 < s \le 1$, see row 1 in table 4.6.

1. *Put $\langle a,b,c,d \rangle = \langle 1,0,0,0 \rangle$ and $\langle a',b',c',d' \rangle = \langle 1,0,0,D \rangle$ where D is an integer number such that $D > \frac{1-s}{s}$. Then we have*

   - $a' \ge a \wedge b' \le b \wedge c' \le c$
   - $F_{\odot_s}(a,b,c,d) = 1$ *because of* $\frac{a}{a+b+c+d} = \frac{1}{1+0+0+0} \ge s$
   - $F_{\odot_s}(a',b',c',d') = 0$ *because of* $\frac{a'}{a'+b'+c'+d'} = \frac{1}{1+0+0+D} < \frac{1}{1+0+0+\frac{1-s}{s}} = \frac{1}{\frac{1}{s}} = s$

   which means that $\odot_s$ is not double implicational.
   Let be $0 < s \le 1$ a real number, $F_{\odot_s}(a,b,c,d) = 1$, $a' \ge a \wedge b' \le b \wedge c' \le c$ and $a+b+c+d = a'+b'+c'+d'$. Then we have $\frac{a}{a+b+c+d} \ge s$ and

   $$\frac{a'}{a'+b'+c'+d'} = \frac{a'}{a+b+c+d} \ge \frac{a}{a+b+c+d} \ge s$$

   which means that $F_{\odot_s}(a',b',c',d') = 1$ and thus $\odot_s$ is weakly double implicational. This finishes the proof of point 1.

2. The proof is analogous to the proof of point 1.

This finishes the proof. □

Lemma 8.6. *Let $\approx_1$ be a double implicational 4ft-quantifier with associated function $F_{\approx_1}$ and let $\approx_2$ be a weakly double implicational quantifier with associated function $F_{\approx_2}$ which is not double implicational. Then*

1. the compound 4ft-quantifier $\approx_1 \wedge \approx_2$ is a weakly double implicational quantifier
2. if there are 4ft-tables $\langle a_0, b_0, c_0, d_0 \rangle$ and $\langle a_0', b_0', c_0', d_0' \rangle$ such that

$$a_0' \geq a_0 \wedge b_0' \leq b_0 \wedge c_0' \leq c_0, \quad F_{\approx_1}(a_0, b_0, c_0, d_0) = 1,$$

$$F_{\approx_2}(a_0, b_0, c_0, d_0) = 1, \text{ and } F_{\approx_2}(a_0', b_0', c_0', d_0') = 0$$

then the compound 4ft-quantifier $\approx_1 \wedge \approx_2$ is not a double implicational quantifier.

Proof

1. We have to prove that if $F_{\approx_1 \wedge \approx_2}(a, b, c, d) = 1$, $a' \geq a \wedge b' \leq b \wedge c' \leq c$, and $a + b + c + d = a' + b' + c' + d'$ then $F_{\approx_1 \wedge \approx_2}(a', b', c', d') = 1$. Assumption $F_{\approx_1 \wedge \approx_2}(a, b, c, d) = 1$ means $F_{\approx_1}(a, b, c, d) \times F_{\approx_2}(a, b, c, d) = 1$, see definition 4.3. This implies both $F_{\approx_1}(a, b, c, d)$ and $F_{\approx_2}(a, b, c, d) = 1$. 4ft-quantifier $\approx_1$ is double implicational, thus the assumption $a' \geq a \wedge b' \leq b \wedge c' \leq c$ means that $F_{\approx_1}(a', b', c', d') = 1$. 4ft-quantifier $\approx_2$ is weakly double implicational and thus $a' \geq a \wedge b' \leq b \wedge c' \leq c$ and $a + b + c + d = a' + b' + c' + d'$ means that $F_{\approx_2}(a', b', c', d') = 1$. We can conclude $F_{\approx_1}(a', b', c', d') \times F_{\approx_2}(a', b', c', d') = 1$. This finishes the proof of point 1.

2. The assumptions $F_{\approx_1}(a_0, b_0, c_0, d_0) = 1$ and $F_{\approx_2}(a_0, b_0, c_0, d_0) = 1$ imply $F_{\approx_1}(a_0, b_0, c_0, d_0) \times F_{\approx_2}(a_0, b_0, c_0, d_0) = 1$ and thus $F_{\approx_1 \wedge \approx_2}(a_0, b_0, c_0, d_0) = 1$. In addition, $F_{\approx_2}(a_0', b_0', c_0', d_0') = 0$, thus $F_{\approx_1}(a_0', b_0', c_0', d_0') \times F_{\approx_2}(a_0', b_0', c_0', d_0') = 0$ and this means that $F_{\approx_1 \wedge \approx_2}(a_0', b_0', c_0', d_0') = 0$. However, it holds $a_0' \geq a_0 \wedge b_0' \leq b_0 \wedge c_0' \leq c_0$ and we can conclude that compound 4ft-quantifier $\approx_1 \wedge \approx_2$ is not a double implicational quantifier. This finishes the proof. □

Lemma 8.7. *Let $\approx_1$ be a Σ-double implicational 4ft-quantifier with associated function $F_{\approx_1}$ and let $\approx_2$ be a weakly Σ-double implicational quantifier with associated function $F_{\approx_2}$ which is not Σ-double implicational. Then*

1. the compound 4ft-quantifier $\approx_1 \wedge \approx_2$ is a weakly Σ-double implicational quantifier
2. if there are 4ft-tables $\langle a_0, b_0, c_0, d_0 \rangle$ and $\langle a_0', b_0', c_0', d_0' \rangle$ such that

$$a_0' \geq a_0 \wedge b_0' + c_0' \leq b_0 + c_0, \quad F_{\approx_1}(a_0, b_0, c_0, d_0) = 1,$$

$$F_{\approx_2}(a_0, b_0, c_0, d_0) = 1, \text{ and } F_{\approx_2}(a_0', b_0', c_0', d_0') = 0$$

then the compound 4ft-quantifier $\approx_1 \wedge \approx_2$ is not a Σ-double implicational quantifier.

Proof. The proof is analogous to the proof of lemma 8.6. $\square$

Theorem 8.7. *The following 4ft-quantifiers are weakly Σ-double implicational but they are neither double implicational nor Σ-double implicational:*

1. *Supported double p-implication $\leftrightarrow_{p,s}$ defined as $\Leftrightarrow_p \wedge \odot_s$*
2. *Supported lower critical double implication $\leftrightarrow^!_{p,\alpha,s}$ defined as $\Leftrightarrow^!_{p,\alpha} \wedge \odot_s$*
3. *Supported upper critical double implication $\leftrightarrow^?_{p,\alpha,s}$ defined as $\Leftrightarrow^?_{p,\alpha} \wedge \odot_s$.*

Proof. We prove the theorem in three steps:

A) *We prove that all 4ft-quantifiers in question are weakly Σ-double implicational quantifiers.*

B) *We prove for each of 4ft-quantifiers in question that it is not double implicational.*

C) *We prove for each of 4ft-quantifiers in question that it is not Σ-double implicational.*

ad A): According to theorem 8.2, all 4ft-quantifiers $\Leftrightarrow_p$, $\Leftrightarrow^!_{p,\alpha}$, and $\Leftrightarrow^?_{p,\alpha}$ are Σ-double implicational quantifiers. Thus, it follows from lemma 8.2 that they are also weakly Σ-double implicational. The 4ft-quantifier $\odot_s$ is weakly Σ-double implicational, see point 2 of lemma 8.5. Point 1 of lemma 8.7 implies that all the above defined supported quantifiers are weakly Σ-double implicational quantifiers.

ad B): Theorem 8.2 says that all 4ft-quantifiers $\Leftrightarrow_p$, $\Leftrightarrow^!_{p,\alpha}$, and $\Leftrightarrow^?_{p,\alpha}$ are Σ-double implicational quantifiers and thus theorem 8.1 implies that all of them are double implicational. In addition, the 4ft-quantifier $\odot_s$ is weakly double implicational but it is not double implicational, see point 1 of lemma 8.5.

We use point 2 of lemma 8.6 to prove for each 4ft-quantifier $\approx$ of 4ft-quantifiers in question that it is not double implicational. We assume $\approx = \approx_1 \wedge \odot$. We have to find 4ft-tables $\langle a_0, b_0, c_0, d_0 \rangle$ and $\langle a'_0, b'_0, c'_0, d'_0 \rangle$ such that $a'_0 \geq a_0 \wedge b'_0 \leq b_0 \wedge c'_0 \leq c_0$, $F_{\approx_1}(a_0, b_0, c_0, d_0) = 1$, $F_{\odot}(a_0, b_0, c_0, d_0) = 1$, and $F_{\odot}(a'_0, b'_0, c'_0, d'_0) = 0$.

Let $\langle a_0, b_0, c_0, d_0 \rangle = \langle 1, 0, 0, 0 \rangle$ and $\langle a'_0, b'_0, c'_0, d'_0 \rangle = \langle 1, 0, 0, D \rangle$ where D is an integer number such that $D > \frac{1-s}{s}$. Then for $0 < s \leq 1$ we have:

- $a'_0 \geq a_0 \wedge b'_0 \leq b_0 \wedge c'_0 \leq c_0$
- $F_{\odot_s}(a_0, b_0, c_0, d_0) = 1$ *because of* $\frac{1}{1+0+0+0} \geq s$
- $F_{\odot_s}(a'_0, b'_0, c'_0, d'_0) = 0$ *because of* $\frac{1}{1+0+0+D} < \frac{1}{1+0+0+\frac{1-s}{s}} = \frac{1}{\frac{1}{s}} = s$.

In addition we have:

- $F_{\Leftrightarrow_p}(a_0, b_0, c_0, d_0) = 1$ *because* $\frac{1}{1+0+0} \geq p$ *and thus the supported double p-implication $\leftrightarrow_{p,s}$ is not double implicational, see point 1 of the theorem.*
- $F_{\Leftrightarrow^?_{p,\alpha,s}}(a_0, b_0, c_0, d_0) = 1$ *for $0 < p < 1$ and $0 < \alpha \leq 0.5$ because*

$$\sum_{i=0}^{1} \binom{1+0+0}{i} p^i (1-p)^{1+0+0-i} = \binom{1}{0} p^0 (1-p)^{1-0} + \binom{1}{1} p^1 (1-p)^{1-1} =$$
$$= 1 - p + p > \alpha \text{ and thus supported upper critical double implication } \rightarrow^?_{p,\alpha,s} \text{ is}$$

not double implicational, see point 3 of the theorem.

It remains to prove that the supported 4ft-quantifier $\leftrightarrow^!_{p,\alpha,s}$ (see point 2 of the theorem) is not double implicational. Let $\langle a_0, b_0, c_0, d_0 \rangle = \langle A, 0, 0, 0 \rangle$ such that $p^A \leq \alpha$ and let $\langle a'_0, b'_0, c'_0, d'_0 \rangle = \langle A, 0, 0, D \rangle$ where D is an integer number such that $D > \frac{A - sA}{s}$. Then for $0 < s \leq 1$ we have:

- $a'_0 \geq a_0 \wedge b'_0 \leq b_0 \wedge c'_0 \leq c_0$
- $F_{\odot Base}(a_0, b_0, c_0, d_0) = 1$ *because* $\frac{A}{A+0+0} \geq s$
- $F_{\odot Base}(a'_0, b'_0, c'_0, d'_0) = 0$ *because* $\frac{A}{A+0+0+D} < \frac{A}{A+0+0+\frac{A-sA}{s}} = \frac{A}{\frac{A}{s}} = s$
- $F_{\leftrightarrow^!_{p,\alpha}}(a_0, b_0, c_0, d_0) = 1$ *for* $0 < p < 1$ *and* $0 < \alpha \leq 0.5$ *because*

  $\sum_{i=A}^{A+0+0} \binom{A+0+0}{i} p^i (1-p)^{A+0+0-i} = \binom{A}{A} p^A (1-p)^{A-A} = p^A \leq \alpha$. *Thus the supported lower critical double implication $\to^!_{p,\alpha,s}$ is not double implicational.*

This finishes the step B).

ad C): We have to prove for each 4ft-quantifier $\approx$ of the 4ft-quantifiers in question that $\approx$ is not Σ-double implicational. We have just proven that $\approx$ is not double implicational. Thus, according to theorem 8.1, $\approx$ is not Σ-double implicational.

This finishes the proof of the theorem. □

8.4 Interesting Σ-double Implicational Quantifiers

There exist important theoretical results for specific Σ-double implicational quantifiers, which we call interesting Σ-double implicational quantifiers, similarly as was the case for implicational quantifiers. First we define (b+c)-dependent Σ-double implicational quantifiers, see also definition 7.4. Please remember that for a double implicational quantifier $\leftrightarrow^*$ we can write $F_{\leftrightarrow^*}(a,b,c)$ instead of $F_{\leftrightarrow^*}(a,b,c,d)$, see note 6.2.

Definition 8.8. Let $\approx$ be a 4ft-quantifier. Then $\approx$ is *(b+c)-dependent* if there are non-negative integers a, b, c, d, b', c' such that

$$a + b + c + d > 0 \wedge a + b' + c' + d > 0 \wedge b + c \neq b' + c'$$

and

$$F_{\approx}(a,b,c,d) \neq F_{\approx}(a,b',c',d) .$$

Definition 8.9. *Σ-double implicational quantifier $\leftrightarrow^*$ is interesting* if it satisfies:

- $\leftrightarrow^*$ is a-dependent (see definition 7.4)
- $\leftrightarrow^*$ is (b+c)-dependent
- $F_{\leftrightarrow^*}(0,0,0) = 0$.

We are going to prove that 4ft-quantifiers $\leftrightarrow_p$ and $\leftrightarrow^!_{p,\alpha}$ are interesting Σ-double implicational rules. In addition, we prove that 4ft-quantifier $\leftrightarrow^?_{p,\alpha}$ is a-dependent and (b+c)-dependent but it does not satisfy $F_{\leftrightarrow^?_{p,\alpha}}(0,0,0) = 0$.

Theorem 8.8. *The following 4ft-quantifiers are interesting Σ-double implicational quantifiers:*

1. *Double p-implication $\Leftrightarrow_p$ (i.e. Jaccard), see rows 4 in Tables 4.2 and 4.4 and rows 12 in Tables 4.5 and 4.6*
2. *Lower critical double implication $\Leftrightarrow'_{p,\alpha}$, see rows 5 in Tables 4.2 and 4.4*

Proof. Both 4ft-quantifiers are Σ-double implicational according to theorem 8.2. We have to prove the following facts D_a), D_{b+c}, $D_{0,0,0}$) for each of them:

D_a)   *there are non-negative integers a, a', b, c, d such that*

$$a+b+c+d>0 \wedge a'+b+c+d>0 \wedge F_{\Leftrightarrow^*}(a,b,c,d) \neq F_{\Leftrightarrow^*}(a',b,c,d)$$

D_{b+c})   *there are non-negative integers a, b, b', c, c', d such that*

$$a+b+c+d>0 \wedge a+b'+c'+d>0 \wedge b+c \neq b'+c'$$

and $F_{\Leftrightarrow^*}(a,b,c,d) \neq F_{\Leftrightarrow^*}(a,b',c,d)$.
$D_{0,0,0}$) $F_{\Leftrightarrow^*}(0,0,0) = 0.$

1. *$F_{\Leftrightarrow_p}(a,b,c,d) = 1$ if and only if $\frac{a}{a+b+c} \geq p \wedge a+b+c>0$ where $0<p\leq 1$, which means:*

    D_a)   *$F_{\Leftrightarrow_p}(1,0,0,1) = 1$ and $F_{\Leftrightarrow_p}(0,0,0,1) = 0$ (see point G_3 in definition 4.1); note that for $0<p<1$ we also have $F_{\Leftrightarrow_p}(A,1,0,0) = 1$ for $A \geq \frac{p}{p+1}$ and $F_{\Leftrightarrow_p}(0,1,0,0) = 0$*

    D_{b+c})   *$F_{\Leftrightarrow_p}(1,0,0,0) = 1$ and for $b > \frac{1-p}{p}$ also $F_{\Leftrightarrow_p}(1,b,0,0) = 0$ because of*
    $$\frac{1}{1+b} < \frac{1}{1+\frac{1-p}{p}} = \frac{p}{p+1-p} = p$$
    $D_{0,0,0}$)   *$F_{\Leftrightarrow_p}(0,0,0) = 0$, see point G_3 in definition 4.1.*

    This finishes the proof for 4ft-quantifier $\Leftrightarrow_p$.

2. *$F_{\Leftrightarrow'_{p,\alpha}}(a,b,c,d) = 1$ if and only if $\sum_{i=a}^{a+b+c} \binom{a+b+c}{i} p^i(1-p)^{a+b+c-i} \leq \alpha$, we assume $0<p<1$ and $0<\alpha\leq 0.5$, which means:*

    D_a)   *for a > 0 satisfying $p^a < \alpha$ it holds $F_{\Leftrightarrow'_{p,\alpha}}(a,0,0,1) = 1$ because*
    $\sum_{i=a}^{a+0+0} \binom{a+0+0}{i} p^i(1-p)^{a+0+0-i} = \binom{a+0+0}{a} p^a(1-p)^{a+0+0-a} = p^a < \alpha$
    and $F_{\Leftrightarrow'_{p,\alpha}}(0,0,0,1) = 0$ (see point G_4 in definition 4.1)

    D_{b+c})   *for a > 0 satisfying $p^a < \alpha$ it holds $F_{\Leftrightarrow'_{p,\alpha}}(a,0,0,0) = 1$, see above; according to point 1 in lemma 5.16 we have*
    $\lim_{b\to\infty} \sum_{i=a}^{a+b+0} \binom{a+b+0}{i} p^i(1-p)^{a+b+0-i} = 1$, *thus there is b such that* $\sum_{i=a}^{a+b+0} \binom{a+b+0}{i} p^i(1-p)^{a+b+0-i} > \alpha$, *which means that there are b,c such that $b+c>0$ and $F_{\Leftrightarrow'_{p,\alpha}}(a,b,0,0) = 0$*

    $D_{0,0,0}$)   *$F_{\Leftrightarrow'_{p,\alpha}}(0,0,0) = 0$, see point G_4 in definition 4.1.*

    This finishes the proof for 4ft-quantifier $\Leftrightarrow'_{p,\alpha}$.

This finishes the proof.   □

Theorem 8.9. *The 4ft-quantifier $\Leftrightarrow^?_{p,\alpha}$ of suspicious critical double implication (i.e. upper critical double implication), see rows 6 in Tables 4.2 and 4.4, is both a-dependent and (b+c)-dependent for $0 < p < 1$ and $0 < \alpha \leq 0.5$, but it does not satisfy the condition $F_{\Leftrightarrow^?_{p,\alpha}}(0,0,0) = 0$.*

Proof. It holds $F_{\Leftrightarrow^?_{p,\alpha}}(a,b,c,d) = 1$ if and only if $\sum_{i=0}^{a} \binom{a+b+c}{i} p^i (1-p)^{a+b+c-i} > \alpha$, we assume $0 < \alpha \leq 0.5$, which means (we use the same notation $D_a)$, $D_{b+c})$, and $D_{0,0,0})$ as in theorem 8.8):

$D_a)$ $\sum_{i=0}^{0} \binom{0+b+0}{i} p^i (1-p)^{0+b+0-i} = (1-p)^b$ *and thus there is $b > 0$ such that $(1-p)^b < \alpha$, which means $F_{\Leftrightarrow^?_{p,\alpha}}(0,b,0,0) = 0$; according to point 3 of lemma 5.16 there is a' such that $\sum_{i=0}^{a'} \binom{a'+b+0}{i} p^i (1-p)^{a'+b+0-i} > \alpha$ which means $F_{\Leftrightarrow^?_{p,\alpha}}(a',b,0,1) = 1$*

$D_{b+c})$ $F_{\Leftrightarrow^?_{p,\alpha}}(0,0,0,1) = 1$ *because $\sum_{i=0}^{0} \binom{0+0+0}{i} p^i (1-p)^{0+0+0-i} = 1$; and there is also $b' > 0$ such that $F_{\Leftrightarrow^?_{p,\alpha}}(0,b',0,1) = 0$ (see $D_a)$)*

$D_{0,0,0})$ $\sum_{i=0}^{0} \binom{0+0+0}{i} p^i (1-p)^{0+0+0-i} = 1 > \alpha$ *for $0 < \alpha \leq 0.5$ and thus we have $F_{\Leftrightarrow^?_{p,\alpha}}(0,0,0,d) = 1$ for all $d > 0$, which means that condition $F_{\Leftrightarrow^*}(0,0,0) = 0$ is not satisfied.*

This finishes the proof.   □

Additionally we prove that the founded Σ-double implicational quantifiers from theorem 8.5 are interesting Σ-double implicational quantifiers.

Theorem 8.10. *The following founded 4ft-quantifiers are interesting Σ-double implicational quantifiers.*

1. *Founded double p-implication $\Leftrightarrow_{p,Base}$ defined as $\Leftrightarrow_p \wedge \oplus_{Base}$*
2. *Founded lower critical double implication $\Leftrightarrow^!_{p,\alpha,Base}$ defined as $\Leftrightarrow^!_{p,\alpha} \wedge \oplus_{Base}$*
3. *Founded upper critical double implication $\Leftrightarrow^?_{p,\alpha,Base}$ defined as $\Leftrightarrow^?_{p,\alpha} \wedge \oplus_{Base}$*

Proof. All the above three 4ft-quantifiers are Σ-double implicational according to theorem 8.5. We need to prove that the following $D_a)$, $D_{b+c})$ and $D_{0,0,0})$ hold for all three of them:

$D_a)$ *there are non-negative integers a, a', b, c, d such that*

$$a+b+c+d > 0 \wedge a'+b+c+d > 0 \wedge F_{\Leftrightarrow^*}(a,b,c,d) \neq F_{\Leftrightarrow^*}(a',b,c,d)$$

$D_{b+c})$ *there are non-negative integers a, b, b', c, c', d such that*

$$a+b+c+d > 0 \wedge a+b'+c'+d > 0 \wedge b+c \neq b'+c'$$

and $F_{\Leftrightarrow^*}(a,b,c,d) \neq F_{\Leftrightarrow^*}(a,b',c,d)$.
$D_{0,0,0})$ $F_{\Leftrightarrow^*}(0,0,0) = 0$.

All these quantifiers are founded, meaning that $F_{\Leftrightarrow^*}(0,0,0) = 0$ for all of them because $Base > 0$ and thus $F_{\oplus_{Base}}(0,0,0,d) = 0$ which means $F_{\Leftrightarrow^*}(0,0,0) = 0$. Therefore it is sufficient to prove only $D_a)$ and $D_{b+c})$ for each of the above three 4ft-quantifiers. In most cases we can use the proofs of theorems 8.8 and 8.9. Proofs for particular 4ft-quantifiers follow.

1. $F_{\Leftrightarrow_{p,Base}}(a,b,c,d) = 1$ if and only if $\frac{a}{a+b+c} \geq p \wedge a \geq Base$ where $0 < p \leq 1$, which means:

   $D_a)$ $F_{\Leftrightarrow_{p,Base}}(Base,0,0,1) = 1$ and $F_{\Leftrightarrow_{p,Base}}(0,0,0,1) = 0$.

   $D_{b+c})$ $F_{\Leftrightarrow_{p,Base}}(Base,0,0,0) = 1$ and for $b > Base\frac{1-p}{p}$ also
   $F_{\Leftrightarrow_{p,Base}}(Base,b,0,0) = 0$ because of $\frac{Base}{Base+b+0} < \frac{Base}{Base+Base\frac{1-p}{p}} = p$.

2. $F_{\Leftrightarrow^!_{p,\alpha,Base}}(a,b,c,d) = 1$ if and only if
   $\sum_{i=a}^{a+b+c} \binom{a+b+c}{i} p^i (1-p)^{a+b+c-i} \leq \alpha \wedge a \geq Base$, we assume $0 < p < 1$ and $0 < \alpha \leq 0.5$, which means:

   $D_a)$ for $a > Base$ satisfying $p^a < \alpha$ it holds $F_{\Leftrightarrow^!_{p,\alpha}}(a,0,0,1) = 1$ because
   $\sum_{i=a}^{a+0+0} \binom{a+0+0}{i} p^i (1-p)^{a+0-i} = \binom{a+0+0}{a} p^a (1-p)^{a+0+0-a} = p^a < \alpha$
   and $F_{\Leftrightarrow^!_{p,\alpha,Base}}(0,0,0,1) = 0$

   $D_{b+c})$ for $a > Base$ satisfying $p^a < \alpha$ it holds $F_{\Leftrightarrow^!_{p,\alpha,Base}}(a,0,0,0) = 1$, see
   above; point 1 in lemma 5.16 implies $\lim_{b\to\infty} \sum_{i=a}^{a+b+0} \binom{a+b+0}{i} p^i (1-p)^{a+b+0-i}$
   $= 1$, thus there exists b such that $\sum_{i=a}^{a+b+0} \binom{a+b+0}{i} p^i (1-p)^{a+b+0-i} > \alpha$ and
   this means $F_{\Leftrightarrow^!_{p,\alpha}}(a,b,0,0) = 0$.

3. $F_{\Leftrightarrow^?_{p,\alpha,Base}}(a,b,c,d) = 1$ if and only if
   $\sum_{i=0}^{a} \binom{a+b+c}{i} p^i (1-p)^{a+b+c-i} > \alpha \wedge a \geq Base$, we assume $0 < p < 1$ and $0 < \alpha \leq 0.5$, which means:

   $D_a)$ $F_{\Leftrightarrow^?_{p,\alpha,Base}}(Base,0,0,1) = 1$ because
   $\sum_{i=0}^{Base} \binom{Base+0+0}{i} p^i (1-p)^{Base+0+0-i} = 1$ and $F_{\Leftrightarrow^?_{p,\alpha,Base}}(0,0,0,1) = 0$.

   $D_{b+c})$ $F_{\Leftrightarrow^?_{p,\alpha,Base}}(Base,0,0,1) = 1$, see $D_a)$; according to point 1 in lemma
   5.15 we have $\lim_{b\to\infty} \sum_{i=0}^{Base} \binom{Base+b+0}{i} p^i (1-p)^{Base+b+0-i} = 0$ and thus
   there is b such that $\sum_{i=0}^{Base} \binom{Base+b+0}{i} p^i (1-p)^{Base+b+0-i} < \alpha$ and
   $F_{\Leftrightarrow^?_{p,\alpha,Base}}(Base,b,0,0) = 0$.

This finishes the proof. $\square$

8.5   Tables of Critical Frequencies for Double Implicational Quantifiers

We defined the table $Tb_{\Rightarrow^*}$ of maximal b for implicational quantifier $\Rightarrow^*$ in Sect. 7.5. It can be used to avoid complex computations when evaluating the associated function $F_{\Rightarrow^*}$ of $\Rightarrow^*$ as well as to achieve theoretical results. Table $Tb_{\Rightarrow^*}$ of maximal b is a table of critical frequencies for implicational quantifiers.

We define similar tables of critical frequencies also for double imlicational and Σ-double imlicational quantifiers. Then we prove two simple theorems showing that these tables can be also used to avoid complex computations when evaluating the associated function of a given double imlicational or Σ-double imlicational quantifier.

In the following definitions and theorems we again use the fact that a value $F_{\Leftrightarrow^*}(a,b,c,d)$ of the associated function of double implicational quantifier $\Leftrightarrow^*$ does not depend on d, see note 6.2 and lemma 6.6. Thus we write only $F_{\Leftrightarrow^*}(a,b,c)$ instead of $F_{\Leftrightarrow^*}(a,b,c,d)$ for the double implicational quantifier $\Leftrightarrow^*$.

Please remember definition 7.6 where the set $\mathcal{N}^+$ is defined as a union $\mathcal{N}^+ = \mathcal{N} \cup \{\infty\}$, it holds also $n < \infty$, $n + \infty = \infty$, $\infty - n = \infty$ for each $n \in \mathcal{N}$, $\max(\mathcal{N}) = \infty$, and $\min(\mathcal{N}^+) = \infty$.

Definition 8.10. Let $\Leftrightarrow^*$ be a double implicational quantifier. Then

1. a *table $Tb_{c,\Leftrightarrow^*}$ of maximal c for* $\Leftrightarrow^*$ is a function assigning to each couple $\langle a,b\rangle$ of non-negative integer numbers $a,b \in \mathcal{N}$ a value $Tb_{c,\Leftrightarrow^*}(a,b)$ from $\mathcal{N}^+$ such that
$$Tb_{c,\Leftrightarrow^*}(a,b) = \min\{e|F_{\Leftrightarrow^*}(a,b,e) = 0\}$$

2. a *table $Tb_{c,\Leftrightarrow^*,b}$ of maximal c for* $\Leftrightarrow^*$ *and frequency b* is a function assigning to each integer $a \geq 0$ a value
$$Tb_{c,\Leftrightarrow^*,b}(a) = Tb_{c,\Leftrightarrow^*}(a,b)$$

3. a *table $Tb_{b,\Leftrightarrow^*}$ of maximal b for* $\Leftrightarrow^*$ is a function assigning to each couple $\langle a,c\rangle$ of non-negative integer numbers $a,c \in \mathcal{N}$ a value $Tb_{b,\Leftrightarrow^*}(a,c)$ from $\mathcal{N}^+$ such that
$$Tb_{b,\Leftrightarrow^*}(a,c) = \min\{e|F_{\Leftrightarrow^*}(a,e,c) = 0\}$$

4. a *table $Tb_{b,\Leftrightarrow^*,c}$ of maximal b for* $\Leftrightarrow^*$ *and frequency c* is a function assigning to each integer $a \geq 0$ value
$$Tb_{b,\Leftrightarrow^*,c}(a) = Tb_{b,\Leftrightarrow^*}(a,c) \,.$$

Theorem 8.11. *Let* $\Leftrightarrow^*$ *be a double implicational quantifier with associated function* $F_{\Leftrightarrow^*}$. *Then*

1. $F_{\Leftrightarrow^*}(a,b,c) = 1$ *if and only if* $c < Tb_{c,\Leftrightarrow^*}(a,b)$ *where* $Tb_{c,\Leftrightarrow^*}$ *is a table of maximal c for* $\Leftrightarrow^*$
2. *table* $Tb_{c,\Leftrightarrow^*,b}(a)$ *of maximal c for* $\Leftrightarrow^*$ *and frequency b is a nondecreasing function of a*
3. $F_{\Leftrightarrow^*}(a,b,c) = 1$ *if and only if* $b < Tb_{b,\Leftrightarrow^*}(a,c)$ *where* $Tb_{b,\Leftrightarrow^*}$ *is a table of maximal b for* $\Leftrightarrow^*$
4. *table* $Tb_{b,\Leftrightarrow^*,c}(a)$ *of maximal b for* $\Leftrightarrow^*$ *and frequency c is a nondecreasing function of a.*

Proof

1. *We use the fact that* $F_{\Leftrightarrow^*}(a,b,Tb_{c,\Leftrightarrow^*}(a,b)) = 0$. *It follows from the definition that* $Tb_{c,\Leftrightarrow^*}(a,b) = \min\{e|F_{\Leftrightarrow^*}(a,b,e) = 0\}$.

   - *Let* $F_{\Leftrightarrow^*}(a,b,c) = 1$, *we have to show that* $c < Tb_{c,\Leftrightarrow^*}(a,b)$. *It cannot happen that* $Tb_{c,\Leftrightarrow^*}(a,b) \leq c$ *because* $F_{\Leftrightarrow^*}(a,b,c) = 1$ *and* $Tb_{c,\Leftrightarrow^*}(a,b) \leq c$ *implies* $F_{\Leftrightarrow^*}(a,b,Tb_{c,\Leftrightarrow^*}(a,b)) = 1$ *due to the assumption that* $\Leftrightarrow^*$ *is a double implicational quantifier. Thus it must be the case that* $c < Tb_{c,\Leftrightarrow^*}(a,b)$.
   - *Let* $F_{\Leftrightarrow^*}(a,b,c) = 0$. *Then* $Tb_{c,\Leftrightarrow^*}(a,b) \leq c$ *because of definition* $Tb_{c,\Leftrightarrow^*}(a,b) = \min\{e|F_{\Leftrightarrow^*}(a,b,e) = 0\}$. *Thus it is not true that* $c < Tb_{c,\Leftrightarrow^*}(a,b)$.

   This finishes the proof of point 1.

2. *Let* $a' > a$. *We have to prove* $Tb_{c,\Leftrightarrow^*,b}(a') \geq Tb_{c,\Leftrightarrow^*,b}(a)$ *which is equivalent to* $Tb_{c,\Leftrightarrow^*}(a',b) \geq Tb_{c,\Leftrightarrow^*}(a,b)$. *Let us assume* $Tb_{c,\Leftrightarrow^*}(a',b) < Tb_{c,\Leftrightarrow^*}(a,b)$. *This means* $F_{\Leftrightarrow^*}(a,b,Tb_{c,\Leftrightarrow^*}(a',b)) = 1$ *because of point 1 (we substitute* $c = Tb_{c,\Leftrightarrow^*}(a',b)$). *The 4ft-quantifier* $\Leftrightarrow^*$ *is double implicational, we assume* $a' > a$ *and thus we have* $F_{\Leftrightarrow^*}(a',b,Tb_{c,\Leftrightarrow^*}(a',b)) = 1$. *However, this contradicts* $F_{\Leftrightarrow^*}(a',b,Tb_{c,\Leftrightarrow^*}(a',b)) = 0$. *Thus it must* $Tb_{c,\Leftrightarrow^*}(a',b) \geq Tb_{c,\Leftrightarrow^*}(a',b)$ *i.e.* $Tb_{c,\Leftrightarrow^*,b}(a') \geq Tb_{c,\Leftrightarrow^*,b}(a)$. *This finishes the proof of point 2.*

3. *The proof is similar to that of point 1.*
4. *The proof is similar to that of point 2.*

This finishes the proof.   □

Definition 8.11. Let $\Leftrightarrow^*$ be a Σ-double imlicational quantifier. Then a *table* $Tb_{\Sigma,\Leftrightarrow^*}$ *of maximal* $b + c$ for $\Leftrightarrow^*$ is a function $Tb_{\Sigma,\Leftrightarrow^*}$ assigning to each non-negative integer number $a \in \mathcal{N}$ a value $Tb_{\Sigma,\Leftrightarrow^*}(a)$ from $\mathcal{N}^+$ such that

$$Tb_{\Sigma,\Leftrightarrow^*}(a) = \min\{b + c|F_{\Leftrightarrow^*}(a,b,c) = 0\} \,.$$

Theorem 8.12. *Let* $Tb_{\Sigma,\Leftrightarrow^*}$ *be a table of maximal* $b + c$ *for* Σ-double implicational quantifier $\Leftrightarrow^*$ *with associated function* $F_{\Leftrightarrow^*}$. *Then we have*

1. $F_{\Leftrightarrow^*}(a,b,c) = 1$ *if and only if* $b + c < Tb_{\Sigma,\Leftrightarrow^*}(a)$
2. $Tb_{\Sigma,\Leftrightarrow^*}(a)$ *is a nondecreasing function of a.*

Proof. We use the fact that $F_{\Leftrightarrow^*}(a,b,c) = 0$ *for each* $b \geq 0$, $c \geq 0$ *such that* $b + c = Tb_{\Sigma,\Leftrightarrow^*}(a)$. *Keep in mind that* $Tb_{\Sigma,\Leftrightarrow^*}(a) = \min\{b + c|F_{\Rightarrow^*}(a,b,c) = 0\}$.

1. • *Let* $F_{\Leftrightarrow^*}(a,b,c) = 1$, *we have to prove* $b + c < Tb_{\Sigma,\Leftrightarrow^*}(a)$. *We cannot have* $Tb_{\Sigma,\Leftrightarrow^*}(a) \leq b + c$ *because* $F_{\Leftrightarrow^*}(a,b,c) = 1$ *and* $Tb_{\Sigma,\Leftrightarrow^*}(a) \leq b + c$ *implies* $F_{\Leftrightarrow^*}(a,Tb_{\Sigma,\Leftrightarrow^*}(a),0) = 1$ *due to the assumption that* $\Leftrightarrow^*$ *is a* Σ-double implicational quantifier. However, we know that $F_{\Leftrightarrow^*}(a,Tb_{\Sigma,\Leftrightarrow^*}(a),0) = 0$ *and thus it must hold that* $b + c < Tb_{\Sigma,\Leftrightarrow^*}(a)$.
   • *Let* $F_{\Leftrightarrow^*}(a,b,c) = 0$, *we have to prove that* $b + c \geq Tb_{\Sigma,\Leftrightarrow^*}(a)$. *We have* $Tb_{\Sigma,\Leftrightarrow^*}(a) \leq b + c$ *because* $Tb_{\Sigma,\Leftrightarrow^*}(a) = \min\{b + c|F_{\Leftrightarrow^*}(a,b,c) = 0\}$.
   This finishes the proof of point 1.

2. *Let $a' > a$. We have to prove $Tb_{\Sigma,\Leftrightarrow^*}(a') \geq Tb_{\Sigma,\Leftrightarrow^*}(a)$. Let us assume $Tb_{\Sigma,\Leftrightarrow^*}(a') < Tb_{\Sigma,\Leftrightarrow^*}(a)$. This means $F_{\Leftrightarrow^*}(a, Tb_{\Sigma,\Leftrightarrow^*}(a'), 0) = 1$ because of point 1 (we substitute $b = Tb_{\Sigma,\Leftrightarrow^*}(a')$ and $c = 0$). 4ft-quantifier $\Leftrightarrow^*$ is Σ-double implicational and $a' > a$, thus $F_{\Leftrightarrow^*}(a', Tb_{\Sigma,\Leftrightarrow^*}(a'), 0) = 1$ which is a contradiction with $F_{\Leftrightarrow^*}(a', Tb_{\Sigma,\Leftrightarrow^*}(a'), 0) = 0$. Thus it must be $Tb_{\Sigma,\Leftrightarrow^*}(a') \geq Tb_{\Sigma,\Leftrightarrow^*}(a)$.*

This finishes the proof. □

8.6 Typical Σ-double Implicational Quantifiers

We are going to define a typical Σ-double implicational quantifier. We show also that quantifiers $\Leftrightarrow_p$, $\Leftrightarrow^!_{p,\alpha}$, and $\Leftrightarrow^?_{p,\alpha}$ are typical Σ-double implicational quantifier for most values of their parameters.

Definition 8.12. Let $\Leftrightarrow^*$ be a Σ-double implicational quantifier and let $Tb_{\Sigma,\Leftrightarrow^*}$ be its table of maximal $b + c$. We say that the quantifier $\Leftrightarrow^*$ is a *typical Σ-double implicational* if there is an integer A such that $1 < Tb_{\Sigma,\Leftrightarrow}(A) < \infty$.

We say that a Σ-double implicational quantifier $\Leftrightarrow^*$ is *untypical Σ-double implicational* if it is not typical Σ-double implicational.

Theorem 8.13. *Let $0 < p < 1$ and $0 < \alpha \leq 0.5$ be real numbers. Then the quantifiers $\Leftrightarrow_p$ of double p-implication, $\Leftrightarrow^!_{p,\alpha}$ of lower critical double implication, and $\Leftrightarrow^?_{p,\alpha}$ of upper critical double implication are typical Σ-double implicational.*

Proof. To prove that a Σ-double implicational quantifier $\Leftrightarrow$ is typical we have to find integer A such that $1 < Tb_{\Sigma,\Leftrightarrow}(A) < \infty$. If we find integer $A > 0, B, C, B', C'$, such that $B + C \geq 1$, $F_\Leftrightarrow(A, B, C) = 1$ and $F_\Leftrightarrow(A, B', C') = 0$ then it holds

$$1 \leq B + C < Tb_{\Sigma,\Leftrightarrow}(A) = \min\{b + c | F_\Leftrightarrow(A, b, c) = 0\} \leq B' + C'$$

which finishes the proof that the Σ-double implicational quantifier $\Leftrightarrow$ is typical. We are going to apply this approach to each of the 4ft-quantifiers $\Leftrightarrow_p$, $\Leftrightarrow^!_{p,\alpha}$, and $\Leftrightarrow^?_{p,\alpha}$. We write only $F_\Leftrightarrow(A, b, c)$ etc. in accordance with lemma 6.6.

1. The 4ft-quantifier $\Leftrightarrow_p$ of double p-implication is defined such that

$$F_{\Leftrightarrow_p}(a, b, c, d) = \begin{cases} 1 & \text{if } \frac{a}{a+b+c} \geq p \\ 0 & \text{otherwise.} \end{cases}$$

Let $B = 1$, $C = 0$. We assume $0 < p < 1$ thus there surely exists A such that $\frac{A}{A+1} \geq p$. This means $F_{\Leftrightarrow_p}(A, B, C) = 1$. In addition let be $C' = 0$. Then there surely is an integer $B' > 0$ such that $\frac{A}{A+B'} < p$ which means $F_{\Leftrightarrow_p}(A, B', C') = 0$. This finishes the proof for 4ft-quantifier $\Leftrightarrow_p$.

2. The 4ft-quantifier $\Leftrightarrow^!_{p,\alpha}$ of lower critical double implication is defined for $0 < p < 1$ and $0 < \alpha \leq 0.5$ as follows:

$$F_{\Leftrightarrow^!_{p,\alpha}}(a, b, c, d) = \begin{cases} 1 & \text{if } \sum_{i=a}^{a+b+c} \binom{a+b+c}{i} p^i (1-p)^{a+b+c-i} \leq \alpha \\ 0 & \text{otherwise.} \end{cases}$$

Let $B = 1$ and $C = 0$. We assume $0 < p < 1$, according to lemma 5.15, point 2) it holds

$$\lim_{N \to \infty} \sum_{i=N-1}^{N} \binom{N}{i} p^i (1-p)^{N-i} = 0$$

and thus there is an integer $Q > 1$ such that

$$\sum_{i=Q-1}^{Q} \binom{Q}{i} p^i (1-p)^{Q-i} \leq \alpha$$

which means $\Leftrightarrow_{p,\alpha}^! (Q-1,1,0) = 1$. In addition, according to lemma 5.16, point 1) it holds

$$\lim_{N \to \infty} \sum_{i=Q-1}^{Q-1+N} \binom{Q-1+N}{i} p^i (1-p)^{Q-1+N-i} = 1$$

and thus there is an integer B', such that

$$\sum_{i=Q-1}^{Q-1+B'} \binom{Q-1+B'}{i} p^i (1-p)^{Q-1+B'-i} > \alpha$$

which means $\Leftrightarrow_{p,\alpha}^! (Q, B', 0) = 0$. This finishes the proof for 4ft-quantifier $\Leftrightarrow_{p \cdot \alpha}^!$.

3. The 4ft-quantifier $\Leftrightarrow_{p,\alpha}^?$ of upper critical double implication is defined for $0 < p \leq 1$ and $0 < \alpha \leq 0.5$ as follows:

$$\Leftrightarrow_{p,\alpha}^? (a,b,c,d) = \begin{cases} 1 \text{ if } \sum_{i=0}^{a} \binom{a+b+c}{i} p^i (1-p)^{a+b+c-i} > \alpha \\ 0 \text{ otherwise.} \end{cases}$$

Let $B = 1$, $C = 0$. We assume $0 < p < 1$ and according to lemma 5.16, point 2) it holds

$$\lim_{N \to \infty} \sum_{i=0}^{N-1} \binom{N}{i} p^i (1-p)^{N-i} = 1$$

thus there is an integer $Q > 1$ such that

$$\sum_{i=0}^{Q-1} \binom{Q}{i} p^i (1-p)^{Q-i} > \alpha$$

which means $\Leftrightarrow_{p,\alpha}^? (Q-1,1,0) = 1$. In addition, according to lemma 5.15, point 1) it holds

$$\lim_{N \to \infty} \sum_{i=0}^{A} \binom{A+N}{i} p^i (1-p)^{A+N-i} = 0$$

and thus there is an integer B' such that

$$\sum_{i=0}^{Q-1+B'} \binom{Q-1+B'}{i} p^i (1-p)^{Q-1+B'-i} \le \alpha$$

which means $\Leftrightarrow_{p,\alpha}^{!} (Q-1, B', 0) = 0.$

This finishes the proof.

8.7 Pure Double Implicational Quantifiers

The truth preservation condition $TPC_{\Leftrightarrow}$ for double implicational rules originates from trying to express the relation of equivalence of attributes φ and ψ using a "double implicational" 4ft-quantifier $\Leftrightarrow^*$ such that $\varphi \Leftrightarrow^* \psi$ if and only if both $\varphi \Rightarrow^* \psi$ and $\psi \Rightarrow^* \varphi$, where $\Rightarrow^*$ is a suitable implicational quantifier, see Sect. 6.3. We defined $TPC_{\Leftrightarrow}$ in definition 6.3 as a combination of the truth preservation conditions for implicational quantifier $\Rightarrow^*$ both in rule $\varphi \Rightarrow^* \psi$ and in rule $\psi \Rightarrow^* \varphi$.

Thus it is natural to ask what about the double implicational 4ft-quantifier $\Leftrightarrow^*$ satisfying condition

$$\varphi \Leftrightarrow^* \psi \text{ if and only if } \varphi \Rightarrow^* \psi \text{ and } \psi \Rightarrow^* \varphi, \qquad (8.2)$$

where $\Rightarrow^*$ is a suitable implicational quantifier.

We show in this section that there is no such suitable implicational quantifier for most Σ-double implicational quantifiers defined in Sect. 8.2. First we define a class of pure double implicational quantifiers using condition 8.2 and we show that each pure double implicational quantifier is double implicational. Then we find a condition equivalent to the fact that a double implicational quantifier is a pure double implicational quantifier. We show that a Σ-double implicational quantifier is pure double implicational if and only if it is not typical and we use this result to show that most practically important Σ-double implicational quantifiers are not pure double implicational.

Definition 8.13. Let $\approx$ be a 4ft-quantifier with associated function $F_{\approx}$ such that there is an implicational quantifier $\Rightarrow^*$ with associated function $F_{\Rightarrow^*}$ satisfying

$$F_{\approx}(a,b,c,d) = 1 \text{ if and only if both } F_{\Rightarrow^*}(a,b) = 1 \text{ and } F_{\Rightarrow^*}(a,c) = 1.$$

Then we say:

1. $\approx$ is a *pure double implicational quantifier*
2. *implicational quantifier* $\Rightarrow^*$ *is associated to pure double implicational quantifier* $\approx$,
3. *pure double implicational quantifier* $\approx$ *is associated to implicational quantifier* $\Rightarrow^*$.

The following simple theorem is true:

Theorem 8.14. *Each pure double implicational quantifier is double implicational.*

Proof. Let $\approx$ be a pure double implicational quantifier with associated function $F_\approx$. We have to prove that if $F_\approx(a,b,c,d) = 1$ and $a' \geq a \wedge b' \leq b \wedge c' \leq c$, then also $F_\approx(a',b',c',d) = 1$, see definition 6.3.

$\approx$ is the pure double implicational quantifier, thus there is an implicational quantifier $\Rightarrow^*$ with associated function $F_{\Rightarrow^*}$ such that

$$F_\approx(a,b,c,d) = 1 \text{ if and only if both } F_{\Rightarrow^*}(a,b) = 1 \text{ and } F_{\Rightarrow^*}(a,c) = 1 \,.$$

We assume $F_\approx(a,b,c,d) = 1$, which means $F_{\Rightarrow^*}(a,b) = 1$ and $F_{\Rightarrow^*}(a,c) = 1$. We also assume $a' \geq a \wedge b' \leq b \wedge c' \leq c$ which implies both $F_{\Rightarrow^*}(a',b') = 1$ and $F_{\Rightarrow^*}(a',c') = 1$. This however means $F_\approx(a',b',c',d) = 1$ according to definition 8.13. This finishes the proof. □

We show that we can use a suitable table of critical frequencies to define pure double implicational quantifiers, see also definition 7.6.

Theorem 8.15. *The 4ft-quantifier $\approx$ with associated function $F_\approx$ is pure double implicational if and only if there is a non-negative nondecreasing function $T_\approx$ assigning to each integer $a \geq 0$ a value from $\mathcal{N}^+$ such that*

$$F_\approx(a,b,c,d) = 1 \text{ if and only if both } b < T_\approx(a) \text{ and } c < T_\approx(a) \,.$$

Proof. 1. Let $\approx$ be pure double implicational. Then there is an implicational quantifier $\Rightarrow^*$ with associated function $F_{\Rightarrow^*}$ such that

$$F_\approx(a,b,c,d) = 1 \text{ if and only if both } F_{\Rightarrow^*}(a,b) = 1 \text{ and } F_{\Rightarrow^*}(a,c) = 1 \,.$$

Let us define $T_\approx(a) = Tb_{\Rightarrow^*}(a)$ where $Tb_{\Rightarrow^*}$ is a table of maximal b for quantifier $\Rightarrow^*$, see definition 7.7. According to theorem 7.8 it holds $F_{\Rightarrow^*}(a,b) = 1$ if and only if $b < Tb_{\Rightarrow^*}(a)$, $F_{\Rightarrow^*}(a,c) = 1$ if and only if $c < Tb_{\Rightarrow^*}(a)$, and $Tb_{\Rightarrow^*}$ is nondecreasing. Thus we have

$$F_\approx(a,b,c,d) = 1 \text{ if and only if both } b < Tb_{\Rightarrow^*}(a) \text{ and } c < Tb_{\Rightarrow^*}(a) \,.$$

2. Let $T_\approx$ be a non-negative nondecreasing function assigning to each integer non-negative number a value from $\mathcal{N}^+$ such that

$$F_\approx(a,b,c,d) = 1 \text{ if and only if both } b < T_\approx(a) \text{ and } c < T_\approx(a) \,.$$

We need an implicational quantifier $\Rightarrow^*$ with associated function $F_{\Rightarrow}^*$ satisfying

$$F_\approx(a,b,c,d) = 1 \text{ if and only if both } F_{\Rightarrow^*}(a,b) = 1 \text{ and } F_{\Rightarrow^*}(a,c) = 1 \,.$$

Let us define 4ft-quantifier $\Rightarrow^*$ such that

$$F_{\Rightarrow^*}(a,b,c,d) = 1 \text{ if and only if } b < T_\approx(a) \,.$$

Then, according to theorem 7.9, the quantifier $\Rightarrow^*$ is implicational and function $T_\approx(a)$ is its table of maximal b. According to theorem 7.8 it holds $F_{\Rightarrow^*}(a,b) = 1$

if and only if $b < T_{\approx}(a)$ and also $F_{\Rightarrow^*}(a,c) = 1$ if and only if $c < T_{\approx}(a)$. This means

$$F_{\approx}(a,b,c,d) = 1 \text{ if and only if both } F_{\Rightarrow^*}(a,b) = 1 \text{ and } F_{\Rightarrow^*}(d,c) = 1$$

which finishes the proof.   □

The following two theorems deal with the relation of Σ-double implicational quantifiers and pure double implicational quantifiers. The final theorem 8.18 of this section shows that most Σ-double implicational quantifiers defined in Sect. 8.2 are not pure double implicational.

Theorem 8.16. *A Σ-double implicational quantifer is pure double implicational if and only if it is not typical.*

Proof. Recall that a Σ-double implicational quantifier $\Leftrightarrow^*$ is typical Σ-double implicational if there is A such that $1 < Tb_{\Sigma,\Leftrightarrow^*}(A) < \infty$ where $Tb_{\Sigma,\Leftrightarrow^*}$ is a table of maximal $b+c$ of $\Leftrightarrow^*$, see definition 8.12. Thus, if quantifier $\Leftrightarrow^*$ is not typical, then $Tb_{\Sigma,\Leftrightarrow^*}$ only includes values 0, 1 or ∞.

According to theorem 8.12, the function $Tb_{\Sigma,\Leftrightarrow^*}$ is nondecreasing, thus if quantifier $\Leftrightarrow^*$ is not typical then one of the following possibilities a) – g) occurs:

a)   For each $a \geq 0$ it holds $Tb_{\Sigma,\Leftrightarrow^*}(a) = 0$.
b)   For each $a \geq 0$ it holds $Tb_{\Sigma,\Leftrightarrow^*}(a) = 1$.
c)   For each $a \geq 0$ it holds $Tb_{\Sigma,\Leftrightarrow^*}(a) = \infty$.
d)   For each $a \geq 0$ it holds $Tb_{\Sigma,\Leftrightarrow^*}(a) \in \{0,1\}$ and there are a_0 and a_1 such that $Tb_{\Sigma,\Leftrightarrow^*}(a_0) = 0$ and $Tb_{\Sigma,\Leftrightarrow^*}(a_1) = 1$. The function $Tb_{\Sigma,\Leftrightarrow^*}(a)$ is nondecreasing and thus there is $0 \leq Z_1$ such that for $a \leq Z_1$ we have $Tb_{\Sigma,\Leftrightarrow^*}(a) = 0$ and for $a > Z_1$ we have $Tb_{\Sigma,\Leftrightarrow^*}(a) = 1$.
e)   For each $a \geq 0$ it holds $Tb_{\Sigma,\Leftrightarrow^*}(a) \in \{0,\infty\}$ and there are a_0 and a_2 such that $Tb_{\Sigma,\Leftrightarrow^*}(a_0) = 0$ and $Tb_{\Sigma,\Leftrightarrow^*}(a_2) = \infty$ The function $Tb_{\Sigma,\Leftrightarrow^*}(a)$ is nondecreasing and thus there is $0 \leq Z_2$ such that for $a \leq Z_2$ we have $Tb_{\Sigma,\Leftrightarrow^*}(a) = 0$ and for $a > Z_2$ we have $Tb_{\Sigma,\Leftrightarrow^*}(a) = \infty$.
f)   For each $a \geq 0$ it holds $Tb_{\Sigma,\Leftrightarrow^*}(a) \in \{1,\infty\}$ and there are a_1 and a_2 such that $Tb_{\Sigma,\Leftrightarrow^*}(a_1) = 1$ and $Tb_{\Sigma,\Leftrightarrow^*}(a_2) = \infty$. The function $Tb_{\Sigma,\Leftrightarrow^*}(a)$ is nondecreasing and thus there is $0 \leq Z_2$ such that for $a \leq Z_2$ we have $Tb_{\Sigma,\Leftrightarrow^*}(a) = 1$ and for $a > Z_2$ we have $Tb_{\Sigma,\Leftrightarrow^*}(a) = \infty$.
g)   For each $a \geq 0$ it holds $Tb_{\Sigma,\Leftrightarrow^*}(a) \in \{0,1,\infty\}$ and there are a_0, a_1, and a_2 such that $Tb_{\Sigma,\Leftrightarrow^*}(a_0) = 0$, $Tb_{\Sigma,\Leftrightarrow^*}(a_1) = 1$, and $Tb_{\Sigma,\Leftrightarrow^*}(a_2) = \infty$. The function $Tb_{\Sigma,\Leftrightarrow^*}(a)$ is nondecreasing and thus there are $0 \leq Z_1 \leq Z_2$ such that

• for $0 \leq a \leq Z_1$ we have $Tb_{\Sigma,\Leftrightarrow^*}(a) = 0$,
• for $Z_1 < a \leq Z_2$ we have $Tb_{\Sigma,\Leftrightarrow^*}(a) = 1$,
• for $Z_2 < a$ we have $Tb_{\Sigma,\Leftrightarrow^*}(a) = \infty$.

We are going to prove the theorem in steps 1) – 5). First we prove that if Σ-double implicational quantifier $\Leftrightarrow^*$ is typical then it is not pure double implicational, see step 1.

Then we prove that if a Σ-double implicational quantifier ⇔* is not typical then it is a pure double implicational quantifer. We use the above shown fact that if Σ-double implicational quantifier ⇔* is not typical then one of possibilities a) – g) occurs. We will find an associated implicational quantifier ⇒* for Σ-double implicational quantifier ⇔* (see definition 8.13) for possibilities a), b), c) and g) in steps 2), 3), 4), and 5) respectively. The proof for possibilities d), e), and f) is analogous to the proof for possibility g).

Recall that the table of maximal b + c for Σ-double implicational quantifier ⇔* is defined such that

$$Tb_{\Sigma,\Leftrightarrow^*}(a) = \min\{b+c|F_{\Leftrightarrow^*}(a,b,c) = 0\},$$

see definition 8.11.

1. *Let Σ-double implicational quantifier ⇔* be typical. Then there is a non-negative A such that $1 < Tb_{\Sigma,\Leftrightarrow^*}(A) < \infty$. We have to proof that there is no implicational quantifier ⇒* which satisfies*

$$F_{\Leftrightarrow^*}(a,b,c) = 1 \text{ if and only if both } F_{\Rightarrow^*}(a,b) = 1 \text{ and } F_{\Rightarrow^*}(a,c). \qquad (8.3)$$

So, let us assume there is such quantifier ⇒* and denote $\sigma = Tb_{\Sigma,\Leftrightarrow^*}(A) - 1$. This means $F_{\Leftrightarrow^*}(A,\sigma,0) = 1$ which implies $F_{\Rightarrow^*}(A,\sigma) = 1$ because of (8.3). However, $F_{\Rightarrow^*}(A,\sigma) = 1$ and (8.3) implies also $F_{\Leftrightarrow^*}(A,\sigma,\sigma) = 1$.

We defined $\sigma = Tb_{\Sigma,\Leftrightarrow^*}(A) - 1$ and we assume $Tb_{\Sigma,\Leftrightarrow^*}(A) > 1$ which means $Tb_{\Sigma,\Leftrightarrow^*}(A) \geq 2$. Thus we have

$$\sigma + \sigma = Tb_{\Sigma,\Leftrightarrow^*}(A) - 1 + Tb_{\Sigma,\Leftrightarrow^*}(A) - 1 = Tb_{\Sigma,\Leftrightarrow}(A) + (Tb_{\Sigma,\Leftrightarrow^*}(A) - 2),$$

which implies

$$\sigma + \sigma \geq Tb_{\Sigma,\Leftrightarrow^*}(A).$$

According to definition 8.11 we have

$$Tb_{\Sigma,\Leftrightarrow^*}(A) = \min\{b+c| \Leftrightarrow^* (A,b,c) = 0\},$$

and because of $\sigma + \sigma \geq Tb_{\Sigma,\Leftrightarrow^*}(A)$ it must be $F_{\Leftrightarrow^*}(A,\sigma,\sigma) = 0$. This is however a contradiction with $F_{\Leftrightarrow^*}(A,\sigma,\sigma) = 1$ as proved above. This finishes the proof that if ⇔* is typical Σ-double implicational then it is not pure double implicational.

2. *In case a) we assume $Tb_{\Sigma,\Leftrightarrow^*}(a) = 0$ for each $a \geq 0$. It cannot be the case that $b+c < 0$ and thus it holds $F_{\Leftrightarrow^*}(a,b,c) = 0$ for all $a \geq 0$, $b \geq 0$, and $c \geq 0$. Let us define quantifier ⇒* such that $F_{\Rightarrow^*}(a,b,c,d) = 0$ for all non-negative integers a, b, c, and d satisfying $a+b+c+d > 0$. This quantifier is certainly implicational and it holds*

$$F_{\Leftrightarrow^*}(a,b,c) = 1 \text{ if and only if both } F_{\Rightarrow^*}(a,b) = 1 \text{ and } F_{\Rightarrow^*}(a,c) = 1.$$

This means that ⇔* is pure double implicational in case a).

3. *In case b) we assume $Tb_{\Sigma,\Leftrightarrow^*}(a) = 1$ for each $a \geq 0$. This means that for each $a \geq 0$ it holds $F_{\Leftrightarrow^*}(a,b,c) = 1$ if and only if $b = c = 0$. Let us define quantifier $\Rightarrow^*$ such that for all non-negative integers a, b, c, and d satisfying $a + b + c + d > 0$ it holds $F_{\Rightarrow^*}(a,b,c,d) = 1$ if and only if $b = 0$, otherwise $F_{\Rightarrow^*}(a,b,c,d) = 0$. This quantifier is certainly implicational and*

$$F_{\Leftrightarrow^*}(a,b,c) = 1 \text{ if and only if both } F_{\Rightarrow^*}(a,b) = 1 \text{ and } F_{\Rightarrow^*}(a,c) = 1 \,.$$

Thus quantifier $\Leftrightarrow^*$ is pure double implicational in case b).

4. *In case c) we assume $Tb_{\Sigma,\Leftrightarrow^*}(a) = \infty$ for each $a \geq 0$. This means that for all non-negative integers a, b, c, d satisfying $a + b + c + d > 0$ we have $F_{\Leftrightarrow^*}(a,b,c) = 1$. Let us define quantifier $\Rightarrow^*$ such that for all non-negative integers a, b, c, and d satisfying $a + b + c + d > 0$ it holds $F_{\Rightarrow^*}(a,b,c,d) = 1$. This quantifier is certainly implicational and*

$$F_{\Leftrightarrow^*}(a,b,c) = 1 \text{ if and only if both } F_{\Rightarrow^*}(a,b) = 1 \text{ and } F_{\Rightarrow^*}(a,c) = 1 \,,$$

thus quantifier $\Leftrightarrow^*$ is pure double implicational in case c).

5. *In case g) we assume there are $0 \leq Z_1 \leq Z_2$ such that:*

- *for $0 \leq a \leq Z_1$ we have $Tb_{\Sigma,\Leftrightarrow^*}(a) = 0$ which means $F_{\Leftrightarrow^*}(a,b,c) = 0$ for all $b \geq 0$ and $c \geq 0$*
- *for $Z_1 < a \leq Z_2$ we have $Tb_{\Sigma,\Leftrightarrow^*}(a) = 1$ which means $F_{\Leftrightarrow^*}(a,b,c) = 1$ for $b + c = 0$ and $F_{\Leftrightarrow^*}(a,b,c) = 0$ otherwise*
- *for $Z_2 < a$ we have $Tb_{\Sigma,\Leftrightarrow^*}(a) = \infty$ which means $F_{\Leftrightarrow^*}(a,b,c) = 1$ for all $a \geq 0$, $b \geq 0$, and $c \geq 0$.*

Let us define 4ft-quantifier $\Rightarrow^*$ such that

- *for $0 \leq a \leq Z_1$ we have $F_{\Rightarrow^*}(a,b,c,d) = 0$ for all non-negative integers b, c, and d satisfying $a + b + c + d > 0$*
- *for $Z_1 < a \leq Z_2$ we have $F_{\Rightarrow^*}(a,b,c,d) = 1$ if and only if $b = 0$, otherwise $F_{\Rightarrow^*}(a,b,c,d) = 0$*
- *for $Z_2 < a$ we have $F_{\Rightarrow^*}(a,b,c,d) = 1$ for all b, c, d.*

This 4ft-quantifier is certainly implicational and it holds

$$F_{\Leftrightarrow^*}(a,b,c) = 1 \text{ if and only if both } F_{\Rightarrow^*}(a,b) = 1 \text{ and } F_{\Rightarrow^*}(a,c) = 1 \,.$$

Thus, 4ft-quantifier $\Leftrightarrow^*$ is pure double implicational in case g).

The proof for possibilities d), e), and f) is analogous to the proof for possibility g). This finishes the proof. □

Theorem 8.17. *Let $\Leftrightarrow_\Sigma$ be a Σ-double implicational quantifier and $\Rightarrow^*$ is an implicational quantifier such that*

$$F_{\Leftrightarrow_\Sigma}(a,b,c) = 1 \text{ if and only if both } F_{\Rightarrow^*}(a,b) = 1 \text{ and } F_{\Rightarrow^*}(a,c) = 1 \,.$$

Then exactly one of the following possibilities a) – g) occur.

a) $F_{\Rightarrow^*}(a,b) = 0$ for all a, b
b) $F_{\Rightarrow^*}(a,b) = 1$ if and only if $b = 0$, otherwise $F_{\Rightarrow^*}(a,b) = 0$.
c) $F_{\Rightarrow^*}(a,b) = 1$ for all a, b.
d)   There exists $0 \leq Z_1$ such that

- for $0 \leq a \leq Z_1$ it holds $F_{\Rightarrow^*}(a,b) = 0$ for each b.
- for $Z_1 < a$ it holds $F_{\Rightarrow^*}(a,b) = 1$ if and only if $b = 0$, otherwise $F_{\Rightarrow^*}(a,b) = 0$.

e)   There exists $0 \leq Z_2$ such that

- for $0 \leq a \leq Z_2$ it holds $F_{\Rightarrow^*}(a,b) = 0$ for each b
- for $Z_2 < a$ it holds $F_{\Rightarrow^*} = 1$ for each b.

f)   There exists $0 \leq Z_2$ such that

- for $0 \leq a \leq Z_2$ it holds $F_{\Rightarrow^*}(a,b) = 1$ if and only if $b = 0$, otherwise $F_{\Rightarrow^*}(a,b) = 0$
- for $Z_2 < a$ it holds $F_{\Rightarrow^*}(a,b) = 1$ for each b.

g)   There exists $0 \leq Z_1 \leq Z_2$ such that

- for $0 \leq a \leq Z_1$ it holds $F_{\Rightarrow^*}(a,b) = 0$ for each b
- for $Z_1 < a \leq Z_2$ it holds $F_{\Rightarrow^*}(a,b) = 1$ if and only if $b = 0$, otherwise $F_{\Rightarrow^*}(a,b) = 0$
- for $Z_2 < a$ it holds $F_{\Rightarrow^*}(a,b) = 1$ for each b.

Proof. It follows from the proof of theorem 8.16.   □

Theorem 8.18. *Let $0 < p < 1$ and $0 < \alpha \leq 0.5$ be real numbers. Then the quantifiers $\Leftrightarrow_p$ of double p-implication, $\Leftrightarrow_{p,\alpha}^!$ of lower critical double implication, and $\Leftrightarrow_{p,\alpha}^?$ of upper critical double implication $\Leftrightarrow_{p,\alpha}^?$ are not pure double implicational.*

Proof. This is a direct consequence of theorems 8.13 and 8.16.   □

8.8 Strong Double Implicational Quantifiers

A natural generalization of pure double implicational quanifiers are "strong double imlicational quantifiers". Informally speaking, $\Leftrightarrow^*$ is a strong double implicational quantifier if there are two implicational quantifiers $\Rightarrow_1^*$ and $\Rightarrow_2^*$ such that

$$\varphi \Leftrightarrow^* \psi \text{ if and only if both } \varphi \Rightarrow_1^* \psi \text{ and } \psi \Rightarrow_2^* \varphi .$$

Strong double implicational quantifiers are studied in this section. After their formal definition we show that strong double implicational quantifiers have properties similar to properties of pure double implicational quantifiers.

Definition 8.14. Let $\approx^*$ be a quantifier with associated function $F_{\approx}$ and let there be two implicational quantifiers $\Rightarrow_1$ and $\Rightarrow_2$ with associated functions $F_{\Rightarrow_1}$ and $F_{\Rightarrow_2}$ respectively such that

$$F_{\approx}(a,b,c,d) = 1 \text{ if and only if both } F_{\Rightarrow_1}(a,b) = 1 \text{ and } F_{\Rightarrow_2}(a,c) = 1 .$$

Then we say:

1. $\Leftrightarrow^*$ is a *strong double implicational quantifier* ,
2. *implicational quantifier* $\Rightarrow_1$ *is a left associated quantifier to strong double implicational quantifier* $\Leftrightarrow^*$ *and implicational quantifier* $\Rightarrow_2$ *is a right associated quantifier to strong double implicational quantifier* $\Leftrightarrow^*$
3. *strong double implicational quantifier* $\Leftrightarrow^*$ *is associated to implicational quantifiers* $\Rightarrow_1$ *and* $\Rightarrow_2$.

Theorem 8.19

1. *Each strong double implicational quantifier is double implicational.*
2. *Each pure double implicational quantifier is strong double implicational.*

Proof. Proof of point 1) is analogous to the proof of theorem 8.14. Point 2) directly follows from definitions 8.13 and 8.14. □

The following theorem shows that strong double implicational quantifiers can also be defined using suitable tables of critical frequencies.

Theorem 8.20. *4ft-quantifier* $\approx$ *with associated function* $F_{\approx}$ *is strong double implicational if and only if there are non-negative nondecreasing functions* $T_{1,\approx}$ *and* $T_{2,\approx}$ *assigning to each integer* $a \geq 0$ *a value from* $\mathcal{N}^+$ *such that*

$$F_{\approx}(a,b,c) = 1 \text{ if and only if both } b < T_{1,\approx}(a) \text{ and } c < T_{2,\approx}(a) .$$

Proof. Proof is analogous to the proof of theorem 8.15. □

Theorem 8.21 concerns relation of Σ-double implicational quantifiers and strong implicational quantifiers. It is analogous to theorem 8.16. Theorem 8.22 is analogous to theorem 8.18 and shows that most of Σ-double implicational quantifiers defined in Sect. 8.2 are not strong double implicational.

Theorem 8.21. *A Σ-double implicational quantifier is a strong double implicational quantifier if and only if it is not typical.*

Proof. The proof is similar to the proof of theorem 8.16. It partially differs only what concerns step 1) which we give here in a shortened form.

Let us assume that $\Leftrightarrow^*$ *is a typical Σ-double implicational quantifier i.e. that there is an integer A such that* $1 < Tb_{\Sigma,\Leftrightarrow^*}(A) < \infty$ *where* $Tb_{\Sigma,\Leftrightarrow^*}$ *is a table of maximal* $b + c$ *of* $\Leftrightarrow^*$. *We have to prove that there are not two implicational quantifiers* $\Rightarrow_1$ *and* $\Rightarrow_2$ *with associated functions* $F_{\Rightarrow_1}$ *and* $F_{\Rightarrow_2}$ *such that*

$$F_{\Leftrightarrow^*}(a,b,c) = 1 \text{ if and only if both } F_{\Rightarrow_1}(a,b) = 1 \text{ and } F_{\Rightarrow_2}(a,c) .$$

Let us assume there are such two 4ft-quantifiers $\Rightarrow_1$ *and* $\Rightarrow_2$. *Let us denote* $\sigma = Tb_{\Sigma,\Leftrightarrow^*}(A) - 1$. *We assume*

$$F_{\Leftrightarrow^*}(a,b,c) = 1 \text{ if and only if both } F_{\Rightarrow_1}(a,b) = 1 \text{ and } F_{\Rightarrow_2}(a,c).$$

$\sigma = Tb_{\Sigma,\Leftrightarrow^*}(A) - 1$ implies $F_{\Leftrightarrow^*}(A,\sigma,0) = 1$ which means $F_{\Rightarrow_1}(A,\sigma) = 1$. Analogously it holds $F_{\Leftrightarrow^*}(A,0,\sigma) = 1$ which means also $F_{\Rightarrow_2}(A,\sigma) = 1$. This, however, also means $F_{\Leftrightarrow^*}(A,\sigma,\sigma) = 1$.

According to definition 8.11 we have $Tb_{\Sigma,\Leftrightarrow^*}(a) = \min\{b + c | F_{\Leftrightarrow^*}(a,b,c) = 0\}$, thus it must be $F_{\Leftrightarrow^*}(A,\sigma,\sigma) = 0$ because of

$$\sigma + \sigma \geq Tb_{\Sigma,\Leftrightarrow^*}(A).$$

This is however a contradiction with the above proved $F_{\Leftrightarrow^*}(A,\sigma,\sigma) = 1$. We have proved that if $\Leftrightarrow^*$ is typical Σ-double implicational then it is not strong implicational. The rest of the proof is the same as in theorem 8.16. This finishes the proof. □

Theorem 8.22. Let $0 < p < 1$ and $0 < \alpha \leq 0.5$ be real numbers. Then the quantifiers $\Leftrightarrow_p$ of double p-implication, $\Leftrightarrow_{p,\alpha}^{!}$ of lower critical double implication, and $\Leftrightarrow_{p,\alpha}^{?}$ of upper critical double implication $\Leftrightarrow_{p,\alpha}^{?}$ are not strong double implicational

Proof. This is a direct consequence of theorems 8.13 and 8.21. □

8.9 Relations among Subclasses of Double Implicational Rules

In this section we summarize relations among the subclasses of double implicational quantifiers introduced above. First we prove the following theorem.

Theorem 8.23. There is a double iplicational quantifier which is neither Σ-double implicational nor strong double implicational.

Proof. Let us define 4ft-quantifier $\Leftrightarrow_{0.9,\omega}^{*}$ for $0 < \omega$ such that

$$F_{\Leftrightarrow_{0.9,\omega}^{*}}(a,b,c,d) = \begin{cases} 1 & \text{if } \frac{a}{a+b+\omega c} \geq 0.9 \text{ and } a+b+c > 0 \\ 0 & \text{otherwise.} \end{cases}$$

Then we have:

1. Let both $F_{\Leftrightarrow_{0.9,\omega}^{*}}(a,b,c,d) = 1$ and $a' \geq a$, $b' \leq b$, $c' \leq c$. Then $a' + b' + c' > 0$ and also

$$\frac{a'}{a'+b'+\omega c'} \geq \frac{a'}{a'+b+\omega c} \geq \frac{a}{a+b+\omega c} \geq 0.9.$$

   This means that the quantifier $\Leftrightarrow_{0.9,\omega}^{*}$ is double implicational.
2. Let be $a = 90$, $b = 9$, $c = 2$ and $d = 0$ and assume $0 < \omega < 0.5$. Then $b + \omega c < 10$ which means

$$\frac{a}{a+b+\omega c} = \frac{90}{90+b+\omega c} > \frac{90}{90+10} = 0.9, \text{ thus } F_{\Leftrightarrow_{0.9,\omega}^{*}}(90,9,2,0) = 1.$$

Let us assume that $\Leftrightarrow^*_{0.9,\omega}$ is Σ-double implicational. Then it must also hold $F_{\Leftrightarrow^*_{0.9,\omega}}(90,9+2,0,0) = 1$. However, we have

$$\frac{90}{90+9+2+\omega*0} = \frac{90}{90+11} < 0.9, \text{ thus } F_{\Leftrightarrow^*_{0.9,\omega}}(90,9,2,0) = 0.$$

This is a contradiction and means that $\Leftrightarrow^*_{0.9,\omega}$ is not Σ-double implicational for $0 < \omega < 0.5$.

3. *Let us assume that $\Leftrightarrow^*_{0.9,\omega}$ is for $0 < \omega < 1$ strong double implicational. This means that there are implicational quantifiers $\Rightarrow^1_{0.9,\omega}$ and $\Rightarrow^2_{0.9,\omega}$ such that*

$$F_{\Leftrightarrow^*_{0.9,\omega}}(a,b,c,d) = 1 \text{ if and only if both } F_{\Rightarrow^1_{0.9,\omega}}(a,b) = 1 \text{ and } F_{\Rightarrow^2_{0.9,\omega}}(a,c) = 1.$$

It is $F_{\Leftrightarrow^*_{0.9,\omega}}(90,10,0,0) = 1$, thus it must hold

$$\text{both } F_{\Rightarrow^1_{0.9,\omega}}(90,10) = 1 \text{ and } F_{\Rightarrow^2_{0.9,\omega}}(90,0) = 1.$$

It is also $F_{\Leftrightarrow^*_{0.9,\omega}}(90,9,1,0) = 1$ which means

$$\text{both } F_{\Rightarrow^1_{0.9,\omega}}(90,9) = 1 \text{ and } F_{\Rightarrow^2_{0.9,\omega}}(90,1) = 1.$$

We have $F_{\Rightarrow^1_{0.9,\omega}}(90,10) = 1$ and also $F_{\Rightarrow^2_{0.9,\omega}}(90,1) = 1$. The assumption

$$F_{\Leftrightarrow^*_{0.9,\omega}}(a,b,c,d) = 1 \text{ if and only if both } F_{\Rightarrow^1_{0.9,\omega}}(a,b) = 1 \text{ and } F_{\Rightarrow^2_{0.9,\omega}}(a,c) = 1$$

implies that also $F_{\Leftrightarrow^*_{0.9,\omega}}(90,10,1,0) = 1$. It however holds that $F_{\Leftrightarrow^*_{0.9,\omega}}(90,10,1,0) = 0$ which is a contradiction. We can conclude that quantifier $\Leftrightarrow^*_{0.9,\omega}$ is not strong double implicational.

This means that each quantifier $\Leftrightarrow^*_{0.9,\omega}$ is double implicational for $0 < \omega < 0.5$ but it is neither Σ-double implicational nor strong double implicational. This finishes the proof.   □

We can summarize relations of classes of rules studied in this chapter in the following way:

1. The following quantifiers are double implicational

   - Σ-double implicational, see theorem 8.1
   - pure double implicational, see theorem 8.14
   - strong double implicational, see theorem 8.19

2. Each pure double implicational quantifier is strong double implicational, see theorem 8.19.

3. It is reasonable to distinguish typical and untypical Σ-double implicational quantifiers, see definition 8.12. We have the following results

- Quantifiers of double p-implication $\Leftrightarrow_p$, double lower critical implication $\Leftrightarrow_{p,\alpha}^{!}$, and double upper critical implication $\Leftrightarrow_{p,\alpha}^{?}$ with parameters $0 < p < 1$ and $0 < \alpha \leq 0.5$. are typical Σ-double implicational quantifiers, see theorem 8.13.
- A typical Σ-double implicational quantifier is neither pure double implicational nor strong double implicational, see theorems 8.16 and 8.21.
- Each untypical Σ-double implicational quantifier is both pure double implicational and strong double implicational, see theorems 8.16 and 8.21.

4. There are double implicational quantifiers which are not Σ-double implicational, see theorem 8.23.
5. There are double implicational quantifiers which are not pure double implicational. Examples of such quantifiers are:

- strong double implicational quantifiers associated to two distinct implicational quantifiers
- suitably defined double implicational quantifiers which are not strong implicational, see theorem 8.23,
- typical Σ-double implicational quantifiers, see theorem 8.16.

6. There are double implicational quantifiers which are not strong double implicational. Examples of such quantifiers are:

- suitably defined double implicational quantifiers, see theorem 8.23,
- typical Σ-double implicational quantifiers, see theorem 8.21.

Chapter 9
Equivalence Rules

Equivalence quantifiers and rules are introduced in a way similar to the way the double implicational quantifiers and rules are introduced. There is an equivalent structure of important or at least interesting subclasses of equivalence quantifiers. This means that it is again easy to see that equivalence quantifiers are $\mathcal{M}$-independent and it is also reasonable to introduce weakly equivalence quantifiers.

Important equivalence quantifiers deal with the sum $a + d$ of frequencies of 4ft-table $\langle a, b, c, d \rangle$. This leads to definition of Σ-equivalence quantifiers and weakly Σ-equivalence quantifiers. We define and study two important subclasses of Σ-equivalence quantifiers – interesting Σ-equivalence quantifiers and typical Σ-equivalence quantifiers. Tables of critical frequencies are an important tool for the study of 4ft-quantifiers There is only one type of tables of critical frequencies for equivalence quantifiers and it is used to define typical equivalence quantifiers.

The definition of equivalence quantifiers can be seen as a combination of two definitions of implicational quantifiers. This results in definitions of additional important or at least interesting subclasses of equivalence quantifiers, such as pure equivalence quantifiers and strong equivalence quantifiers.

Weakly equivalence quantifiers, Σ-equivalence quantifiers and weakly Σ-equivalence quantifiers are introduced in Sect. 9.1. There are several practically important equivalence quantifiers defined in Tables 4.4 and 4.6, some of which are Σ-equivalence quantifiers, see Sect. 9.2. It is also possible to define and study particular important weakly equivalence quantifiers, see Sect. 9.3. The class of interesting Σ-equivalence quantifiers is defined and studied in Sect. 9.4. The table of critical frequencies for Σ-equivalence quantifiers is introduced in Sect. 9.5, which is then used to define typical Σ-equivalence quantifiers introduced in Sect. 9.6. Pure and strong equivalence quantifiers are defined and studied in Sects. 9.7 and 9.8. Relations among particular subclasses of equivalence quantifiers are summarized in Sect. 9.9.

9.1 Equivalence and Weakly Equivalence Quantifiers

Equivalence quantifiers were introduced in [18] under the name associational rules. Here we use the name equivalence rules due to reason listed in Sect. 6.4. The

J. Rauch: *Observational Calculi and Association Rules*, SCI 469, pp. 127–147.
DOI: 10.1007/978-3-642-11737-4_9 © Springer-Verlag Berlin Heidelberg 2013

equivalence quantifiers are defined using truth preservation condition $TPC_{\equiv}$ in definition 6.3. Truth preservation condition $TPC_{\equiv}$ is $\mathscr{M}$-independent as well as the equivalence quantifiers, see following lemma.

Lemma 9.1

1. *Truth preservation condition $TPC_{\equiv}$ is $\mathscr{M}$-independent.*
2. *Equivalence quantifier is $\mathscr{M}$-independent.*

Proof

1. *According to point 4) of definition 6.11 we have to prove that there are 4ft-tables $\langle a,b,c,d \rangle$ and $\langle a',b',c',d' \rangle$ such that*

$$TPC_{\equiv}(a,b,c,d,a',b',c',d') = 1 \text{ and } a+b+c+d \neq a'+b'+c'+d'.$$

4ft-tables $\langle 10,10,10,10 \rangle$ and $\langle 99,9,9,99 \rangle$ satisfy this condition, it holds that

$$TPC_{\equiv}(10,10,10,10,99,9,9,99) = 1 \text{ and } 10+10+10+10 \neq 99+9+9+99.$$

This finishes the proof of point 1.
2. *This follows from just proved point 1) and point 6) of definition 6.11.*

This finishes the proof.   □

We use the notion of equivalence quantifier to define the notion of an *equivalence rule*.

Definition 9.1. Let $\mathscr{LC}$ be a logical calculus of association rules and let $\varphi \approx \psi$ be an association rule of $\mathscr{LC}$. If quantifier $\approx$ is equivalence then the rule $\varphi \approx \psi$ is an *equivalence rule*.

We derive a new $\mathscr{M}$-dependent class of 4ft-quantifiers from the class of equivalence 4ft-quantifiers in the way given by definition 6.12. We will call the 4ft-quantifiers belonging to this class weakly equivalence 4ft-quantifiers. Consequently, we define weakly equivalence rules. Please note that the class of weakly equivalence quantifiers is introduced in [74] as a class of $\mathscr{M}$-dependent equivalence quantifiers. We will use the name *weakly equivalence quantifiers* instead of $\mathscr{M}$-*dependent equivalence quantifiers* for the same reason we use the name *weakly implicational quantifiers* instead of $\mathscr{M}$-*dependent implicational quantifiers*, see note 7.1.

Definition 9.2. The *truth preservation condition* $TPC_{\equiv}^{\mathscr{M}}$ *for weakly equivalence quantifiers* is a $\{0,1\}$-valued function $TPC_{\equiv}^{\mathscr{M}}(a,b,c,d,a',b',c',d')$ defined such that

$$TPC_{\equiv}^{\mathscr{M}}(a,b,c,d,a',b',c',d') = \begin{cases} 1 & \text{if } a' \geq a \wedge b' \leq b \wedge c' \leq c \wedge d' \geq d \text{ and} \\ & a+b+c+d = a'+b'+c'+d' \\ 0 & \text{otherwise.} \end{cases}$$

Class $4ft[TPC_{\equiv}^{\mathscr{M}}]$ is a *class of weakly equivalence quantifiers*. If quantifier $\approx$ belongs to $4ft[TPC_{\equiv}^{\mathscr{M}}]$ then it is a *weakly equivalence quantifier*.

Definition 9.3. Let $\mathscr{LC}$ be a logical calculus of association rules and let $\varphi \approx \psi$ be an association rule of $\mathscr{LC}$. If quantifier $\approx$ is weakly equivalence then the rule $\varphi \approx \psi$ is a *weakly equivalence rule*.

Associated function $F_{\equiv_p}$ of 4ft-quantifier of p-equivalence $\equiv_p$ is defined such that $F_{\equiv_p} = 1$ if and only if $\frac{a+d}{a+b+c+d} \geq p$. The sum $a+d$ plays an important role in this definition. The sum $a+d$ plays a similarly important role in the definition of associated functions of 4ft-quantifiers $\equiv^!_{p,\alpha}$ and $\equiv^?_{p,\alpha}$, see Table 4.4. We use the sum $a+d$ in the definition of Σ-equivalence quantifiers. To make sure the defined quantifiers are equivalence quantifiers, we have to somehow limit frequencies b and c, which is done by a restriction on the sum $b+c$. This leads to definition of Σ-equivalence quantifiers and weakly Σ-equivalence quantifiers and rules.

Definition 9.4. The *truth preservation condition* $TPC_{\Sigma,\equiv}$ *for* Σ*-equivalence quantifiers* is a $\{0,1\}$-valued function $TPC_{\Sigma,\equiv}(a,b,c,d,a',b',c',d')$ defined for all 8-tuples $\langle a,b,c,d,a',b',c',d' \rangle$ of non-negative integer numbers such that

$$TPC_{\Sigma,\equiv}(a,b,c,d,a',b',c',d') = \begin{cases} 1 & \text{if } a'+d' \geq a+d \wedge b'+c' \leq b+c \\ 0 & \text{otherwise.} \end{cases}$$

The *class of* Σ*-equivalence quantifiers* is the class $4ft[TPC_{\Sigma,\equiv}]$ defined by truth preservation condition $TPC_{\Sigma,\equiv}$. We say that 4ft-quantifier $\approx$ is Σ*-equivalence* if it belongs to $4ft[TPC_{\Sigma,\equiv}]$.

The condition $a'+d' \geq a+d \wedge b'+c' \leq b+c$ is called the *simple form of* $TPC_{\Sigma,\equiv}$.

Definition 9.5. Let $\mathscr{LC}$ be a logical calculus of association rules and let $\varphi \approx \psi$ be an association rule of $\mathscr{LC}$. If quantifier $\approx$ is Σ-equivalence then the rule $\varphi \approx \psi$ is Σ*-equivalence rule*.

Please note that $a'+d' \geq a+d$ and $a+b+c+d = a'+b'+c'+d'$ implies $b'+c' \leq b+c$. This makes it possible to use the condition

$$a'+d' \geq a+d \text{ and } a+b+c+d = a'+b'+c'+d'$$

instead of

$$a'+d' \geq a+d \wedge b'+c' \leq b+c \text{ and } a+b+c+d = a'+b'+c'+d'$$

in the following definition of weakly Σ-equivalence quantifiers.

Definition 9.6. The *truth preservation condition* $TPC^{\mathscr{M}}_{\Sigma,\equiv}$ *for weakly* Σ*-equivalence quantifiers* is $\{0,1\}$-valued function $TPC^{\mathscr{M}}_{\Sigma,\equiv}(a,b,c,d,a',b',c',d')$ defined such that

$$TPC^{\mathscr{M}}_{\Sigma,\equiv}(a,b,c,d,a',b',c',d') = \begin{cases} 1 & \text{if } a'+d' \geq a+d \text{ and} \\ & a+b+c+d = a'+b'+c'+d' \\ 0 & \text{otherwise.} \end{cases}$$

Class $4ft[TPC^{\mathcal{M}}_{\Sigma,\equiv}]$ is a *class of weakly Σ-equivalence quantifiers*. If quantifier $\approx$ belongs to $4ft[TPC^{\mathcal{M}}_{\Sigma,\equiv}]$ then it is a *weakly Σ-equivalence quantifier*.

Definition 9.7. Let $\mathcal{LC}$ be a logical calculus of association rules and let $\varphi \approx \psi$ be an association rule of $\mathcal{LC}$. If quantifier $\approx$ is weakly Σ-equivalence then the rule $\varphi \approx \psi$ is a *weakly Σ-equivalence rule*.

It is easy to prove that each Σ-equivalence quantifier is an equivalence quantifier:

Theorem 9.1. *Each Σ-equivalence quantifier is an equivalence quantifier.*

Proof. We have to prove that if quantifier $\approx$ belongs to class $4ft[TPC\Sigma,\equiv]$ then it belongs to class $4ft[TPC_\equiv]$. According to lemma 6.1 it is enough to prove that

$$TPC_\equiv(a,b,c,d,a',b',c',d') = 1 \text{ implies } TPC_{\Sigma,\equiv}(a,b,c,d,a',b',c',d') = 1 .$$

This means that we have to prove that

$$a' \geq a \wedge b' \leq b \wedge c' \leq c \wedge d' \geq d \text{ implies } a'+d' \geq a+d \wedge b'+c' \leq b+c$$

which is obvious. This finishes the proof. □

The following lemma says that each equivalence quantifier is also a weakly equivalence quantifier and that each Σ-equivalence quantifier is also a weakly Σ-equivalence quantifier.

Lemma 9.2

1. If $\equiv^*$ is an equivalence quantifier then $\equiv^*$ is also a weakly equivalence quantifier.
2. If $\equiv^*$ is a Σ-equivalence quantifier then $\equiv^*$ is also a weakly Σ-equivalence quantifier.

Proof. The lemma is a direct consequence of lemma 6.15. □

9.2 Selected Equivalence Quantifiers

Several of the 4ft-quantifiers introduced in Tables 4.4 and 4.6 are Σ-equivalence or at least equivalence, see the following theorems.

Theorem 9.2. *The following quantifiers are Σ-equivalence quantifiers.*

1. *p-equivalence $\equiv_p$, see rows 7 in Tables 4.2 and 4.4 and rows 7 in Tables 4.5 and 4.6*
2. *Lower critical equivalence $\equiv^!_{p,\alpha}$, see rows 8 in Tables 4.2 and 4.4*
3. *Upper critical equivalence $\equiv^?_{p,\alpha}$, see rows 9 in Tables 4.2 and 4.4.*

Proof. For each of the 4ft-quantifiers listed above, we have to prove:

if $F_{\approx}(a,b,c,d) = 1$ and $a'+d' \geq a+d \wedge b'+c' \leq b+c$ then $F_{\approx}(a',b',c',d') = 1$

where $F_{\approx}(a,b,c,d)$ is the associated function of the 4ft-quantifier $\approx$, see definition 9.4.

1. *It holds that $F_{\equiv_p}(a,b,c,d) = 1$ if and only if $\frac{a+d}{a+b+c+d} \geq p$. If $F_{\equiv_p}(a,b,c,d) = 1$ and $a'+d' \geq a+d \wedge b'+c' \leq b+c$ then we have*

$$\frac{a'+d'}{a'+b'+c'+d'} \geq \frac{a+d}{a+b'+c'+d} \geq \frac{a+d}{a+b+c+d} \geq p, \qquad (9.1)$$

and this finishes the proof for $\equiv_p$.

2. *It holds that $F_{\equiv^!_{p,\alpha}}(a,b,c,d) = 1$ if and only if $\sum_{i=a+d}^{n} \binom{n}{i} p^i (1-p)^{n-i} \leq \alpha$ where $n = a+b+c+d$. We have to prove that if $F_{\equiv^!_{p,\alpha}}(a,b,c,d) = 1$ and $a'+d' \geq a+d \wedge b'+c' \leq b+c$ then*

$$\sum_{i=a'+d'}^{a'+b'+c'+d'} \binom{a'+b'+c'+d'}{i} p^i (1-p)^{a'+b'+c'+d'-i} \leq \alpha.$$

This however directly follows from theorem 5.1 if we use $u = a+d$, $v = b+c$, $u' = a'+d'$, and $v' = b'+c'$.

3. *It holds that $F_{\equiv^?_{p,\alpha}}(a,b,c,d) = 1$ if and only if $\sum_{i=0}^{a+d} \binom{n}{i} p^i (1-p)^{n-i} > \alpha$ where $n = a+b+c+d$. We have to prove that if $F_{\equiv^?_{p,\alpha}}(a,b,c,d) = 1$ and $a'+d' \geq a+d \wedge b'+c' \leq b+c$ then*

$$\sum_{i=0}^{a'+d'} \binom{a'+b'+c'+d'}{i} p^i (1-p)^{a'+b'+c'+d'-i} > \alpha.$$

This however directly follows from theorem 5.2 if we use $u = a+d$, $v = b+c$, $u' = a'+d'$, and $v' = b'+c'$.

This finishes the proof. □

Theorem 9.3. *The following quantifiers are equivalence quantifiers.*

1. *p-equivalence $\equiv_p$, see rows 7 in Tables 4.2 and 4.4 and rows 7 in Table 4.5 and 4.6*
2. *Lower critical equivalence $\equiv^!_{p,\alpha}$, see rows 8 in Tables 4.2 and 4.4*
3. *Upper critical equivalence $\equiv^?_{p,\alpha}$, see rows 9 in Tables 4.2 and 4.4.*

Proof. The theorem follows from theorems 9.1 and 9.2. □

Theorem 9.4. *The following quantifiers are equivalence quantifiers but not Σ-equivalence quantifiers.*

1. Simple deviation $\sim_q$, see rows 10 in Tables 4.2 and 4.4
2. Fisher's quantifier $\sim_\alpha^1$, see rows 11 in Tables 4.2 and 4.4
3. χ^2-quantifier $\sim_\alpha^2$, see rows 12 in Tables 4.2 and 4.4
4. Base $\oplus_{Base}$, see rows 18 in Tables 4.2 and 4.4.

Proof. For each of the 4ft-quantifiers listed above, we have to prove:

A)   *If $F_\approx(a,b,c,d) = 1$ and $a' \geq a \wedge b' \leq b \wedge c' \leq c \wedge d' \geq d$ then $F_\approx(a',b',c',d') = 1$*
B)   *There are 4ft-tables $\langle a,b,c,d \rangle$ and $\langle a',b',c',d' \rangle$ such that $a' + d' \geq a + d$, $b' + c' \leq b + c$, $F_\approx(a,b,c,d) = 1$, and $F_\approx(a',b',c',d) = 0$.*

Here $F_\approx(a,b,c,d)$ is the associated function of the 4ft-quantifier $\approx$, see also definitions 6.4 and 9.4.

1. It holds that $F_{\sim_q}(a,b,c,d) = 1$ if and only if $ad > e^q bc$.

   A)   *We have to prove that if $a' \geq a \wedge b' \leq b \wedge c' \leq c \wedge d' \geq d$ and $ad > e^q bc$ then $a'd' > e^q b'c'$. It is, however, obvious that $a'd' \geq ad > e^q bc \geq e^q b'c'$.*
   B)   *Le m be a positive integer number and let $\langle a,b,c,d \rangle = \langle m,1,0,1 \rangle$ and $\langle a',b',c',d' \rangle = \langle m+1,1,0,0 \rangle$ be 4ft-tables. Then we have $a' + d' \geq a + d$, $b' + c' \leq b + c$, $F_{\sim_q}(a,b,c,d) = 1$ and $F_{\sim_q}(a',b',c',d') = 0$.*

2. It holds that $F_{\sim_\alpha^1}(a,b,c,d) = 1$ if and only if $\sum_{i=a}^{\min(r,k)} \frac{\binom{k}{i}\binom{n-k}{r-i}}{\binom{n}{r}} \leq \alpha \wedge ad > bc$ where $r = a + b$, $k = a + c$, and $n = a + b + c + d$.

   A)   *We have to prove that if $a' \geq a \wedge b' \leq b \wedge c' \leq c \wedge d' \geq d$ and $\sum_{i=a}^{\min(r,k)} \frac{\binom{k}{i}\binom{n-k}{r-i}}{\binom{n}{r}} \leq \alpha \wedge ad > bc$ then $\sum_{i=a'}^{\min(r',k')} \frac{\binom{k'}{i}\binom{n'-k'}{r'-i}}{\binom{n'}{r'}} \leq \alpha \wedge a'd' > b'c'$ where $r' = a' + b'$, $k' = a' + c'$, and $n' = a' + b' + c' + d'$. This however directly follows from theorem 5.3.*
   B)   *There must be an integer $m > 0$ such that $\frac{1}{\binom{2m}{m}} \leq \alpha$, i.e. an m satisfying $\frac{m!m!}{(2m)!} = \frac{1}{(m+1)(m+2)...2m} \leq \alpha$. Let us define $\langle a,b,c,d \rangle = \langle m,0,0,m \rangle$ and $\langle a',b',c',d' \rangle = \langle 2m,0,0,0 \rangle$. Then we have $a' + d' \geq a + d$, $b' + c' \leq b + c$ and it also holds that:*

    • *$F_{\sim_\alpha^1}(m,0,0,m) = 1$ because*
$$\sum_{i=m}^{\min(m,m)} \frac{\binom{m}{i}\binom{2m-m}{m-m}}{\binom{2m}{m}} = \frac{\binom{m}{m}\binom{m}{0}}{\binom{2m}{m}} = \frac{1}{\binom{2m}{m}} \leq \alpha \wedge m.m > 0.0$$
    • *$F_{\sim_\alpha^1}(2m,0,0,0) = 1$ because $2m.0 \not> 0.0$.*

3. It holds that $F_{\sim_\alpha^2}(a,b,c,d) = 1$ if and only if $\frac{(ad-bc)^2}{rkls}n \geq \chi_\alpha^2 \wedge ad > bc$ where $r = a + b$, $s = c + d$, $k = a + c$, $l = b + d$, and $n = a + b + c + d$.

   A)   *We have to prove that if $a' \geq a \wedge b' \leq b \wedge c' \leq c \wedge d' \geq d$ and $\frac{(ad-bc)^2}{rkls}n \geq \chi_\alpha^2 \wedge ad > bc$ then $\frac{(a'd'-b'c')^2}{r'k'l's'}n' \geq \chi_\alpha^2 \wedge a'd' > b'c'$ where $r' = a' + b'$, $s' = c' + d'$, $k' = a' + c'$, $l' = b' + d'$, and $n' = a' + b' + c' + d'$. It however directly follows from theorem 5.5.*

B) *There is an integer m such that $2m \geq \chi_\alpha^2$. Let us define $\langle a,b,c,d \rangle = \langle m,0,0,m \rangle$ and $\langle a',b',c',d' \rangle = \langle 2m,0,0,0 \rangle$. Then we have $a' + d' \geq a + d$, $b' + c' \leq b + c$ and it also holds that:*

- $F_{\sim_\alpha^2}(m,0,0,m) = 1$ *because* $\frac{(ad-bc)^2}{rkls}n = \frac{(mm)^2}{m^4}2m = 2m \geq \chi_\alpha^2$.
- $F_{\sim_\alpha^2}(2m,0,0,0) = 1$ *because* $2m.0 \not> 0$.

4. *It holds that $F_{\oplus_{Base}}(a,b,c,d) = 1$ if and only if $a \geq Base$.*

A) *It is obvious that if $a' \geq a \wedge b' \leq b \wedge c' \leq c \wedge d' \geq d$ and $a \geq Base$ then also $a' \geq Base$.*

B) *Let us define $\langle a,b,c,d \rangle = \langle Base,0,0,0 \rangle$ and $\langle a',b',c',d' \rangle = \langle 0,0,0,Base \rangle$. Then $a' + d' \geq a + d$, $b' + c' \leq b + c$ and it holds $F_{\oplus_{Base}}(Base,0,0,0) = 1$ and $F_{\oplus_{Base}}(0,0,0,Base) = 0$.*

This finishes the proof. □

We have derived several founded implicational 4ft-quantifiers from important implicational 4ft-quantifiers in Sect. 7.2, see theorems 7.2 and 7.3 and several founded Σ-double implicational 4ft-quantifiers in Sect. 8.2, see theorems 8.4 and 8.5. We apply a similar approach to equivalence quantifiers.

Theorem 9.5. *Let $\mathcal{LC}$ be a logical calculus of association rules with 4ft-quantifiers $\approx_1, \ldots \approx_Q$ with associated functions $F_{\approx_1}, \ldots, F_{\approx_Q}$ respectively. Then*

1. *the compound 4ft-quantifier $\approx_{i_1} \wedge \ldots \wedge \approx_{i_k}$ is equivalence if and only if all 4ft-quantifiers $\approx_{i_1}, \ldots, \approx_{i_k}$ are equivalence.*
2. *the compound 4ft-quantifier $\approx_{i_1} \wedge \ldots \wedge \approx_{i_k}$ is Σ-equivalence if and only if all 4ft-quantifiers $\approx_{i_1}, \ldots, \approx_{i_k}$ are Σ-equivalence.*
3. *the compound 4ft-quantifier $\approx_{i_1} \wedge \ldots \wedge \approx_{i_k}$ is weakly equivalence if and only if all 4ft-quantifiers $\approx_{i_1}, \ldots, \approx_{i_k}$ are weakly equivalence.*
4. *the compound 4ft-quantifier $\approx_{i_1} \wedge \ldots \wedge \approx_{i_k}$ is weakly Σ-equivalence if and only if all 4ft-quantifiers $\approx_{i_1}, \ldots, \approx_{i_k}$ are weakly Σ-equivalence.*

Proof. The proof is analogous to the proof of theorem 8.4. □

Theorem 9.6. *The following quantifiers are equivalence quantifiers but not Σ-equivalence quantifiers.*

1. *Founded p-equivalence $\equiv_{p,Base}$ defined as $\equiv_p \wedge \oplus_{Base}$*
2. *Founded lower critical equivalence $\equiv_{p,\alpha,Base}^!$ defined as $\equiv_{p,\alpha}^! \wedge \oplus_{Base}$*
3. *Founded upper critical equivalence $\equiv_{p,\alpha,Base}^?$ defined as $\equiv_{p,\alpha}^? \wedge \oplus_{Base}$*
4. *Founded simple deviation $\sim_{q,Base}$ defined as $\sim_q \wedge \oplus_{Base}$*
5. *Founded Fisher's quantifier $\sim_{\alpha,Base}^1$ defined as $\sim_\alpha^1 \wedge \oplus_{Base}$*
6. *Founded χ^2-quantifier $\sim_{\alpha,Base}^2$ defined as $\sim_\alpha^2 \wedge \oplus_{Base}$*

Proof. Directly follows from theorems 9.3, 9.4, and 9.5. □

All the above introduced equivalence quantifiers are also symmetrical, see the next theorem. Please remember that 4ft-quantifier $\approx$ is symmetrical if its associated function $F_\approx$ satisfies

$$F_\approx(a,b,c,d) = F_\approx(a,c,b,d) ,$$

see definitions 6.5 and 6.6 which are equivalent, see lemma 6.9.

Theorem 9.7. *The following 4ft-quantifiers are symmetrical:*

1. *p-equivalence $\equiv_p$ and founded p-equivalence $\equiv_{p,Base}$*
2. *Lower critical equivalence $\equiv^!_{p,\alpha}$ and founded lower critical equivalence $\equiv^!_{p,\alpha,Base}$*
3. *Upper critical equivalence $\equiv^?_{p,\alpha}$ and founded upper critical equivalence $\equiv^?_{p,\alpha,Base}$*
4. *Simple deviation $\sim_q$ and founded simple deviation $\sim_{q,Base}$*
5. *Fisher's quantifier $\sim^1_\alpha$ and founded Fisher's quantifier $\sim^1_{\alpha,Base}$*
6. *χ^2-quantifier $\sim^2_\alpha$ and founded χ^2-quantifier $\sim^2_{\alpha,Base}$.*

Proof. Proof for points 1 - 4 directly follows from definitions of these quantifiers and from definition 6.5. The proof for point 5 follows from lemma 5.9 and the proof for point 5 follows from lemma 5.12. This finishes the proof. □

9.3 Selected Weakly Equivalence Quantifiers

We have derived several supported weakly implicational 4ft-quantifiers from important implicational 4ft-quantifiers and quantifier $\odot_s$ of support, see Sect. 7.3. We have also derived several supported weakly Σ-double implicational 4ft-quantifiers from important Σ-double implicational 4ft-quantifiers and quantifier $\odot_s$ of support, see Sect. 8.3. A similar approach can be applied to equivalence quantifiers. The results are provided in theorem 9.8, however first we will need lemmas 9.3 and 9.4.

Lemma 9.3. *The following is true for 4ft-quantifier $\odot_s$ of support with $0 < s \leq 1$:*

1. *$\odot_s$ is not equivalence but it is weakly equivalence.*
2. *$\odot_s$ is neither Σ-equivalence nor weakly Σ-equivalence.*

Proof. Remember that $\odot_s(a,b,c,d) = 1$ if and only if $\frac{a}{a+b+c+d} \geq s$ for a real number $0 < s \leq 1$, see row 1 in table 4.6.

ad 1.: Let be $0 < s \leq 1$ a real number and $F_{\odot_s}$ be the associated function of $\odot_s$. Put $\langle a,b,c,d \rangle = \langle 1,0,0,0 \rangle$ and $\langle a',b',c',d' \rangle = \langle 1,0,0,D \rangle$ where D is an integer number such that $D > \frac{1-s}{s}$. Then

- $a' \geq a \wedge b' \leq b \wedge c' \leq c \wedge d' \geq d$
- $F_{\odot_s}(a,b,c,d) = 1$ because $\frac{a}{a+b+c+d} = \frac{1}{1+0+0+0} \geq s$
- $F_{\odot_s}(a',b',c',d') = 0$ because $\frac{a'}{a'+b'+c'+d'} = \frac{1}{1+0+0+D} < \frac{1}{1+0+0+\frac{1-s}{s}} = \frac{1}{\frac{1}{s}} = s$

which means that $\odot_s$ is not equivalence.

Let $0 < s \leq 1$ be a real number, $F_{\odot_s}(a,b,c,d) = 1$, $a' \geq a \wedge b' \leq b \wedge c' \leq c \wedge d' \geq d$ and $a+b+c+d = a'+b'+c'+d'$. Then we have $\frac{a}{a+b+c+d} \geq s$ and

$$\frac{a'}{a'+b'+c'+d'} = \frac{a'}{a+b+c+d} \geq \frac{a}{a+b+c+d} \geq s$$

which means that $F_{\odot_s}(a',b',c',d') = 1$ and thus $\odot_s$ is weakly equivalence. This finishes the proof for point 1.

ad 2.: We prove first that $\odot_s$ is not weakly Σ-equivalence. Put $\langle a,b,c,d\rangle = \langle 1,0,0,0\rangle$ and $\langle a',b',c',d'\rangle = \langle 0,0,0,1\rangle$. Then

- $a'+d' \geq a+d$ and $a+b+c+d = a'+b'+c'+d'$
- $F_{\odot_s}(a,b,c,d) = 1$ because $\frac{a}{a+b+c+d} = 1 \geq s$
- $F_{\odot_s}(a',b',c',d') = 0$ because $\frac{a'}{a'+b'+c'+d'} = 0 < s$

and this means that $\odot_s$ is not weakly Σ-equivalence. Thus, according to lemma 6.15, 4ft-quantifier $\odot_s$ is not Σ-equivalence.

This finishes the proof.   □

Lemma 9.4. *Let $\approx_1$ be an equivalence 4ft-quantifier with associated function $F_{\approx_1}$ and let $\approx_2$ be a weakly equivalence quantifier with associated function $F_{\approx_2}$ which is not equivalence. Then*

1. compound 4ft-quantifier $\approx_1 \wedge \approx_2$ is weakly equivalence quantifier
2. if there are 4ft-tables $\langle a_0,b_0,c_0,d_0\rangle$ and $\langle a'_0,b'_0,c'_0,d'_0\rangle$ such that

$$a'_0 \geq a_0 \wedge b'_0 \leq b_0 \wedge c'_0 \leq c_0 \wedge d'_0 \geq d_0, \quad F_{\approx_1}(a_0,b_0,c_0,d_0) = 1,$$

$$F_{\approx_2}(a_0,b_0,c_0,d_0) = 1, \text{ and } F_{\approx_2}(a'_0,b'_0,c'_0,d'_0) = 0$$

then compound 4ft-quantifier $\approx_1 \wedge \approx_2$ is not equivalence quantifier.

Proof. The proof is analogous to the proof of lemma 8.6.   □

Theorem 9.8. *All the compound 4ft-quantifiers*

1. Supported p-equivalence $\equiv^{\odot}_{p,s}$ defined as $\equiv_p \wedge \odot_s$
2. Supported lower critical equivalence $\equiv^{\odot!}_{p,\alpha,s}$ defined as $\equiv^!_{p,\alpha} \wedge \odot_s$
3. Supported upper critical equivalence $\equiv^{\odot?}_{p,\alpha,s}$ defined as $\equiv^?_{p,\alpha} \wedge \odot_s$
4. Supported simple deviation $\sim^{\odot}_{q,s}$ defined as $\sim_q \wedge \odot_s$
5. Supported Fisher's quantifier $\sim^{\odot1}_{\alpha,s}$ defined as $\sim^1_{\alpha} \wedge \odot_s$
6. Supported χ^2-quantifier $\sim^{\odot2}_{\alpha,s}$ defined as $\sim^2_{\alpha} \wedge \odot_s$

satisfy both I) and II):

I)   is weakly equivalence but not equivalence
II) is not weakly Σ-equivalence nor Σ-equivalence.

Proof. We prove the theorem in four steps:

A)   We prove that all 4ft-quantifiers in question are weakly equivalence quantifiers.
B)   We prove for each of the 4ft-quantifiers in question that it is not equivalence.
C)   We prove for each of the 4ft-quantifiers in question that it is not Σ-equivalence.

D)   We prove for each of the 4ft-quantifiers in question that it is not weakly Σ-equivalence.

Ad A): All 4ft-quantifiers $\equiv_p$, $\equiv_{p,\alpha}^!$, $\equiv_{p,\alpha}^?$, $\sim_q$, $\sim_\alpha^1$, and $\sim_\alpha^2$ are equivalence according to theorems 9.3 and 9.4. Point 1 of lemma 9.3 says that 4ft-quantifier $\odot_s$ is weakly equivalence. Thus, according to point 1 of lemma 9.4, all 4ft-quantifiers in question are weakly equivalence quantifiers.

Ad B): Point 1 of lemma 9.3 says that 4ft-quantifier $\odot_s$ is not equivalence. Thus, according to point 1 of theorem 9.5 it is true for each 4ft-quantifier $\approx$ of 4ft-quantifiers in question that $\approx$ is not equivalence.

Ad C): Point 2 of lemma 9.3 says that 4ft-quantifier $\odot_s$ is not Σ-equivalence. Thus, according to point 2 of theorem 9.5 it is true for each 4ft-quantifier $\approx$ of 4ft-quantifiers in question that $\approx$ is not Σ-equivalence.

Ad D): Point 2 of lemma 9.3 says that 4ft-quantifier $\odot_s$ is not weakly Σ-equivalence. Thus, according to point 4 of theorem 9.5 it is true for each 4ft-quantifier $\approx$ of 4ft-quantifiers in question that $\approx$ is not weakly Σ-equivalence.

This finishes the proof.   □

9.4   Interesting Σ-equivalence Quantifiers

We introduced interesting implicational quantifiers in Sect. 7.4 and interesting Σ-double implicational quantifiers in Sect. 8.4 because there are interesting theoretical results related to them. The same reason leads to the definition of interesting Σ-equivalence quantifiers. In addition, we define also interesting weakly Σ-equivalence quantifiers. We start with definition of (a+d)-dependent Σ-equivalence quantifiers, see also definitions 7.4 and 8.8.

Definition 9.8. Let $\approx$ be a 4ft-quantifier. Then $\approx$ is *(a+d)-dependent* if there are non-negative integers a, b, c, d, a', d' such that

$$a+b+c+d > 0 \ \wedge \ a'+b+c+d' > 0 \ \wedge \ a+d \neq a'+d'$$

and

$$F_\approx(a,b,c,d) \ \neq \ F_\approx(a',b,c,d') \,.$$

Definition 9.9. Let $\equiv^*$ be a Σ-equivalence quantifier or a weakly Σ-equivalence quantifier. Then we say that $\equiv^*$ is an *interesting Σ-equivalence quantifier* or *interesting weakly Σ-equivalence quantifier* respectively if it satisfies:

• $\equiv^*$ is (a+d)-dependent
• $F_{\equiv^*}(0,b,c,0) = 0$ for $b+c > 0$.

We prove that 4ft-quantifiers $\equiv_p$, $\equiv_{p,\alpha}^!$ and $\equiv_{p,\alpha}^?$ are interesting Σ-equivalence for most of the values of their parameters.

Note 9.1. Please note that we do not introduce a reasonable weakly Σ-equivalence quantifier which is not Σ-equivalence. Finding such a 4ft-quantifier is a challenge, see also Sect. 16.9. However, there is a weakly Σ-equivalence quantifier which is not Σ-equivalence, see lemma 11.3.

Theorem 9.9. *The following 4ft-quantifiers are interesting Σ-equivalence quantifiers:*

1. *4ft-quantifier $\equiv_p$ of p-equivalence for $0 < p < 1$*
2. *4ft-quantifier $\equiv^!_{p,\alpha}$ of lower critical equivalence for $0 < p < 1$ and $0 < \alpha \leq 0.5$.*
3. *4ft-quantifier $\equiv^?_{p,\alpha}$ of upper critical equivalence for $0 < p < 1$ and $0 < \alpha \leq 0.5$ such that $1 - p \leq \alpha$.*

Proof. *According to theorem 9.2 all these 4ft-quantifiers are Σ-equivalence. It remains to prove for each 4ft-quantifier $\equiv^*$ from the 4ft-quantifiers in question that $\equiv^*$ is (a+d)-dependent and that $F_{\equiv^*}(0,b,c,0) = 0$ for $b + c > 0$. We prove for each 4ft-quantifier $\equiv^*$:*

A) $F_{\equiv^*}(0,b,c,0) = 0$ for $b + c > 0$
B) *for each $b + c > 0$ there are $a \geq 0$ and $d \geq 0$ such that $a + d > 0$ and $F_{\equiv^*}(a,b,c,d) = 1$.*

Let us assume $a' = 0$ and $d' = 0$, then A) and B) implies that for $b + c > 0$ it holds that $a + b + c + d > 0 \wedge a' + b + c + d' > 0$, $a + d \neq a' + d'$ and $F_{\approx}(a,b,c,d) \neq F_{\approx}(a',b,c,d')$. In other words, A) and B) implies that $\equiv^*$ is (a+d)-dependent and that $F_{\equiv^*}(0,b,c,0) = 0$ for $b + c > 0$.

1. *4ft-quantifier $\equiv_p$ of founded equivalence is defined such that*

$$F_{\equiv_p}(a,b,c,d) = \begin{cases} 1 \text{ if } \frac{a+d}{a+b+c+d} \geq p \\ 0 \text{ otherwise.} \end{cases}$$

A) $F_{\equiv_p}(0,b,c,0) = 0$ for $b + c > 0$ because $\frac{0+0}{0+b+c+0} = 0$.
B) *We assume $0 < p < 1$ and thus there must be $a > 0$ for all b,c satisfying $b + c > 0$ such that $\frac{a}{a+b+c+0} \geq p$. This means $F_{\equiv_p}(a,b,c,0) = 1$.*

2. *4ft-quantifier $\equiv^!_{p,\alpha}$ of lower critical equivalence is defined such that*

$$F_{\equiv^!_{p,\alpha}}(a,b,c,d) = \begin{cases} 1 \text{ if } \sum_{i=a+d}^{n} \binom{n}{i} p^i (1-p)^{n-i} \leq \alpha \\ 0 \text{ otherwise,} \end{cases}$$

where $n = a + b + c + d$.

A) $F_{\equiv^!_{p,\alpha}}(0,b,c,0) = 0$ for $b + c > 0$ because $\sum_{i=0}^{b+c} \binom{b+c}{i} p^i (1-p)^{b+c-i} = 1$.
B) *It holds that $\lim_{N \to \infty} \sum_{i=N-A}^{N} \binom{N}{i} p^i (1-p)^{N-i} = 0$ for integer $A \geq 0$, see point 2 of lemma 5.15. This also means that $\lim_{a \to \infty} \sum_{i=a}^{a+b+c} \binom{a+b+c}{i} p^i (1-p)^{a+b+c-i} = 0$ for $b + c > 0$. Thus there is $a > 0$ such that $\sum_{i=a}^{a+b+c} \binom{a+b+c}{i} p^i (1-p)^{a+b+c-i} \leq \alpha$ which implies $F_{\equiv^!_{p,\alpha}}(a,b,c,0) = 1$.*

3. *4ft-quantifier* $\equiv^?_{p,\alpha}$ *of upper critical equivalence is defined such that*

$$\equiv^?_{p,\alpha} (a,b,c,d) = \begin{cases} 1 & \text{if } \sum_{i=0}^{a+d} \binom{n}{i} p^i (1-p)^{n-i} > \alpha \\ 0 & \text{otherwise,} \end{cases}$$

where $n = a+b+c+d$.

A) $F_{\equiv^!_{p,\alpha}}(0,b,c,0) = 0$ *for* $b+c > 0$ *because of*
    $\sum_{i=0}^{0} \binom{b+c}{0} p^0 (1-p)^{b+c} = (1-p)^{b+c} \leq 1 - p$ *and we assume* $1 - p \leq \alpha$.

B) *It holds that* $\lim_{N \to \infty} \sum_{i=0}^{N} \binom{N+A}{i} p^i (1-p)^{N-i} = 1$ *for integer* $A \geq 0$,
    see point 3 of lemma 5.16 and thus there is $a > 0$ *such that*
    $\sum_{i=0}^{a} \binom{a+b+c}{i} p^i (1-p)^{a+b+c-i} > \alpha$ *which means that* $F_{\equiv^?_{p,\alpha}}(a,b,c,0) = 1$.

This finishes the proof.   □

9.5   Table of Critical Frequencies for Σ-equivalence Quantifiers

We defined tables of critical frequencies for implicational quantifiers and for double implicational quantifiers in Sects. 7.5 and 8.5 We define similar tables of critical frequencies also for Σ-equivalence quantifiers.

Please remember definition 7.6 where the set $\mathcal{N}^+$ is defined as a union $\mathcal{N}^+ = \mathcal{N} \cup \{\infty\}$, and additionally $n < \infty$, $n + \infty = \infty$, $\infty - n = \infty$ for each $n \in \mathcal{N}$, $\max(\mathcal{N}) = \infty$, and $\min(\mathcal{N}^+) = \infty$.

Definition 9.10. Let $\equiv^*$ be a Σ-equivalence quantifier. Then a *table* $Tb_{E,\equiv^*}$ *of maximal* $b+c$ *for* $\equiv^*$ is a function assigning to each non-negative integer number A a value $Tb_{E,\equiv^*}(A)$ from $\mathcal{N}^+$ such that

$$Tb_{E,\equiv^*}(A) = \min\{b+c \mid F_{\equiv^*}(a,b,c,d) = 0 \wedge a+d = A\} .$$

There is a useful theorem concerning the tables of maximal $b+c$ for Σ-equivalence quantifiers.

Theorem 9.10. *Let* $Tb_{E,\equiv^*}$ *be a table of maximal* $b+c$ *for* Σ-*equivalence quantifier* $\equiv^*$ *with associated function* $F_{\equiv^*}$. *Then we have*

1. $F_{\equiv^*}(a,b,c,d) = 1$ *if and only if* $b+c < Tb_{E,\equiv^*}(a+d)$
2. $Tb_{E,\equiv^*}(A)$ *is a nondecreasing function of* A.

Proof. The proof is similar to the proof of theorem 8.12 and we present it in a bit shortened form.

1. *The fact that* $F_{\equiv^*}(a,b,c,d) = 1$ *if and only if* $b + c < Tb_{E,\equiv^*}(a+d)$ *directly follows from the simple form of truth preservation condition* $TPC_{\Sigma,\equiv}$ *for* Σ-*equivalence quantifiers:*

$$a' + d' \geq a+d \wedge b' + c' \leq b+c,$$

see definition 9.4.

2. Let us assume

$$\text{both } A' > A \text{ and } Tb_{E,\equiv^*}(A') < Tb_{E,\equiv^*}(A).$$

Let us denote $u = Tb_{E,\equiv^*}(A')$. *This means* $u < Tb_{\equiv^*}(A)$ *and thus* $F_{\equiv^*}(A,u,0,0) = 1$. *Quantifier* $\equiv^*$ *is Σ-equivalence and* $A' > A$, *and this also means* $F_{\equiv^*}(A',u,0,0) = 1$. *However, according to definition 9.10 we have*

$$u = Tb_{E,\equiv^*}(A') = \min\{b+c|F_{\equiv^*}(a,b,c,d) = 0 \wedge a+d = A'\},$$

and thus also $F_{\equiv^*}(A',u,0,0) = 0$, *which is a contradiction.*

This finishes the proof. □

9.6 Typical Σ-equivalence Quantifiers

We are going to define typical Σ-equivalence quantifiers used in the next sections. We prove a simple lemma and then we show that 4ft-quantifiers $\equiv_p$, $\equiv_{p,\alpha}^!$, and $\equiv_{p,\alpha}^?$ are typical Σ-equivalence quantifiers for most of values of their parameters.

Definition 9.11. Let $\equiv^*$ be a Σ-equivalence quantifier and let $Tb_{E,\equiv^*}$ be its table of maximal $b+c$. We say that the quantifier $\equiv^*$ is a *typical Σ-equivalence quantifier* if there is an integer $A \geq 0$ such that $1 < Tb_{E,\equiv^*}(A) < \infty$.

We say that a Σ-equivalence quantifier $\equiv^*$ is *untypical Σ-equivalence* if it is not typical Σ-equivalence.

Lemma 9.5. *Let* $\equiv^*$ *be a Σ-equivalence quantifier and let* $Tb_{E,\equiv^*}$ *be its table of maximal* $b+c$. *If there are integers P, Q, R such that* $P \geq 1$ *and* $1 \leq Q < R$ *satisfying* $F_{\equiv^*}(P,Q,0,0) = 1$ *and* $F_{\equiv^*}(P,R,0,0) = 0$ *then* $\equiv^*$ *is a typical Σ-equivalence quantifier.*

Proof. It holds that $Tb_{E,\equiv^*}(P) = \min\{b+c \mid F_{\equiv^*}(a,b,c,d) = 0 \wedge a+d = P\}$ *and thus assumptions* $1 \leq Q < R$ *and* $F_{\equiv^*}(P,Q,0,0) = 1$ *imply* $Tb_{E,\equiv^*}(P) > 1$ *(see definition 9.10 and theorem 9.10). In addition, assumptions* $1 \leq Q < R$ *and* $F_{\equiv^*}(P,R,0,0) = 0$ *imply* $Tb_{E,\equiv^*}(P) < \infty$.

This finishes the proof. □

Theorem 9.11. *Let be* $0 < p < 1$ *and* $0 < \alpha \leq 0.5$. *Then the 4ft-quantifiers* $\equiv_p$ *of p-equivalence,* $\equiv_{p,\alpha}^!$ *of lower critical equivalence, and* $\equiv_{p,\alpha}^?$ *of upper critical equivalence are typical Σ-equivalence quantifiers.*

Proof. All the quantifiers in question are Σ-equivalence quantifiers according to theorem 9.2. For each quantifier $\equiv^*$ *from the quantifiers* $\equiv_p$, $\equiv_{p,\alpha}^!$, *and* $\equiv_{p,\alpha}^?$ *we find integer numbers* P,Q,R *satisfying* $P \geq 1$ *and* $1 \leq Q < R$ *such that* $F_{\equiv^*}(P,Q,0,0) = 1$ *and* $F_{\equiv^*}(P,R,0,0) = 0$. *Then, lemma 9.5 implies that all of these 4ft-quantifiers are typical Σ-equivalence quantifiers.*

1. 4ft-quantifier $\equiv_p$ of p-equivalence is defined such that

$$F_{\equiv_p}(a,b,c,d) = \begin{cases} 1 \text{ if } \frac{a+d}{a+b+c+d} \geq p \\ 0 \text{ otherwise.} \end{cases}$$

It holds that $0 < p < 1$ and thus there is $P > 1$ such that $\frac{P}{P+1} \geq p$ i.e. $F_{\equiv_p}(P,Q,0,0) = 1$ for $Q = 1$. There is also $R > 1$ such that $\frac{P}{P+R} < p$ i.e. $F_{\equiv_p}(P,R,0,0) = 0$. This finishes the proof for 4ft-quantifier $\equiv_p$.

2. 4ft-quantifier $\equiv_{p,\alpha}^!$ of lower critical double implication is for $0 < p \leq 1$ and $0 < \alpha \leq 0.5$ defined such that

$$F_{\equiv_{p,\alpha}^!}(a,b,c,d) = \begin{cases} 1 \text{ if } \Sigma_{i=a+d}^{n} \binom{n}{i} p^i (1-p)^{n-i} \leq \alpha \\ 0 \text{ otherwise} \end{cases}$$

where $n = a+b+c+d$. We assume $0 < p < 1$, and according to lemma 5.15 point 2 we have

$$\lim_{N \to \infty} \sum_{i=N-A}^{N} \binom{N}{i} p^i (1-p)^{N-i} = 0$$

and thus there is an integer $P > 1$ such that

$$\sum_{i=P-1}^{P} \binom{P}{i} p^i (1-p)^{P-i} \leq \alpha$$

which means $\equiv_{p,\alpha}^! (P,Q,0,0) = 1$ for $Q = 1$.
In addition, according to lemma 5.16 point 1 we have

$$\lim_{N \to \infty} \sum_{i=A}^{A+N} \binom{A+N}{i} p^i (1-p)^{A+N-i} = 1$$

and thus there is an integer R, such that

$$\sum_{i=P}^{P+R} \binom{P+R}{i} p^i (1-p)^{P+R-i} > \alpha$$

which means $\equiv_{p,\alpha}^! (P,R,0,0) = 0$. This finishes the proof for 4ft-quantifier $\equiv_p^!$.

3. 4ft-quantifier $\equiv_{p,\alpha}^?$ of upper critical double implication is for $0 < p \leq 1$ and $0 < \alpha \leq 0.5$ defined such that

$$\equiv_{p,\alpha}^? (a,b,c,d) = \begin{cases} 1 \text{ if } \Sigma_{i=0}^{a+d} \binom{n}{i} p^i (1-p)^{n-i} > \alpha \\ 0 \text{ otherwise} \end{cases}$$

where $n = a+b+c+d$. We assume $0 < p < 1$ and according to lemma 5.16 point 2 we have

$$\lim_{N \to \infty} \sum_{i=0}^{N-A} \binom{N}{i} p^i (1-p)^{N-i} = 1$$

thus there is an integer $P > 1$ such that

$$\sum_{i=0}^{P-1} \binom{P}{i} p^i (1-p)^{P-i} > \alpha$$

which means $\equiv^?_{p,\alpha} (P,Q,0,0) = 1$ for $Q = 1$.
In addition, according to lemma 5.15 point 1 we have

$$\lim_{N \to \infty} \sum_{i=0}^{A} \binom{A+N}{i} p^i (1-p)^{A+N-i} = 0$$

and thus there is an integer R such that

$$\sum_{i=0}^{P+R} \binom{P+R}{i} p^i (1-p)^{P+R-i} \le \alpha$$

which means $\equiv^?_{p,\alpha} (P,R,0,0) = 0$.

This finishes the proof. □

9.7 Pure Equivalence Quantifiers

Truth preservation condition $TPC_{\equiv}$ for equivalence rules originates from an attempt to express equivalence relation of attributes φ and ψ using an "equivalence" 4ft-quantifier $\equiv^*$ such that $\varphi \equiv^* \psi$ if and only if both $\varphi \Rightarrow^* \psi$ and $\neg\psi \Rightarrow^* \neg\varphi$, where $\Rightarrow^*$ is a suitable implicational quantifier, see Sect. 6.4.

$TPC_{\equiv}$ is introduced in definition 6.4 as a combination of truth preservation conditions for implicational quantifier $\Rightarrow^*$ both in rule $\varphi \Rightarrow^* \psi$ and in rule $\neg\psi \Rightarrow^* \neg\varphi$. Thus, it is natural to investigate the equivalence 4ft-quantifier $\equiv^*$ satisfying condition

$$\varphi \equiv^* \psi \quad \text{if and only if} \quad \varphi \Rightarrow^* \psi \text{ and } \neg\psi \Rightarrow^* \neg\varphi, \tag{9.2}$$

where $\Rightarrow^*$ is a suitable implicational quantifier. We call such 4ft-quantifier as pure equivalence quantifiers.

First we define the class of pure equivalence quantifiers using condition 9.2 and we show that each pure equivalence quantifier is equivalence. Then we find a condition equivalent to the fact that an equivalence quantifier is pure equivalence.

Definition 9.12. Let $\approx$ be a 4ft-quantifier with associated function $F_{\approx}$ such that there is an implicational quantifier $\Rightarrow^*$ with associated function $F_{\Rightarrow^*}$ satisfying

$$F_{\approx}(a,b,c,d) = 1 \quad \text{if and only if both} \quad F_{\Rightarrow^*}(a,b) = 1 \text{ and } F_{\Rightarrow^*}(d,c) = 1 .$$

Then we say:

1. $\approx$ is a *pure equivalence quantifier*
2. *implicational quantifier* $\Rightarrow^*$ *is associated to pure equivalence quantifier* $\approx$,
3. *pure equivalence quantifier* $\approx$ *is associated to implicational quantifier* $\Rightarrow^*$.

Theorem 9.12. *Each pure equivalence quantifier is equivalence.*

Proof. Let $\approx$ be a pure equivalence quantifier with associated function $F_\approx$. We have to prove that if $F_\approx(a,b,c,d) = 1$ and $a' \geq a \wedge b' \leq b \wedge c' \leq c \wedge d' \geq d$, then also $F_\approx(a',b',c',d) = 1$, see definition 6.4.

$\approx$ is a pure equivalence quantifier, thus there is an implicational quantifier $\Rightarrow^*$ with associated function $F_{\Rightarrow^*}$ such that

$$F_\approx(a,b,c,d) = 1 \quad \text{if and only if both} \quad F_{\Rightarrow^*}(a,b) = 1 \text{ and } F_{\Rightarrow^*}(d,c) = 1 \,.$$

We assume $F_\approx(a,b,c,d) = 1$, which means $F_{\Rightarrow^*}(a,b) = 1$ and $F_{\Rightarrow^*}(d,c) = 1$. We also assume $a' \geq a \wedge b' \leq b \wedge c' \leq c \wedge d' \geq d$ which implies both $F_{\Rightarrow^*}(a',b') = 1$ and $F_{\Rightarrow^*}(d',c') = 1$. This however means $F_\approx(a',b',c',d) = 1$ according to definition 9.12. This finishes the proof. $\qquad\qquad\square$

We show also that we can use a suitable table of critical frequencies to define pure equivalence quantifiers. Please remember definition 7.6 of $\mathcal{N}^+$.

Theorem 9.13. *4ft-quantifier $\approx$ with associated function $F_\approx$ is pure equivalence if and only if there is a non-negative nondecreasing function $T_\approx$ assigning to each integer $a \geq 0$ a value from $\mathcal{N}^+$ such that*

$$F_\approx(a,b,c,d) = 1 \text{ if and only if both } b < T_\approx(a) \text{ and } c < T_\approx(d) \,.$$

Proof

1. *Let $\approx$ be pure equivalence. Then there is an implicational quantifier $\Rightarrow^*$ with associated function $F_{\Rightarrow^*}$ such that*

$$F_\approx(a,b,c,d) = 1 \text{ if and only if both } F_{\Rightarrow^*}(a,b) = 1 \text{ and } F_{\Rightarrow^*}(d,c) = 1 \,.$$

Let us define $T_\approx(a) = Tb_{\Rightarrow^*}(a)$ where $Tb_{\Rightarrow^*}$ is a table of maximal b for quantifier $\Rightarrow^*$, see definition 7.7. According to theorem 7.8 we have $F_{\Rightarrow^*}(a,b) = 1$ if and only if $b < Tb_{\Rightarrow^*}(a)$, $F_{\Rightarrow^*}(d,c) = 1$ if and only if $c < Tb_{\Rightarrow^*}(d)$, and $Tb_{\Rightarrow^*}$ is nondecreasing. Thus we have

$$F_\approx(a,b,c,d) = 1 \text{ if and only if both } b < Tb_{\Rightarrow^*}(a) \text{ and } c < Tb_{\Rightarrow^*}(d) \,.$$

2. *Let $T_\approx$ be a non-negative nondecreasing function assigning to each integer non-negative number a value from $\mathcal{N}^+$ such that*

$$F_\approx(a,b,c,d) = 1 \text{ if and only if both } b < T_\approx(a) \text{ and } c < T_\approx(d) \,.$$

We need an implicational quantifier $\Rightarrow^*$ with associated function $F_{\Rightarrow}^*$ satisfying

$$F_{\approx}(a,b,c,d) = 1 \text{ if and only if both } F_{\Rightarrow^*}(a,b) = 1 \text{ and } F_{\Rightarrow^*}(d,c) = 1 .$$

Let us define 4ft-quantifier $\Rightarrow^*$ such that

$$F_{\Rightarrow^*}(a,b,c,d) = 1 \text{ if and only if } b < T_{\approx}(a) .$$

Then, according to theorem 7.9, quantifier $\Rightarrow^*$ is implicational and function $T_{\approx}(a)$ is its table of maximal b. According to theorem 7.8 it holds $F_{\Rightarrow^*}(a,b) = 1$ if and only if $b < T_{\approx}(a)$ and also $F_{\Rightarrow^*}(d,c) = 1$ if and only if $c < T_{\approx}(d)$. This means

$$F_{\approx}(a,b,c,d) = 1 \text{ if and only if both } F_{\Rightarrow^*}(a,b) = 1 \text{ and } F_{\Rightarrow^*}(d,c) = 1$$

which finishes the proof. □

We prove that a typical Σ-equivalence quantifier is not pure equivalence. We then use this result to show that most practically important Σ-equivalence quantifiers are not pure equivalence.

Theorem 9.14. *If $\equiv^*$ is a typical Σ-equivalence quantifier then $\equiv^*$ is not a pure equivalence quantifier.*

Proof. Let us remember that a Σ-equivalence quantifier $\equiv^*$ is typical Σ-equivalence if there is an integer $A \geq 0$ such that $1 < Tb_{E,\equiv^*}(A) < \infty$ where $Tb_{E,\equiv^*}$ is a table of maximal $b+c$ of $\equiv^*$, see definition 9.11. Table of maximal $b+c$ for Σ-equivalence quantifier $\equiv^*$ is defined such that

$$Tb_{E,\equiv^*}(A) = \min\{b+c \mid F_{\equiv^*}(a,b,c,d) = 0 \wedge a+d = A\} ,$$

see definition 9.10.

Let $\equiv^*$ be a typical Σ-equivalence quantifier. Then there is a non-negative A such that $1 < Tb_{E,\equiv^*}(A) < \infty$. We have to prove that there is no implicational quantifier $\Rightarrow^*$ which satisfies

$$F_{\equiv^*}(a,b,c,d) = 1 \text{ if and only if both } F_{\Rightarrow^*}(a,b) = 1 \text{ and } F_{\Rightarrow^*}(d,c) . \qquad (9.3)$$

Let us assume there is such a quantifier $\Rightarrow^*$ and denote $\sigma = Tb_{E,\equiv^*}(A) - 1$. This means $F_{\equiv^*}(A, \sigma, 0, 0) = 1$ which implies $F_{\Rightarrow^*}(A, \sigma) = 1$ because of (9.3). However, $F_{\Rightarrow^*}(A, \sigma) = 1$ and (9.3) also implies $F_{\equiv^*}(A, \sigma, \sigma, A) = 1$. 4ft-quantifier $\equiv^*$ is a Σ-equivalence quantifier and thus $F_{\equiv^*}(A, \sigma, \sigma, A) = 1$ implies $F_{\equiv^*}(A, \sigma, \sigma, 0) = 1$.

We defined $\sigma = Tb_{E,\equiv^*}(A) - 1$ and we assume $Tb_{E,\equiv^*}(A) > 1$ which means $Tb_{E,\equiv^*}(A) \geq 2$. Thus we have

$$\sigma + \sigma = Tb_{E,\equiv^*}(A) - 1 + Tb_{E,\equiv^*}(A) - 1 = Tb_{E,\equiv^*}(A) + (Tb_{E,\equiv^*}(A) - 2) ,$$

which implies

$$\sigma + \sigma \geq Tb_{E,\equiv^*}(A) .$$

According to definition 9.10 we have

$$Tb_{E,\equiv^*}(A) = \min\{b+c \mid F_{\equiv^*}(a,b,c,d) = 0 \wedge a+d = A\}$$

and because $\sigma + \sigma \geq Tb_{E,\equiv^*}(A)$ *it must be* $F_{\equiv^*}(A,\sigma,\sigma,0) = 0$. *This is however a contradiction with the above proved* $F_{\equiv^*}(A,\sigma,\sigma,0) = 1$.

This finishes the proof. ☐

Note 9.2. The proof of theorem 9.14 is similar to step 1 of the proof of theorem 8.16. However, theorem 8.16 says that a Σ-double implicational quantifier is pure double implicational if and only it is not typical and theorem 9.14 says only that if $\equiv^*$ is a typical Σ-equivalence quantifier then $\equiv^*$ is not pure equivalence quantifier.

A proof of a remaining assertion that pure Σ-equivalence quantifier is not typical (may be for a modified notion of typical Σ-equivalence quantifier) is still a challenge.

Theorem 9.15. *Let be* $0 < p < 1$ *and* $0 < \alpha \leq 0.5$. *Then the quantifiers* $\equiv_p$ *of p-equivalence,* $\equiv^!_{p,\alpha}$ *of lower critical equivalence, and* $\equiv^?_{p,\alpha}$ *of upper critical equivalence are not pure equivalence quantifiers.*

Proof. Directly follows from theorems 9.11 and 9.14. ☐

Finally, we prove that E-quantifier $\equiv^E_p$ (see rows 15 in Tables 4.2 and 4.4) is pure equivalence.

Theorem 9.16. *E-quantifier* $\equiv^E_p$ *is pure equivalence for* $0 < p \leq 1$.

Proof. E-quantifier $\equiv^E_p$ *with parameter p where* $0 < p \leq 1$ *is defined as follows:*

$$F_{\equiv^E_p}(a,b,c,d) = \begin{cases} 1 & \text{if } \max(\frac{b}{a+b}, \frac{c}{d+c}) < p \\ 0 & \text{otherwise.} \end{cases}$$

Let us define 4ft-quantifier $\Rightarrow^E_p$ *such that*

$$F_{\Rightarrow^E_p}(a,b,c,d) = \begin{cases} 1 & \text{if } \frac{a}{a+b} \geq 1-p \\ 0 & \text{otherwise.} \end{cases}$$

4ft-quantifier $\Rightarrow^E_p$ *is implicational and it is*

$$F_{\equiv^E_p}(a,b,c,d) = 1 \text{ if and only if both } \frac{b}{a+b} < p \text{ and } \frac{c}{d+c} < p$$

which means

$$F_{\equiv^E_p}(a,b,c,d) = 1 \text{ if and only if both } \frac{a}{a+b} \geq 1-p \text{ and } \frac{d}{d+c} \geq 1-p$$

which is equivalent to

$$F_{\equiv^E_p}(a,b,c,d) = 1 \text{ if and only if both } F_{\Rightarrow^E_p}(a,b) = 1 \text{ and } F_{\Rightarrow^E_p}(d,c) = 1.$$

This finishes the proof. ☐

9.8 Strong Equivalence Quantifiers

"Strong equivalence quantifiers" are a natural generalization of pure equivalence quantifiers. Informally speaking, $\equiv^*$ is a strong equivalence quantifier if there are two implicational quantifiers $\Rightarrow_1^*$ and $\Rightarrow_1^*$ such that

$$\varphi \equiv^* \psi \text{ if and only if both } \varphi \Rightarrow_1^* \psi \text{ and } \neg\varphi \Rightarrow_2^* \neg\psi .$$

Strong equivalence quantifiers are shortly introduced in this section. We start with a formal definition and then show simple properties of strong equivalence quantifiers.

Definition 9.13. Let $\approx^*$ be a 4ft-quantifier with associated function $F_\approx$ and let there are two implicational quantifiers $\Rightarrow_1$ and $\Rightarrow_2$ with associated functions $F_{\Rightarrow_1}$ and $F_{\Rightarrow_2}$ respectively such that

$$F_\approx(a,b,c,d) = 1 \text{ if and only if both } \Rightarrow_1 (a,b) = 1 \text{ and } \Rightarrow_2 (d,c) = 1 .$$

Then we say:

1. $\equiv^*$ is a *strong equivalence quantifier*,
2. *implicational quantifier* $\Rightarrow_1$ *is a left associated quantifier to strong equivalence quantifier* $\equiv^*$ *and implicational quantifier* $\Rightarrow_2$ *is a right associated quantifier to strong equivalence quantifier* $\equiv^*$
3. *strong equivalence quantifier* $\equiv^*$ *is associated to implicational quantifiers* $\Rightarrow_1$ *and* $\Rightarrow_2$.

Theorem 9.17

1. Each strong equivalence quantifier is an equivalence quantifier.
2. Each pure equivalence quantifier is a strong equivalence quantifier.

Proof. Proof of point 1) is analogous to the proof of theorem 9.12. Point 2) directly follows from definitions 9.12 and 9.13. □

The following theorem says that strong equivalence quantifier can also be defined using suitable tables of critical frequencies. Then we prove that typical Σ-equivalence quantifiers are not strong equivalence.

Theorem 9.18. *4ft-quantifier $\approx$ with associated function $F_\approx$ is strong equivalence if and only if there are non-negative nondecreasing functions functions $T_{1,\approx}$ and $T_{2,\approx}$ assigning to each integer $a \geq 0$ a value from $\mathcal{N}^+$ such that*

$$F_\approx(a,b,c) = 1 \text{ if and only if both } b < T_{1,\approx}(a) \text{ and } c < T_{2,\approx}(d) .$$

Proof. The proof is analogous to the proof of theorem 9.13. □

Theorem 9.19. *If $\equiv^*$ is a typical Σ-equivalence quantifier then $\equiv^*$ is not strong equivalence quantifier.*

Proof. The proof is similar to that of theorem 9.14. Let us remember that a Σ-equivalence quantifier $\equiv^*$ is typical Σ-equivalence if there is an integer $A \geq 0$ such that $1 < Tb_{E,\equiv^*}(A) < \infty$ where $Tb_{E,\equiv^*}$ is a table of maximal $b+c$ of $\equiv^*$, see definition 9.11. The table of maximal $b+c$ for Σ-equivalence quantifier $\equiv^*$ is defined such that

$$Tb_{E,\equiv^*}(A) = \min\{b+c \mid F_{\equiv*}(a,b,c,d) = 0 \wedge a+d = A\},$$

see definition 9.10.

Let $\equiv^*$ be a typical Σ-equivalence quantifier. Then there is a non-negative A satisfying $1 < Tb_{E,\equiv^*}(A) < \infty$. We have to prove that there are not implicational quantifiers $\Rightarrow_1^*$ and $\Rightarrow_2^*$ satisfying

$$F_{\equiv^*}(a,b,c,d) = 1 \text{ if and only if both } F_{\Rightarrow_1^*}(a,b) = 1 \text{ and } F_{\Rightarrow_2^*}(d,c). \tag{9.4}$$

Let us assume there are such quantifiers $\Rightarrow_1^*$, $\Rightarrow_2^*$ and denote $\sigma = Tb_{E,\equiv^*}(A) - 1$. This means $F_{\equiv^*}(A,\sigma,0,0) = 1$ which implies $F_{\Rightarrow_1^*}(A,\sigma) = 1$. In addition, it also means $F_{\equiv^*}(0,0,\sigma,A) = 1$ which implies $F_{\Rightarrow_2^*}(A,\sigma) = 1$ because of (9.4). However, $F_{\Rightarrow_1^*}(A,\sigma) = 1$, $F_{\Rightarrow_2^*}(A,\sigma) = 1$, and (9.4) imply also $F_{\equiv^*}(A,\sigma,\sigma,A) = 1$. 4ft-quantifier $\equiv^*$ is a Σ-equivalence quantifier and thus $F_{\equiv^*}(A,\sigma,\sigma,A) = 1$ implies $F_{\equiv^*}(A,\sigma,\sigma,0) = 1$.

We defined $\sigma = Tb_{E,\equiv^*}(A) - 1$ and we assume $Tb_{E,\equiv^*}(A) > 1$ which means $Tb_{E,\equiv^*}(A) \geq 2$. Thus we have

$$\sigma + \sigma = Tb_{E,\equiv^*}(A) - 1 + Tb_{E,\equiv^*}(A) - 1 = Tb_{E,\equiv^*}(A) + (Tb_{E,\equiv^*}(A) - 2),$$

which implies

$$\sigma + \sigma \geq Tb_{E,\equiv^*}(A).$$

According to definition 9.10 it holds that

$$Tb_{E,\equiv^*}(A) = \min\{b+c \mid F_{\equiv*}(a,b,c,d) = 0 \wedge a+d = A\}$$

and because $\sigma + \sigma \geq Tb_{E,\equiv^*}(A)$ it must be $F_{\equiv^*}(A,\sigma,\sigma,0) = 0$. This is however a contradiction with the $F_{\equiv^*}(A,\sigma,\sigma,0) = 1$ proved above.

This finishes the proof. $\square$

The last theorem of this section shows that important Σ-equivalence quantifiers are not strong equivalence.

Theorem 9.20. *Let $0 < p < 1$ and $0 < \alpha \leq 0.5$. Then the quantifiers $\equiv_p$ of p-equivalence, $\equiv_{p,\alpha}^!$ of lower critical equivalence, and $\equiv_{p,\alpha}^?$ of upper critical equivalence are not strong equivalence quantifiers.*

Proof. Directly follows from theorems 9.11 and 9.19. $\square$

9.9 Relations among Subclasses of Equivalence Quantifiers

In this section we summarize relations among above introduced subclasses of equivalence quantifiers:

1. The following quantifiers belong to equivalence quantifiers:

   - Σ-equivalence quantifiers, see theorem 9.1
   - pure equivalence quantifiers, see theorem 9.12
   - strong equivalence quantifiers, see theorem 9.17.

2. Each pure equivalence quantifier is a strong equivalence quantifier, see theorem 9.17.

3. It is suitable to distinguish typical and untypical Σ-equivalence quantifiers, see definition 9.11. We have the following results

   - Quantifiers $\equiv_p$ of p-equivalence, $\equiv_{p,\alpha}^{!}$ of lower critical equivalence, and $\equiv_{p,\alpha}^{?}$ of upper critical equivalence with parameters $0 < p < 1$ and $0 < \alpha \leq 0.5$ are typical Σ-equivalence quantifiers, see theorem 9.11.
   - A typical Σ-equivalence quantifier is neither pure equivalence nor strong equivalence, see theorems 9.14 and 9.19.

Chapter 10
Rules with F-property

The definition of the class of 4ft-quantifiers with the F-property was inspired by consideration on 4ft-quantifiers with same properties with regards to dealing with missing information as the Fisher's quantifier. This has led to a class which is, in contrast to the classes of implicational, double implicational and equivalence 4ft-quantifiers, $\mathcal{M}$-dependent. This is a simple corollary of the definitions of 4ft-quantifiers with the F-property and $\mathcal{M}$-dependent quantifiers. It is introduced in Sect. 10.1 together with several practically important 4ft-quantifiers with the F-property. Tables of critical frequencies are important tools for the study of 4ft-quantifiers. Useful tables of critical frequencies can also be defined for quantifiers with the F-property. These are introduced in Sect. 10.2.

A way of the definition of the class of 4ft-quantifiers with the F-property implies that it contains no subclasses analogous to the subclasses of pure and strong double implicational quantifiers and subclasses of pure and strong equivalence quantifiers. The theoretical results related to the class of 4ft-quantifiers with the F-property are also not as rich as for classes of 4ft-quantifiers defined in previous chapters. Results concerning dealing with missing information are only presented for the class of 4ft-quantifiers with the F-property. However, several partial results and related challenges are briefly introduced in Chap. 16.

10.1 Selected Quantifiers with F-property

We say that a truth preservation condition $C(a,b,c,d,a',b',c',d')$ is $\mathcal{M}$-dependent if $C(a,b,c,d,a',b',c',d') = 1$ implies $a+b+c+d = a'+b'+c'+d'$, see definition 6.11. The truth preservation condition $TPC_F(a,b,c,d,a',b',c',d')$ for quantifiers with the F-property is defined such that

$$TPC_F(a,b,c,d,a',b',c',d') = \begin{cases} 1 & \text{if } b \geq c-1 \geq 0 \text{ and} \\ & a'=a \wedge b'=b+1 \wedge c'=c-1 \wedge d'=d \\ 1 & \text{if } c \geq b-1 \geq 0 \text{ and} \\ & a'=a \wedge b'=b-1 \wedge c'=c+1 \wedge d'=d \\ 0 & \text{otherwise.} \end{cases}$$

J. Rauch: *Observational Calculi and Association Rules*, SCI 469, pp. 149–154.
DOI: 10.1007/978-3-642-11737-4_10 © Springer-Verlag Berlin Heidelberg 2013

see definition 6.10. This means that each 4ft-quantifier with the F-property is
$\mathcal{M}$-dependent, see the following lemma.

Lemma 10.1. *The truth preservation condition $TPC_F(a,b,c,d,a',b',c',d')$ for
4ft-quantifiers with the F-property is $\mathcal{M}$-dependent. Thus the class of 4ft-quantifiers
with the F-property is $\mathcal{M}$-dependent and each 4ft-quantifier with the F-property is
$\mathcal{M}$-dependent.*

Proof. The lemma is a direct consequence of definitions 6.10 and 6.11.   □

There are important equivalence 4ft-quantifiers of simple deviation $\sim_q$, founded
simple deviation $\sim_{q,Base}$, Fisher's quantifier $\sim_\alpha^1$, founded Fisher's quantifier $\sim_{\alpha,Base}^1$,
χ^2-quantifier $\sim_\alpha^2$ and founded χ^2-quantifier $\sim_{\alpha,Base}^2$, see theorems 9.4 and 9.6. We
prove that all of them have the F-property.

Theorem 10.1. *The following 4ft-quantifiers have the F-property:*

1. *Simple deviation $\sim_q$ and founded simple deviation $\sim_{q,Base}$*
2. *Fisher's quantifier $\sim_\alpha^1$ and founded Fisher's quantifier $\sim_{\alpha,Base}^1$*
3. *χ^2-quantifier $\sim_\alpha^2$ and founded χ^2-quantifier $\sim_{\alpha,Base}^2$.*

Proof. We have to prove for each of the above mentioned quantifiers:

a)   *If $F_\approx(a,b,c,d) = 1$ and $b \geq c-1 \geq 0$ then $F_\approx(a,b+1,c-1,d) = 1$.*
b)   *If $F_\approx(a,b,c,d) = 1$ and $c \geq b-1 \geq 0$ then $F_\approx(a,b-1,c+1,d) = 1$.*

*All the above 4ft-quantifiers are symmetrical according to theorem 9.7. This means
that $F_\approx(a,b,c,d) = F_\approx(a,c,b,d)$.*

*Let us assume that a) is proved, $F_\approx(a,b,c,d) = 1$ and $c \geq b-1 \geq 0$. Then
$F_\approx(a,b,c,d) = 1$ implies $F_\approx(a,c,b,d) = 1$ because of symmetry, we have
$c \geq b-1 \geq 0$ and thus according to a) also $F_\approx(a,c+1,b-1,d) = 1$ and according to symmetry we have $F_\approx(a,b-1,c+1,d) = F_\approx(a,c+1,b-1,d) = 1$ which
means that b) is proved on the basis of a) and symmetry. Thus it is sufficient to prove
only assertion a), assertion b) is a consequence of assertion a).*

1. *It holds that $F_{\sim_q}(a,b,c,d) = 1$ if and only if $ad > e^q bc$, see rows 10 in Tables 4.2
   and 4.4. Note that $(b+1)(c-1) = bc - (b+1-c)$. We assume $b \geq c-1 \geq 0$
   which implies $b+1-c \geq 0$ and thus $bc \geq bc - (b+1-c) = (b+1)(c-1)$.
   This means that if $F_{\sim_q}(a,b,c,d) = 1$ and $b \geq c-1 \geq 0$ then also
   $F_{\sim_q}(a,b+1,c-1,d) = 1$. This finishes the proof for simple deviation $\sim_q$, see
   also lemma 6.13.
   4ft-quantifier $\sim_{q,Base}$ of founded simple deviation is defined as $\sim_q \wedge_\oplus Base$ i.e.
   $F_{\sim_{q,Base}}(a,b,c,d) = 1$ if and only if $ad > e^q bc \wedge a \geq Base$, see theorem 9.6. The
   proof that 4ft-quantifier $\sim_{q,Base}$ has the F-property is thus the same as for 4ft-
   quantifier $\sim_s$.*

2. *It holds that $F_{\sim_\alpha^1}(a,b,c,d) = 1$ if and only if $\sum_{i=a}^{\min(r,k)} \frac{\binom{k}{i}\binom{n-k}{r-i}}{\binom{n}{r}} \leq \alpha \wedge ad > bc$
   where $r = a+b$, $k = a+c$, and $n = a+b+c+d$, see rows 11 in Tables 4.2 and
   4.4. Let us assume $F_{\sim_\alpha^1}(a,b,c,d) = 1$ and $b \geq c-1 \geq 0$. We have proved in the*

previous point that $b \geq c - 1 \geq 0$ implies $ad > (b+1)(c-1)$. Thus it is sufficient to prove that if $b \geq c - 1 \geq 0$ then

$$\sum_{i=a}^{\min(r+1,k-1)} \frac{\binom{k-1}{i}\binom{n-(k-1)}{r+1-i}}{\binom{n}{r+1}} \leq \sum_{i=a}^{\min(r,k)} \frac{\binom{k}{i}\binom{n-k}{r-i}}{\binom{n}{r}}.$$

However, this is proved in theorem 5.4, see also definition 5.1.
Founded Fisher's quantifier $\sim_{\alpha,Base}^{1}$ is defined as $\sim_{\alpha}^{1} \wedge \oplus_{Base}$ which means
$F_{\sim_{\alpha,Base}^{1}}(a,b,c,d) = 1$ *if and only if* $\sum_{i=a}^{\min(r,k)} \frac{\binom{k}{i}\binom{n-k}{r-i}}{\binom{n}{r}} \leq \alpha \wedge ad > bc \wedge a \geq Base$,
see theorem 9.6. The proof that 4ft-quantifier $\sim_{\alpha,Base}^{1}$ has the F-property is thus the same as for 4ft-quantifier $\sim_{\alpha}^{1}$.

3. *It holds that $F_{\sim_{\alpha}^{2}}(a,b,c,d) = 1$ if and only if $\frac{(ad-bc)^2}{rkls}n \geq \chi_{\alpha}^{2} \wedge ad > bc$ where $r = a+b$, $s = c+d$, $k = a+c$, $l = b+d$, and $n = a+b+c+d$, see rows 12 in Tables 4.2 and 4.4 (note that $l = n - k$ and $s = n - r$). Let us assume $F_{\sim_{\alpha}^{2}}(a,b,c,d) = 1$ and $b \geq c - 1 \geq 0$. We know that $b \geq c - 1 \geq 0$ implies $ad > (b+1)(c-1)$. We assume $F_{\sim_{\alpha}^{2}}(a,b,c,d) = 1$ which means $rkls > 0$. In addition we assume $ad > bc$ which implies $a > 0$ and $d > 0$. We also assume $b \geq c - 1 \geq 0$ which implies $k - 1 = a + c - 1 > 0$ and also $s - 1 = d + c - 1 > 0$, thus $(r+1)(k-1)(l+1)(s-1) > 0$. According to theorem 5.6 we have*

$$\frac{(ad - (b+1)(c-1))^2}{(r+1)(k-1)(l+1)(s-1)}n \geq \frac{(ad-bc)^2}{rkls}n$$

which finishes the proof for quantifier $\sim_{\alpha}^{2}$.
Founded χ^2-quantifier $\sim_{\alpha,Base}^{2}$ is defined as the conjunction $\sim_{\alpha}^{2} \wedge \oplus_{Base}$ and thus $F_{\sim_{\alpha,Base}^{2}}(a,b,c,d) = 1$ if and only if $\frac{(ad-bc)^2}{rkls}n \geq \chi_{\alpha}^{2} \wedge ad > bc \wedge a \geq Base$, see theorem 9.6. The proof that 4ft-quantifier $\sim_{\alpha,Base}^{2}$ has the F-property is thus the same as for 4ft-quantifier $\sim_{\alpha}^{2}$.

This finishes the proof.   □

We also prove that both the 4ft-quantifier $\sim_{q}^{+}$ of above average and founded 4ft-quantifier $\sim_{q,Base}^{+}$ of above average are symmetrical and have the F-property.

Theorem 10.2. *Both the 4ft-quantifier $\sim_{q}^{+}$ of above average dependence and founded 4ft-quantifier $\sim_{q,Base}^{+}$ of above average dependence defined as $\sim_{q}^{+} \wedge \oplus_{Base}$ are symmetrical and have the F-property.*

Proof. It holds that $F_{\sim_{q}^{+}}(a,b,c,d) = 1$ if and only if $\frac{a}{a+b} \geq (1+q)\frac{a+c}{a+b+c+d}$, see rows 13 in Tables 4.2 and 4.4. The condition

$$\frac{a}{a+b} \geq (1+q)\frac{a+c}{a+b+c+d}$$

is equivalent to

$$a(a+b+c+d) \geq (1+q)(a+b)(a+c)$$

which means that both $\sim_q^+$ and $\sim_{q,Base}^+$ are symmetrical (remember that $\oplus_{Base}$ is defined by a condition $a \geq Base$, see rows see rows 18 in Tables 4.2 and 4.4).

It is easy to prove that if

$$a(a+b+c+d) \geq (1+q)(a+b)(a+c) \text{ and } b \geq c-1 \geq 0$$

then

$$a(a+(b+1)+(c-1)+d) \geq (1+q)(a+b+1)(a+c-1).$$

This follows from fact that

$$a(a+(b+1)+(c-1)+d) = a(a+b+c+d)$$

and

$$(a+b+1)(a+c-1) = (a+b)(a+c)+c-1-b \leq (a+b)(a+c).$$

The fact that $F_{\sim_q^+}(a,b,c,d) = 1$ as proved above implies $F_{\sim_q^+}(a,b+1,c-1,d) = 1$ and the symmetry of both $\sim_q^+$ and $\sim_{q,Base}^+$ imply that both $\sim_q^+$ and $\sim_{q,Base}^+$ have the F-property.

This finishes the proof. $\square$

Note 10.1. Please note that the 4ft-quantifier $\sim_q^+$ of above average dependence is not equivalence, see theorem 6.3. Thus, it is an example of a 4ft-quantifier which has the F-property but is not equivalence.

10.2 Tables of Critical Frequencies and F-property

We show that there are also reasonable tables of critical frequencies for important quantifiers with the F-property. We start with a lemma.

Lemma 10.2. *Let $\approx$ be a symmetrical 4ft-quantifier with the F-property. Let $\langle a,b,c,d \rangle$ and $\langle a,b',c',d \rangle$ be 4ft-tables such that $b+c = b'+c'$. Then $F_\approx(a,b,c,d) = 1$ and $|b'-c'| \geq |b-c|$ implies $F_\approx(a,b',c',d) = 1$.*

Proof. 4ft-quantifier $\approx$ is symmetrical and thus its associated function $F_\approx(a,b,c,d)$ satisfies $F_\approx(a,b,c,d) = F_\approx(a,c,b,d)$. This means that we can assume $b \geq c$ and $b' \geq c'$ and then $|b'-c'| \geq |b-c|$ means $b'-c' \geq b-c$. We show that this assumption together with $b+c = b'+c'$ implies $b` \geq b \geq c \geq c'$.

Let $c < c'$, then $b+c = b'+c'$ means $b > b'$ which in turn implies $b-c' > b'-c'$. The assumption $c < c'$ also implies $b-c > b-c'$. Together we get $b-c > b-c' > b'-c'$ which is a contradiction to $b'-c' \geq b-c$. Thus it must hold that $c \geq c'$.

Similarly we prove that $b` \geq b$. Let $b' < b$, then $b+c = b'+c'$ means $c' > c$ which in turn implies $b'+c < b+c'$. However, $b'-c' \geq b-c$ means $b'+c \geq b+c'$ which is a contradiction to $b'+c < b+c'$. Thus it must hold that $b' \geq b$.

Let us denote $z = b' - b$, then $z \geq 0$, $b' = b + z$ and it also holds that $c' = c - z$ because $b + c = b' + c'$. If $z = 0$ then there is nothing to prove, thus let $z \geq 1$. We assume that both $\langle a, b, c, d \rangle$ and $\langle a, b', c', d \rangle = \langle a, b + z, c - z, d \rangle$ are 4ft-tables. Thus $c - i - 1 \geq 0$ for $i = 0, \ldots, z - 1$. In addition we assume $b \geq c \geq 0$ which implies $b + i \geq c - i > c - i - 1$ for $i = 0, \ldots, z - 1$. Altogether we have $b + i \geq c - i > c - i - 1 \geq 0$ for $i = 0, \ldots, z - 1$.

4ft-quantifier $\approx$ has the F-property and thus

$$F_{\approx}(a, b + i, c - i, d) = 1 \ \text{ implies } \ F_{\approx}(a, b + i + 1, c - i - 1, d) = 1$$

for $i = 0, \ldots, z - 1$. This means that also $F_{\approx}(a, b, c, d) = 1$ implies $F_{\approx}(a, b', c', d) = 1$. This finishes the proof.   □

Theorem 10.3. *Let $\approx$ be a symmetrical 4ft-quantifier with the F-property. Then there is a function $Tb_{|b-c|, \approx}(a, d, n)$ defined for all integers $a \geq 0$, $d \geq 0$, and $n \geq 0$ satisfying $a + d \leq n$ such that for all $b \geq 0$ and $c \geq 0$ satisfying $a + b + c + d = n$ it holds that*

$$F_{\approx}(a, b, c, d) = 1 \ \text{ if and only if } \ |b - c| \geq Tb_{b-c, \approx}(a, d, n) \ .$$

Proof. Let us define a set $Dfbc_{\approx}(a, d, n)$ for all integers a, d, n satisfying $a \geq 0$, $d \geq 0$, $n > 0$, and $a + d = n$. It is defined such that

$$Dfbc_{\approx}(a, d, n) = \{|b - c| \mid a + b + c + d = n \wedge F_{\approx}(a, b, c, d) = 1\}$$

i.e. $Dfbc_{\approx}(a, d, n)$ is a set of absolute values of differences $b - c$ for all 4ft-tables $\langle a, b, c, d \rangle$ such that $a + b + c + d = n$ and $F_{\approx}(a, b, c, d) = 1$.

Let us define $Tb_{|b-c|, \approx}(a, d, n)$ in the following way:

$$Tb_{|b-c|, \approx}(a, d, nm) = \begin{cases} \min(Dfbc_{\approx}(a, d, n)) & \text{if } Dfbc_{\approx}(a, d, n) \neq \emptyset \\ \\ n + 1 & \text{if } Dfbc_{\approx}(a, d, n) = \emptyset \end{cases}$$

We prove that $F_{\approx}(a, b, c, d) = 1$ if and only if $|b - c| \geq Tb_{|b-c|, \approx}(a, d, n)$ for $b \geq 0$ and $c \geq 0$ satisfying $a + b + c + d = n$.

1. *Let us assume $F_{\approx}(a, b, c, d) = 1$, then $|b - c| \in Dfbc_{\approx}(a, d, n)$ and thus we have $|b - c| \geq Tb_{|b-c|, \approx}(a, d, n)$.*
2. *Let us assume $|b - c| \geq Tb_{b-c, \approx}(a, d, n)$. In addition we assume $a \geq 0$, $b \geq 0$, $c \geq 0$, $d \geq 0$, and $a + b + c + d = n$ which implies $n \geq |b - c|$ and this means $n \geq Tb_{|b-c|, \approx}(a, d, n)$. Thus there are b_m, c_m such that*

$$|b_m - c_m| = \min(\{|b - c| \mid a + b + c + d = n \wedge F_{\approx}(a, b, c, d) = 1\})$$

which means $F_{\approx}(a, b_m, c_m, d) = 1$. We assume $|b - c| \geq |b_m - c_m|$ and according to lemma 10.2 we get $F_{\approx}(a, b, c, d) = 1$.

This finishes the proof.   □

Definition 10.1. Let $\approx$ be a symmetrical 4ft-quantifier with the F-property. Then the function $Tb_{|b-c|,\approx}(a,d,n)$ defined for all integer $a \geq 0$, $d \geq 0$, and $n \geq 0$ satisfying $a+d \leq n$ which is introduced in theorem 10.3 is called a *table of minimal* $|b-c|$ *for* $\approx$.

Remember that 4ft-quantifier $\approx$ is a-d symmetrical if its associated function $F_{\approx}(a,b,c,d)$ satisfies $F_{\approx}(a,b,c,d) = F_{\approx}(d,b,c,a)$, see definition 6.7. The following lemma is an easy consequence of theorem 10.3 and it shows that for a-d symmetrical quantifiers we need to know values of function $Tb_{|b-c|,\approx}(a,d,n)$ only for all $a \leq d \leq n$.

Lemma 10.3. *Let $\approx$ be a symmetrical 4ft-quantifier with the F-property which is also a-d symmetrical. Then*

$$F_{\approx}(a,b,c,d) = 1 \; \text{if and only if} \; |b-c| \geq Tb_{b-c,\approx}(\min(a,d),\max(a,d),n) \,.$$

Proof. 4ft-quantifier $\approx$ is a-d symmetrical which means $F_{\approx}(a,b,c,d) = F_{\approx}(d,b,c,a)$ and thus $F_{\approx}(a,b,c,d) = F_{\approx}(\min(a,d),b,c,\max(a,d))$. This finishes the proof. □

4ft-quantifiers simple deviation $\sim_q$, founded simple deviation $\sim_{q,Base}$, Fisher's quantifier $\sim_{\alpha}^1$, founded Fisher's quantifier $\sim_{\alpha,Base}^1$, χ^2-quantifier $\sim_{\alpha}^2$, and founded χ^2-quantifier $\sim_{\alpha,Base}^2$ are symmetrical according to theorem 9.7 and have the F-property according to theorem 10.1. The following theorem shows that all these quantifiers are also a-d symmetrical. This means that there are useful tables of minimal $|b-c|$ for these quantifiers and that lemma 10.3 can also be applied.

Theorem 10.4. *The following 4ft-quantifiers are a-d symmetrical:*

1. Simple deviation $\sim_q$ and founded simple deviation $\sim_{q,Base}$
2. Fisher's quantifier $\sim_{\alpha}^1$ and founded Fisher's quantifier $\sim_{\alpha,Base}^1$
3. χ^2-quantifier $\sim_{\alpha}^2$ and founded χ^2-quantifier $\sim_{\alpha,Base}^2$.

Proof. 1. It holds that $F_{\sim_q}(a,b,c,d) = 1$ if and only if $ad > e^q bc$, see rows 10 in Tables 4.2 and 4.4. This means that both $\sim_q$ and $\sim_{q,Base}$ are a-d symmetrical.
2. Fisher's measure $\mathscr{F}_F(a,b,c,d)$ (see definition 5.1) satisfies $\mathscr{F}_F(a,b,c,d) = \mathscr{F}_F(d,b,c,a)$, see point 2 of lemma 5.9. This implies that both $\sim_{\alpha}^1$ and $\sim_{\alpha,Base}^1$ are a-d symmetrical.
3. χ^2-measure $\mathscr{F}_{\chi^2}(a,b,c,d)$ (see definition 5.2) satisfies $\mathscr{F}_{\chi^2}(a,b,c,d) = \mathscr{F}_{\chi^2}(d,b,c,a)$, see point 2 of lemma 5.12. This implies that both $\sim_{\alpha}^2$ and $\sim_{\alpha,Base}^2$ are a-d symmetrical.

This finishes the proof. □

Please note that both the 4ft-quantifier $\sim_q^+$ of above average and founded 4ft-quantifier $\sim_{q,Base}^+$ of above average are symmetrical and have the F-property, see theorem 10.2. However, theorem 10.2 says nothing about the a-d symmetry of these two quantifiers.

Part III
Results on Classes of Association Rules

There are four groups of theoretical results concerning 4ft-quantifiers, association rules, and logical calculi of association rules presented in this book. The first group deals with tables of critical frequencies. The tables of critical frequencies are also used to define and study some classes of association rules and that's why they are defined and studied in part II.

The remaining three groups of results concern the classes of association rules introduced in part II. They are presented in this part. The first group of results concerns deduction rules in logical calculi of association rules. Deduction rules in a special simple form are studied and their practical importance is discussed. Results on deduction rules are presented in Chap. 11.

An additional group of results deals with missing information. A special approach to evaluate association rules in data with missing information is developed in the book [18]. Its principle is to consider an association rule true in a given data matrix with missing information if and only if it is true in all possible completions of the given data matrix. Results concerning missing information and logical calculi of association rules are located in Chap. 12.

The last presented group of results is related to the definability of 4ft-quantifiers in classical predicate calculus with equality. It is shown that various 4ft-quantifiers can be expressed by means of classical predicate calculus with equality. A simple criterion of classical definability based on tables of critical frequencies is introduced. Results on definability are presented in Chap. 13.

Chapter 11
Deduction Rules

Practically important deduction rules in calculi of association rules are defined and studied. We are interested in sound deduction rules stating that truthfulness of a given association rule implies truthfulness of an additional rule with the same 4ft-quantifier. Criteria of soundness of such deductions rules are found for important classes of 4ft-quantifiers. It is shown that the question of soundness of a given deduction rule can be converted into a question of whether several propositional formulas are tautologies or not. These propositional formulas are created from an antecedent and a succedent of a given rule in a way depending on the class a 4ft-quantifier in question belongs to.

An introduction to deduction rules in logical calculi of the association rules we are going to study may be found in Sect. 11.1. Propositional formulas used to decide if a given deduction rule is sound or not are created from propositional formulas associated to particular Boolean attributes occurring in the association rules in question. Associated propositional formulas are introduced in Sect. 11.2.

Results presented in this chapter concern association rules with implicational quantifiers, Σ-double implicational quantifiers and Σ-equivalence quantifiers. Criteria of soundness of deduction rules for particular classes of association rules are introduced in Sects. 11.3 – 11.5. Please note that there are similar results on deduction rules for some of 4ft-quantifiers with the F-property. However, their presentation requires some additional effort, see also Sect. 16.3.

11.1 Why and Which Deduction Rules

We are interested in deduction rules of the form

$$\frac{\varphi \approx \psi}{\varphi' \approx \psi'}$$

where $\varphi \approx \psi$ and $\varphi' \approx \psi'$ are association rules. If such a deduction rule is sound then we know that if association rule $\varphi \approx \psi$ is true in a given data matrix $\mathcal{M}$ then also association rule $\varphi' \approx \psi'$ is surely true in data matrix $\mathcal{M}$.

J. Rauch: *Observational Calculi and Association Rules*, SCI 469, pp. 157–180.
DOI: 10.1007/978-3-642-11737-4_11   © Springer-Verlag Berlin Heidelberg 2013

A trivial example of a sound deduction rule is a rule

$$\frac{A(a) \Rightarrow_{0.9} B(b)}{A(a) \Rightarrow_{0.9} B(b) \vee C(c)}$$

which says that if 90 per cent of rows satisfying $A(a)$ satisfy also $B(b)$ then sure 90 per cent of rows satisfying $A(a)$ satisfy also $B(b) \vee C(c)$.

Sound deduction rules in the form $\frac{\varphi \approx \psi}{\varphi' \approx \psi'}$ can be used for example in the following ways:

- *To decrease the number of actually tested association rules:* If the association rule $\varphi \approx \psi$ is true in the analysed data matrix and if $\frac{\varphi \approx \psi}{\varphi' \approx \psi'}$ is a sound deduction rule, then it is not necessary to test $\varphi' \approx \psi'$.
- *To reduce the output of a data mining procedure:* If the association rule $\varphi \approx \psi$ is part of a data mining procedure output (thus it is true in an analysed data matrix) and if $\frac{\varphi \approx \psi}{\varphi' \approx \psi'}$ is a sound deduction rule, then it is not necessary to put the association rule $\varphi' \approx \psi'$ into the output. The used deduction rule must be clear enough from the point of view of the user of the data mining procedure.

An additional possibility of applications of sound deduction rules of the form $\frac{\varphi \approx \psi}{\varphi' \approx \psi'}$ is related to dealing with domain knowledge, see Sect. 15.3.

Thus we are interested in sound deduction rules $\frac{\varphi \approx \psi}{\varphi' \approx \psi'}$ where both $\varphi \approx \psi$ and $\varphi' \approx \psi'$ are association rules belonging to a given logical calculus $\mathscr{LC}_{\mathscr{T}}$ of association rules of type $\mathscr{T} = \langle t_1, \ldots, t_K \rangle$ introduced in definition 3.9.

We show that there is a relatively simple condition equivalent to the fact that the deduction rule $\frac{\varphi \approx \psi}{\varphi' \approx \psi'}$ is sound. We also show that this condition depends on classes of quantifiers defined and studied in Sect. II. An important role is played by propositional formulas associated to Boolean attributes φ, ψ, φ', and ψ', see Sect. 11.2.

11.2 Associated Propositional Formula

Conditions related to soundness of deduction rules of the form $\frac{\varphi \approx \psi}{\varphi' \approx \psi'}$ deal with the fact that a given Boolean attribute ω logically follows from another given Boolean attribute τ. First we define this relation between Boolean attributes.

Definition 11.1. Let $\mathscr{LC}_{\mathscr{T}}$ be a logical calculus of association rules of type $\mathscr{T} = \langle t_1, \ldots, t_K \rangle$ with language $\mathscr{L}_{\mathscr{T}}$ of association rules and let $\mathsf{M}_{\mathscr{T}}$ be a set of all data matrices $\mathscr{M} = \langle M, f_1, \ldots, f_K \rangle$ of type $\mathscr{T}$. Let τ and ω be Boolean attributes of $\mathscr{L}_{\mathscr{T}}$.

We say that ω *logically follows from* τ, symbolically

$$\tau \vdash \omega \, ,$$

if the following holds for each data matrix $\mathscr{M} = \langle M, f_1, \ldots, f_K \rangle$ of the type $\mathscr{T}$:

If $o \in M$ and $f_\tau(o) = 1$, then also $f_\omega(o) = 1$ where f_τ is the interpretation of τ in M and f_ω is the interpretation of ω in M (i.e. if τ is true for $o \in M$ then also ω is true for $o \in M$), see definition 3.5.

We show that it is possible to convert a question of whether the Boolean attribute ω logically follows from the Boolean attribute τ to a question of whether a special formula of a propositional calculus is a tautology. Such a formula of a propositional calculus is presented in theorem 11.1. Propositional formulas of uniqueness and of consistency introduced in definitions 11.3 and 11.4 are used in theorem 11.1. These formulas are built from propositional variables related to a logical calculus of the association rules in question, see definition 11.2.

Definition 11.2. Let $\mathscr{LC}_\mathscr{T}$ be a logical calculus of association rules of type $\mathscr{T} = \langle t_1, \ldots, t_K \rangle$ with language $\mathscr{L}_\mathscr{T}$ of association rules according to definition 3.3. Then

1. *Set* $\mathscr{PV}(\mathscr{LC}_\mathscr{T})$ *of propositional variables related to logical calculus* $\mathscr{LC}_\mathscr{T}$ is a set of all expressions $\overline{A_i(h)}$ where A_i is a basic attribute of the type t_i and $h \in \{1, \ldots, t_i\}$. Each such expression is considered as a propositional variable. (This means it can get values $\{0, 1\}$, i.e. {*false, true*}.)
2. We say that *propositional variable* $\overline{A_i(h)}$ *is related to basic Boolean attribute* $A_i(h)$.
3. An expression $\mathrm{Val}(\overline{A_i(h)})$ denotes a concrete value of the propositional variable $\overline{A_i(h)}$. If $\mathrm{Val}(\overline{A_i(h)}) = 1$ then we say that $\overline{A_i(h)}$ *is true*, if $\mathrm{Val}(\overline{A_i(h)}) = 0$ then we say that $\overline{A_i(h)}$ *is false*.
4. In addition, let $\mathscr{M} = \langle M, f_1, \ldots, f_K \rangle$ be a data matrix of the type $\mathscr{T}$ and let $o \in M$. Then

   • we say that the *value of propositional variable* $\overline{A_i(h)}$ *is given by* $o \in M$ if

$$\mathrm{Val}(\overline{A_i(h)}) = f_{A_i(h)}(o)$$

   where $f_{A_i(h)}$ is the interpretation of $A_i(h)$ in $\mathscr{M}$, see definition 3.5
   • we say that *values of propositional variables* $\mathscr{PV}(\mathscr{LC}_\mathscr{T})$ *related to* $\mathscr{LC}_\mathscr{T}$ *are given by* $o \in M$ if the value of each propositional variable from $\mathscr{PV}(\mathscr{LC}_\mathscr{T})$ is given by $o \in M$.

Definition 11.3. Let $\mathscr{LC}_\mathscr{T}$ be a logical calculus of association rules of type $\mathscr{T} = \langle t_1, \ldots, t_K \rangle$ with language $\mathscr{L}_\mathscr{T}$ of association rules. Furthermore, let A_i be the basic attribute of the type t_i belonging to basic symbols $\mathscr{BS}(\mathscr{L}_\mathscr{T})$ of language $\mathscr{L}_\mathscr{T}$.

1. The *propositional formula* $\kappa_{i,h}(A_i)$ *of uniqueness of the category h of basic attribute* A_i where $h = 1, \ldots, t_i$ is defined as the formula

$$\overline{A_i(h)} \rightarrow \neg\overline{A_i(1)} \wedge \ldots \wedge \neg\overline{A_i(h-1)} \wedge \neg\overline{A_i(h+1)} \wedge \ldots \wedge \neg\overline{A_i(t_i)}$$

where $\overline{A_i(1)}, \ldots, \overline{A_i(t_i)}$ are propositional variables related to basic Boolean attributes $A_i(1), \ldots, A_i(t_i)$ respectively and $\rightarrow$ is a propositional connective of implication.

2. The *propositional formula* $\lambda(A_i)$ *of consistency of basic attribute* A_i is defined as
the formula

$$(\overline{A_i(1)} \vee \ldots \vee \overline{A_i(t_i)}) \wedge \kappa_{i,1}(A_i) \wedge \ldots \wedge \kappa_{i,t_i}(A_i)$$

where $\overline{A_i(1)}, \ldots, \overline{A_i(t_i)}$ are propositional variables related to basic Boolean attributes $A_i(1), \ldots, A_i(t_i)$ respectively and $\kappa_{i,h}(A_i)$ is the propositional formula of
uniqueness of the category h of basic attribute A_i for $h = 1, \ldots, t_i$.

Please note that propositional formula $\kappa_{i,h}(A_i)$

$$\overline{A_i(h)} \rightarrow \neg\overline{A_i(1)} \wedge \ldots \wedge \neg\overline{A_i(h-1)} \wedge \neg\overline{A_i(h+1)} \wedge \ldots \wedge \neg\overline{A_i(t_i)}$$

of uniqueness of the category h of basic attribute A_i states that if the propositional
variable $\overline{A_i(h)}$ is true, then all the other propositional variables $\overline{A_i(j)}$ for $j = 1, \ldots, t_i$
and $j \neq h$ are false. This corresponds to the fact that if a basic Boolean attribute
$A_i(h)$ is true then all the other basic Boolean attributes $A_i(j)$ for $j = 1, \ldots t_i$ and
$j \neq h$ are false.

Similarly, propositional formula $\lambda(A_i)$

$$(\overline{A_i(1)} \vee \ldots \vee \overline{A_i(t_i)}) \wedge \kappa_{i,1}(A_i) \wedge \ldots \wedge \kappa_{i,t_i}(A_i)$$

of the consistency of the basic attribute A_i says that exactly one of the propositional
variables $\overline{A_i(j)}$ for $j = 1, \ldots, t_i$ is true. This corresponds to the fact that exactly one
of the basic Boolean attributes $A_i(j)$ where $j = 1, \ldots t_i$ is true.

There is a simple but useful lemma concerning the truthfulness of the propositional formula of uniqueness and of propositional formula of consistency.

Lemma 11.1. *Let* $\mathcal{LC}_{\mathcal{T}}$ *be a logical calculus of association rules of type*
$\mathcal{T} = \langle t_1, \ldots, t_K \rangle$ *with language* $\mathcal{L}_{\mathcal{T}}$ *of association rules. Let* A_i *be the basic attribute*
of the type t_i *belonging to basic symbols* $\mathcal{BS}(\mathcal{L}_{\mathcal{T}})$ *of language* $\mathcal{L}_{\mathcal{T}}$. *In addition,*
let $\mathcal{M} = \langle M, f_1, \ldots, f_K \rangle$ *be a data matrix of the type* $\mathcal{T}$ *and let* $o \in M$. *If the values*
of propositional variables $\mathcal{PV}(\mathcal{LC}_{\mathcal{T}})$ *related to* $\mathcal{LC}_{\mathcal{T}}$ *are given by* $o \in M$ *then:*

1. *Propositional formula* $\kappa_{i,h}(A_i)$ *of uniqueness of category* h *is true for* $i = 1, \ldots, K$
and $h = 1, \ldots, t_i$.
2. *Propositional formula* $\lambda(A_i)$ *of consistency of the basic attribute* A_i *is true for*
$i = 1, \ldots, K$.

Proof. *Let us assume that values of propositional variables* $\mathcal{PV}(\mathcal{LC}_{\mathcal{T}})$ *related to*
$\mathcal{LC}_{\mathcal{T}}$ *are given by* $o \in M$.

1. *We have to prove for* $i = 1, \ldots, K$ *and* $h = 1, \ldots, t_i$: *if* $\overline{A_i(h)}$ *is true then all* $\overline{A_i(j)}$
for $j = 1, \ldots t_i$ *and* $j \neq h$ *are false. The assumption that* $\overline{A_i(h)}$ *is true means*
that $Val(\overline{A_i(h)}) = f_{A_i(h)}(o) = 1$, *see point 4 of definition 11.2. Thus, according to*
definition 3.5 it holds $f_i(o) \in \mathfrak{I}(h)$ *i.e.* $f_i(o) \in \{h\}$, *see also definition 3.4. This*
implies $f_i(o) \notin \{j\}$ *for* $j = 1, \ldots t_i$ *and* $j \neq h$ *and thus* $f_{A_i(j)}(o) = 0$ *for* $j = 1, \ldots t_i$

and $j \neq h$. We assume that the values of propositional variables $\mathscr{PV}(\mathscr{LC}_{\mathscr{T}})$ related to $\mathscr{LC}_{\mathscr{T}}$ are given by $o \in M$ and this means that

$$Val(\overline{A_i(j)}) = f_{A_i(j)}(o) = 0$$

for $j = 1, \ldots t_i$ and $j \neq h$. This finishes the proof of point 1.

2. *It remains to prove that there is $h \in \{1, \ldots t_i\}$ such that $\overline{A_i(h)}$ is true. It holds that $f_i(o) \in \{1, \ldots t_i\}$ and it is sufficient to lay $h = f_i(o)$. Then $f_{A_i(h)}(o) = 1$ because $f_i(o) \in \{h\} = \{f_i(o)\}$, see point 1 of definition 3.5.*

This finishes the proof. □

Definition 11.4. Let $\mathscr{LC}_{\mathscr{T}}$ be a logical calculus of association rules of type $\mathscr{T} = \langle t_1, \ldots, t_K \rangle$ with language $\mathscr{L}_{\mathscr{T}}$ of association rules.

1. Let A_i be the basic attribute of the type t_i and let $A_i(\alpha)$ be the basic Boolean attribute such that $\alpha = \{h_1, \ldots, h_k\}$ where $\{h_1, \ldots, h_k\} \subset \{1, \ldots, t_i\}$. Then the *propositional disjunction* $\pi(A_i(\alpha))$ *associated with the basic Boolean attribute* $A_i(\alpha)$ is defined such that

$$\pi(A_i(\alpha)) = \overline{A_i(h_1)} \vee \ldots \vee \overline{A_i(h_k)}$$

where $\overline{A_i(h_1)}, \ldots, \overline{A_i(h_k)}$ are propositional variables related to basic Boolean attributes $A_i(h_1), \ldots, A_i(h_k)$ respectively.

2. Let φ be the Boolean attribute from $\mathscr{BA}(\mathscr{L}_{\mathscr{T}})$. Then the *propositional formula* $\pi(\varphi)$ *associated with the Boolean attribute* φ is created from φ such that each basic Boolean attribute $A(\alpha)$ occurring in φ is replaced by an expression $(\pi(A(\alpha)))$ where $\pi(A(\alpha))$ is a propositional disjunction associated with $A(\alpha)$.

Let us give an example of the propositional formula associated with the Boolean attribute.

Example 11.1. The propositional formula

$$\pi(A_1(3,4) \wedge A_7(2,6))$$

associated with Boolean attribute $A_1(3,4) \wedge A_7(2,6)$ is the formula

$$(\overline{A_1(3)} \vee \overline{A_1(4)}) \wedge (\overline{A_7(2)} \vee \overline{A_7(6)})$$

where $\overline{A_1(3)}, \overline{A_1(4)}, \overline{A_7(2)}, \overline{A_7(6)})$ are propositional variables.

There is again a simple but useful lemma concerning the truthfulness of the propositional disjunction $\pi(A(\alpha))$ associated with the basic Boolean attribute $A(\alpha)$ and of the propositional formula $\pi(\varphi)$ associated with the Boolean attribute φ.

Lemma 11.2. *Let $\mathscr{LC}_{\mathscr{T}}$ be a logical calculus of association rules of type $\mathscr{T} = \langle t_1, \ldots, t_K \rangle$ with language $\mathscr{L}_{\mathscr{T}}$ of association rules. In addition, let $\mathscr{M} = \langle M, f_1, \ldots, f_K \rangle$ be a data matrix of the type $\mathscr{T}$ and let $o \in M$. If the values of propositional variables $\mathscr{PV}(\mathscr{LC}_{\mathscr{T}})$ related to $\mathscr{LC}_{\mathscr{T}}$ are given by $o \in M$, then:*

1. *If $A_i(\alpha)$ is the basic Boolean attribute where $\alpha = \{h_1,\ldots,h_k\} \subset \{1,\ldots,t_i\}$ then the propositional disjunction $\pi(A_i(\alpha))$ associated with the basic Boolean attribute $A_i(\alpha)$ is true if and only if $A_i(\alpha)$ is true for $o \in M$.*
2. *If φ is the Boolean attribute from $\mathscr{BA}(\mathscr{L}_{\mathscr{T}})$ then the propositional formula $\pi(\varphi)$ associated with the Boolean attribute φ is true if and only if φ is true for $o \in M$.*

Proof. Let us assume that values of propositional variables $\mathscr{PV}(\mathscr{LC}_{\mathscr{T}})$ related to $\mathscr{LC}_{\mathscr{T}}$ are given by $o \in M$, see point 4 of definition 11.2.

1. *Let $A_i(\alpha)$ be a basic Boolean attribute where $\alpha = \{h_1,\ldots,h_k\} \subset \{1,\ldots,t_i\}$.*
   If $A_i(\alpha)$ is true for $o \in M$ then $f_i(o) \in \{h_1,\ldots,h_k\}$, see definition 3.5. Let us assume $f_i(o) = h_u$, then $f_{A_i(h_u)}(o) = 1$ because $f_i(o) \in \{h_u\} = \{f_i(o)\}$, see point 1 of definition 3.5. We assume $Val(\overline{A_i(h_u)}) = f_{A_i(h_u)}(o)$ which means that $\overline{A_i(h_u)}$ is true and thus the the disjunction $\overline{A_i(h_1)} \vee \ldots \vee \overline{A_i(h_k)}$ is true.
   If $A_i(\alpha)$ is false for $o \in M$ then $f_i(o) \notin \{h_1,\ldots,h_k\}$ i.e. $f_i(o) \notin \{h_j\}$ for $j = 1,\ldots,k$. Thus $f_{A_i(h_j)}(o) = 0$ because $f_i(o) \notin \{h_j\}$ for $j = 1,\ldots,k$, see point 1 of definition 3.5. We assume $Val(\overline{A_i(h_j)}) = f_{A_i(h_j)}(o)$ for $j = 1,\ldots,k$ and this means that the disjunction $\overline{A_i(h_1)} \vee \ldots \vee \overline{A_i(h_k)}$ is false.
   This finishes the proof of point 1.
2. *We prove point 2 by induction on the number $NCon(\varphi)$ of logical connectives $\neg, \wedge, \vee$ in Boolean attribute φ. The just proved point means that the point 2 is true for $NCon(\varphi) = 0$.*
   Let us assume that point 2 is true for all Boolean attributes φ' such that $NCon(\varphi') \leq i$. We prove that the point 2 is true also for all Boolean attributes φ such that $NCon(\varphi) = i+1$.
   There are three possibilities for Boolean attribute φ satisfying $NCon(\varphi) = i+1$: (1): $\varphi = \neg\varphi'$, (2): $\varphi = \varphi' \vee \psi'$, and (3): $\varphi = \varphi' \wedge \psi'$, see definition 3.2. Anyway, we know that $NCon(\varphi') \leq i$ and $NCon(\psi') \leq i$ and thus the propositional formula $\pi(\varphi')$ associated with the Boolean attribute φ' is true if and only if φ' is true for $o \in M$ and similarly $\pi(\psi')$ is true if and only if ψ' is true for $o \in M$.

   (1) *If $\varphi = \neg\varphi'$ then $f_\varphi(o) = 1 - f_{\varphi'}(o)$, see definition 3.5. This means that φ is true for $o \in M$ if and only if φ' is not true for $o \in M$. According to the inductive assumption we know that φ' is true for $o \in M$ if and only if $\pi(\varphi')$ is true and this implies that φ' is not true for $o \in M$ if and only if $\pi(\varphi')$ is not true. We can conclude that φ is true for $o \in M$ if and only if $\pi(\varphi')$ is not true.*
   However, $\varphi = \neg\varphi'$ also means that $\pi(\varphi) = \neg\pi(\varphi')$. According to the properties of propositional connective $\neg$ we know that $\neg\pi(\varphi')$ is true if and only if $\pi(\varphi')$ is not true i.e. $\pi(\varphi)$ is true if and only if $\pi(\varphi')$ is not true. We already know that φ is true for $o \in M$ if and only if $\pi(\varphi')$ is not true and we can conclude that φ is true for $o \in M$ if and if $\pi(\varphi)$ is true. This finishes the induction step for logical connective $\neg$.
   (2) *If $\varphi = \varphi' \vee \psi'$ then $f_\varphi(o) = \max(f_{\varphi'}(o), f_{\psi'}(o))$, see definition 3.5. This means that φ is true for $o \in M$ if and only if φ' is true for $o \in M$ or ψ' is true for $o \in M$. According to the inductive assumption we know that φ' is true for $o \in M$ if and only if $\pi(\varphi')$ is true and that ψ' is true for $o \in M$ if and only if*

$\pi(\psi')$ is true. We can conclude that φ is true for $o \in M$ if and only if $\pi(\varphi')$ is true or $\pi(\psi')$ is true.

However, $\varphi = \varphi' \vee \psi'$ also means that $\pi(\varphi) = \pi(\varphi') \vee \pi(\psi')$ and according to the properties of propositional connective $\vee$ we know that $\pi(\varphi)$ is true if and only if $\pi(\varphi')$ is true or $\pi(\psi')$ is true. We already know that φ is true for $o \in M$ if and only if $\pi(\varphi')$ is true or $\pi(\psi')$ is true. We can conclude that φ is true for $o \in M$ if and only if $\pi(\varphi)$ is true. This finishes the induction step for logical connective $\vee$.

(3) If $\varphi = \varphi' \wedge \psi'$ then $f_\varphi(o) = \min(f_{\varphi'}(o), f_{\psi'}(o))$, see definition 3.5. This means that φ is true for $o \in M$ if and only if both φ' is true for $o \in M$ and ψ' is true for $o \in M$. According to the inductive assumption we know that φ' is true for $o \in M$ if and only if $\pi(\varphi')$ is true and that ψ' is true for $o \in M$ if and only if $\pi(\psi')$ is true. We can conclude that φ is true for $o \in M$ if and only if both $\pi(\varphi')$ is true and $\pi(\psi')$ is true.

However, $\varphi = \varphi' \wedge \psi'$ also means that $\pi(\varphi) = \pi(\varphi') \wedge \pi(\psi')$ and according to the properties of propositional connective $\wedge$ we know that $\pi(\varphi)$ is true if and only if both $\pi(\varphi')$ is true and $\pi(\psi')$ is true. We already know that φ is true for $o \in M$ if and only if both $\pi(\varphi')$ is true and $\pi(\psi')$ is true. We can conclude that φ is true for $o \in M$ if and only if $\pi(\varphi)$ is true. This finishes the induction step for logical connective $\wedge$.

This finishes the proof   □.

The following theorem introduces a criterion converting the question of whether Boolean attribute ω logically follows from Boolean attribute τ to a question of whether a propositional formula created from τ and ω is a tautology.

Theorem 11.1. *Let $\mathscr{LC}_{\mathscr{T}}$ be a logical calculus of association rules of type $\mathscr{T} = \langle t_1, \ldots, t_K \rangle$ with language $\mathscr{L}_{\mathscr{T}}$ of association rules. In addition, let τ and ω be Boolean attributes from $\mathscr{BA}(\mathscr{L}_{\mathscr{T}})$. Then $\tau \vdash \omega$ (i.e. ω logically follows from τ), if and only if the propositional formula*

$$\Lambda(\tau, \omega) \wedge \pi(\tau) \rightarrow \pi(\omega)$$

is a tautology. Here

$$\Lambda(\tau, \omega) = \lambda(A_{i_1}) \wedge \ldots \wedge \lambda(A_{i_J})$$

where $A_{i_1}, \ldots, A_{i_J}$ are all basic attributes occurring in the Boolean attribute τ or in the Boolean attribute ω and $\lambda(A_{i_j})$ is the propositional formula of consistency of the basic attribute A_{i_j} for $j = 1, \ldots, J$. In addition, $\pi(\tau)$ and $\pi(\omega)$ are propositional formulas associated with Boolean attributes τ and ω respectively and $\rightarrow$ is the propositional connective of implication.

Proof. We prove the theorem in steps (1) and (2). In step (1) we prove that if $\Lambda(\tau, \omega) \wedge \pi(\tau) \rightarrow \pi(\omega)$ is a tautology then $\tau \vdash \omega$. In step (2) we prove that if $\tau \vdash \omega$ then $\Lambda(\tau, \omega) \wedge \pi(\tau) \rightarrow \pi(\omega)$ is a tautology.

(1)   We assume that $\Lambda(\tau,\omega) \wedge \pi(\tau) \to \pi(\omega)$ is a tautology. In addition, let $\mathcal{M} = \langle M, f_1, \ldots, f_K \rangle$ be a data matrix of the type $\mathcal{T}$ and let $o \in M$. We have to prove that if τ is true for $o \in M$ then also ω is true for $o \in M$, see definition 11.1. If τ is true for $o \in M$ then the propositional formula $\pi(\tau)$ associated with the Boolean attribute τ is true if values of propositional variables $\mathscr{PV}(\mathscr{LC}_{\mathscr{T}})$ are given by $o \in M$, see point 2 of lemma 11.2. According to point 2 of lemma 11.1 the formulas $\lambda(A_{i_1}), \ldots, \lambda(A_{i_j})$ of consistency of the attributes $A_{i_1}, \ldots, A_{i_j}$ are also true for this setting of values of propositional variables $\mathscr{PV}(\mathscr{LC}_{\mathscr{T}})$. We can conclude that $\Lambda(\tau,\omega) \wedge \pi(\tau)$ is true if the values of propositional variables $\mathscr{PV}(\mathscr{LC}_{\mathscr{T}})$ are given by $o \in M$.
We assume that $\Lambda(\tau,\omega) \wedge \pi(\tau) \to \pi(\omega)$ is a tautology and thus $\pi(\omega)$ is also true if the values of propositional variables $\mathscr{PV}(\mathscr{LC}_{\mathscr{T}})$ are given by $o \in M$. However, according to point 2 of lemma 11.2 this means that ω is true for $o \in M$. This finishes the proof of point (1).

(2)   Please note that all propositional variables occurring in $\pi(\tau)$ or $\pi(\omega)$ occur in $\Lambda(\tau,\omega)$ and that all these propositional variables belong to $\mathscr{PV}(\mathscr{LC}_{\mathscr{T}})$. We assume that $\tau \vdash \omega$ and we have to prove that $\Lambda(\tau,\omega) \wedge \pi(\tau) \to \pi(\omega)$ is a tautology of a propositional calculus. This means that we have to prove that if $\Lambda(\tau,\omega) \wedge \pi(\tau)$ is true for a given setting Ω of values of propositional variables $\mathscr{PV}(\mathscr{LC}_{\mathscr{T}})$ occurring in $\Lambda(\tau,\omega)$ then $\pi(\omega)$ is also true for this setting of propositional variables.
The principle of the proof is the following:

(i)   We construct a data matrix $\mathcal{M}_0 = \langle M_0, f_1, \ldots, f_K \rangle$ such that $M_0 = \{o_0\}$, $\Lambda(\tau,\omega) \wedge \pi(\tau)$ is true for o_0 in M_0 and setting Ω of values of propositional variables $\mathscr{PV}(\mathscr{LC}_{\mathscr{T}})$ is equal to setting of values of propositional variables $\mathscr{PV}(\mathscr{LC}_{\mathscr{T}})$ given by $o_0 \in M_0$ occurring in $\Lambda(\tau,\omega)$.

(ii)   We use point 2 of lemma 11.2 to deduce that τ is true for o_0 in $\mathcal{M}_0$ from the fact that $\pi(\tau)$ is true for setting Ω of values of propositional variables.

(iii)   We use the assumption $\tau \vdash \omega$ to deduce that ω is true for o_0 in $\mathcal{M}_0$.

(iv)   We use point 2 of lemma 11.2 to conclude that $\pi(\omega)$ is true for setting Ω of values of propositional variables.

It remains to construct the data matrix $\mathcal{M}_0 = \langle M_0, f_1, \ldots, f_K \rangle$ according to point (i). We assume that $A_{i_1}, \ldots, A_{i_j}$ are all basic attributes occurring in the Boolean attributes τ or ω. This means that all propositional variables occurring in $\Lambda(\tau,\omega)$ are $\overline{A_{i_j}(u)}$ where $j = 1, \ldots J$ and $u = 1, \ldots t_{i_j}$. We also assume that propositional formula $\Lambda(\tau,\omega) \wedge \pi(\tau)$ is true for setting Ω and this means that all formulas $\lambda(A_{i_1}), \ldots \lambda(A_{i_j})$ are true for this setting.
We have for $j = 1, \ldots J$:

$$\lambda(A_{i_j}) = (\,\overline{A_{i_j}(1)} \vee \ldots \vee \overline{A_{i_j}(t_{i_j})}\,) \wedge \kappa_{i_j,1}(A_{i_j}) \wedge \ldots \wedge \kappa_{i_j,t_{i_j}}(A_{i_j})$$

where

$$\kappa_{i_j,h} = \overline{A_{i_j}(h)} \to \neg\overline{A_{i_j}(1)} \wedge \ldots \wedge \neg\overline{A_{i_j}(h-1)} \wedge \neg\overline{A_{i_j}(h+1)} \wedge \ldots \wedge \neg\overline{A_{i_j}(t_j)}$$

for $h = 1, \ldots, t_{i_j}$. *This means there is exactly one* $z_{i_j} \in \{1, \ldots, t_j\}$ *such that the propositional variable* $\overline{A_{i_j}(z_{i_j})}$ *is true and the propositional variables* $\overline{A_{i_j}(k)}$ *are false for* $k \in \{1, \ldots, t_{i_j}\}$ *and* $k \neq z_{i_j}$.

Let us define $\mathcal{M}_0 = \langle M_0, f_1, \ldots, f_K \rangle$ *such that* M_0 *has exactly one row, let us denote it* o_0 *which means* $M_0 = \{o_0\}$. *In addition, let us define the functions* $f_1, \ldots, f_K$ *such that for* $j = 1, \ldots J$ *it holds* $f_{i_j}(o_0) = z_{i_j}$. *This means that all propositional variables occurring in* $\pi(\tau)$ *have the same truth values as they have in the setting* Ω. *This finishes the proof of step (2).*

This finishes the proof. □

11.3 Deduction Rules for Implicational Quantifiers

We are interested in sound deduction rules of the form $\frac{\varphi \approx \psi}{\varphi' \approx \psi'}$ where both $\varphi \approx \psi$ and $\varphi' \approx \psi'$ are association rules. One problem of interest is deciding whether a given deduction rule is sound or not. There are important results concerning soundness of such deduction rules with interesting implicational quantifiers introduced in definition 7.5.

Theorem 11.2 shows that the question of whether a deduction rule $\frac{\varphi \Rightarrow^* \psi}{\varphi' \Rightarrow^* \psi'}$, where $\Rightarrow^*$ is an interesting implicational quantifier, is a sound deduction rule is equivalent to the question of whether some relations like $\tau \vdash \omega$ are true where ω and τ are Boolean attributes created from φ, ψ, φ', and ψ'. Remember that the question of whether the relation $\tau \vdash \omega$ (i.e. if ω logically follows from τ) is true can be converted into a question of whether the propositional formula $\Lambda(\tau, \omega) \wedge \pi(\tau) \rightarrow \pi(\omega)$ is a tautology or not, see theorem 11.1.

Theorem 11.3 presents a weaker result concerning deduction rules $\frac{\varphi \Rightarrow^* \psi}{\varphi' \Rightarrow^* \psi'}$ where $\Rightarrow^*$ is a weakly implicational quantifier. It is proved that if the same relations $\tau \vdash \omega$ as like in theorem 11.2 are true, then this deduction rule is sound.

Theorem 11.2. *Let* $\mathcal{LC}_{\mathcal{T}}$ *be a logical calculus of association rules of type* $\mathcal{T} = \langle t_1, \ldots, t_K \rangle$ *with a language* $\mathcal{L}_{\mathcal{T}}$ *of association rules. In addition, let* $\varphi \Rightarrow^* \psi$ *and* $\varphi' \Rightarrow^* \psi'$ *be association rules of* $\mathcal{L}_{\mathcal{T}}$. *If* $\Rightarrow^*$ *is an interesting implicational quantifier then deduction rule*

$$\frac{\varphi \Rightarrow^* \psi}{\varphi' \Rightarrow^* \psi'}$$

is sound if and only if at least one of conditions (1) or (2) are satisfied:

(1)   both conditions (1.a) and (1.b) are satisfied

(1.a) $\varphi \wedge \psi \vdash \varphi' \wedge \psi'$
(1.b) $\varphi' \wedge \neg\psi' \vdash \varphi \wedge \neg\psi$

(2) $\varphi \vdash \neg\psi$.

Proof. Let $\mathrm{M}_{\mathcal{T}}$ be a set of all data matrices $\mathcal{M} = \langle M, f_1, \ldots, f_K \rangle$ of type $\mathcal{T}$, see definition 3.9. We have to prove:

I)   If condition (1) is satisfied, $\mathcal{M} \in M_{\mathcal{T}}$ and if $Val(\varphi \Rightarrow^* \psi, \mathcal{M}) = 1$ then also $Val(\varphi' \Rightarrow^* \psi', \mathcal{M}) = 1$.

II)   If condition (2) is satisfied, $\mathcal{M} \in M_{\mathcal{T}}$ and if $Val(\varphi \Rightarrow^* \psi, \mathcal{M}) = 1$ then also $Val(\varphi' \Rightarrow^* \psi', \mathcal{M}) = 1$.

III) If neither (1) nor (2) are satisfied then the deduction rule $\frac{\varphi \Rightarrow^* \psi}{\varphi' \Rightarrow^* \psi'}$ is not sound. This means that we have to find $\mathcal{M} \in M_{\mathcal{T}}$ such that $Val(\varphi \Rightarrow^* \psi, \mathcal{M}) = 1$ and $Val(\varphi' \Rightarrow^* \psi', \mathcal{M}) = 0$.

We assume that $F_{\Rightarrow^*}$ is an associated function of $\Rightarrow^*$, $\langle a,b,c,d \rangle = 4ft(\varphi, \psi, \mathcal{M})$ is 4ft-table of φ and ψ in $\mathcal{M}$ and $\langle a',b',c',d' \rangle = 4ft(\varphi', \psi', \mathcal{M})$ is 4ft-table of φ' and ψ' in $\mathcal{M}$, see Fig. 11.1.

$\mathcal{M}$	ψ	$\neg\psi$
φ	a	b
$\neg\varphi$	c	d

$\mathcal{M}$	ψ'	$\neg\psi'$
φ'	a'	b'
$\neg\varphi'$	c'	d'

4ft-table $\langle a,b,c,d \rangle$ of φ and ψ in $\mathcal{M}$ 4ft-table $\langle a',b',c',d' \rangle$ of φ' and ψ' in $\mathcal{M}$

Fig. 11.1 4ft-tables $\langle a,b,c,d \rangle$ and $\langle a',b',c',d' \rangle$

Remember that $\Rightarrow^*$ is implicational if $F_{\Rightarrow^*}(a,b,c,d) = 1$ and $a' \geq a \wedge b' \leq b$ implies $F_{\Rightarrow^*}(a',b',c',d') = 1$, see definition 6.2 and Sect. 6.2. According to note 6.1 we write only $F_{\Rightarrow^*}(a,b)$ instead of $F_{\Rightarrow^*}(a,b,c,d)$. The proofs for particular points follow.

I): Let us assume $Val(\varphi \Rightarrow^* \psi, \mathcal{M}) = 1$ i.e. $F_{\Rightarrow^*}(a,b) = 1$. Condition (1.a) i.e. $\varphi \wedge \psi \vdash \varphi' \wedge \psi'$ means that if $\varphi \wedge \psi$ is true for $o \in M$ then also $\varphi' \wedge \psi'$ is true for $o \in M$, see definition 11.1. We can conclude that condition (1.a) implies $a' \geq a$. Similarly, condition (1.b) implies $b' \leq b$. We assume $F_{\Rightarrow^*}(a,b) = 1$, we have $a' \geq a$ and $b' \leq b$, thus $F_{\Rightarrow^*}(a',b') = 1$ i.e. $Val(\varphi' \Rightarrow^* \psi', \mathcal{M}) = 1$. This finishes the proof of point I).

II): Condition (2) i.e. $\varphi \vdash \neg\psi$ means that there is no row $o \in M$ satisfying $\varphi \wedge \psi$ and thus $a = 0$. We assume $\Rightarrow^*$ is an interesting implicational quantifier and this means $F_{\Rightarrow^*}(0,0) = 0$, see definition 7.5. 4ft-quantifier $\Rightarrow^*$ is implicational and $F_{\Rightarrow^*}(0,0) = 0$ implies $F_{\Rightarrow^*}(0,b) = 0$ for each $b > 0$. Thus the condition $\varphi \vdash \neg\psi$ means that $Val(\varphi \Rightarrow^* \psi, \mathcal{M}) = 0$ for each $\mathcal{M} \in M_{\mathcal{T}}$. We can conclude that if holds that $Val(\varphi \Rightarrow^* \psi, \mathcal{M}) = 1$ then also $Val(\varphi' \Rightarrow^* \psi', \mathcal{M}) = 1$. This finishes the proof of point II).

III): We assume that neither (1) nor (2) are satisfied. This means that at least one of the following conditions are satisfied:

(III.1)   Neither (1.a) nor (2) are satisfied.

(III.2)   Neither (1.b) nor (2) are satisfied.

We find a data matrix $\mathcal{M}$ both for (III.1) and (III.2) such that $Val(\varphi \Rightarrow^* \psi, \mathcal{M}) = 1$ and $Val(\varphi' \Rightarrow^* \psi', \mathcal{M}) = 0$.

(III.1): Let us assume that neither (1.a) nor (2) are satisfied. The implicational quantifier $\Rightarrow^*$ is interesting which means that it is a-dependent (see definitions 7.5 and 7.4). Thus there are non-negative integers A, B such that $F_{\Rightarrow^*}(A, B) = 1$. This means that $A > 0$ because $F_{\Rightarrow^*}(0, B) = 0$ for each B, see the proof for point II).

Condition (1.a) is not satisfied thus there is a data matrix $\mathscr{M}_1 = \langle M_1, f_1^{(1)}, \ldots, f_K^{(1)} \rangle$ of the type $\mathscr{T}$ where $M_1 = \{o_1^{(1)}, \ldots, o_{n_1}^{(1)}\}$ such that $f_{\varphi \wedge \psi}^{(1)}(o_1^{(1)}) = 1$ and $f_{\varphi' \wedge \psi'}^{(1)}(o_1^{(1)}) = 0$, see Fig. 11.2. Here $f_{\varphi \wedge \psi}^{(1)}$ is the interpretation of $\varphi \wedge \psi$ in $\mathscr{M}_1$, similarly for $f_{\varphi' \wedge \psi'}^{(1)}$.

row	$f_1^{(1)}$	$f_2^{(1)}$	$\ldots$	$f_K^{(1)}$	$f_{\varphi \wedge \psi}^{(1)}$	$f_{\varphi' \wedge \psi'}^{(1)}$
$o_1^{(1)}$	$f_1^{(1)}(o_1^{(1)})$	$f_2^{(1)}(o_1^{(1)})$	$\ldots$	$f_K^{(1)}(o_1^{(1)})$	1	0
$o_2^{(1)}$	$f_1^{(1)}(o_2^{(1)})$	$f_2^{(1)}(o_2^{(1)})$	$\ldots$	$f_K^{(1)}(o_2^{(1)})$	$f_{\varphi \wedge \psi}^{(1)}(o_2^{(1)})$	$f_{\varphi' \wedge \psi'}^{(1)}(o_2^{(1)})$
$\vdots$	$\vdots$	$\vdots$	$\ddots$	$\vdots$	$\vdots$	$\vdots$
$o_{n_1}^{(1)}$	$f_1^{(1)}(o_{n_1}^{(1)})$	$f_2^{(1)}(o_{n_1}^{(1)})$	$\ldots$	$f_K^{(1)}(o_{n_1}^{(1)})$	$f_{\varphi \wedge \psi}^{(1)}(o_{n_1}^{(1)})$	$f_{\varphi' \wedge \psi'}^{(1)}(o_{n_1}^{(1)})$

Fig. 11.2 Data matrix $\mathscr{M}_1$

Let $\mathscr{M}_2 = \langle M_2, f_1^{(2)}, \ldots, f_K^{(2)} \rangle$ be a data matrix such that $M_2 = \{o_1^{(2)}, \ldots, o_A^{(2)}\}$ and $f_i^{(2)}(o_j^{(2)}) = f_i^{(1)}(o_1^{(1)})$ for $i = 1, \ldots K$ and $j = 1, \ldots A$, see Fig. 11.3 where $f_{\varphi \wedge \psi}^{(2)}$ is the interpretation of $\varphi \wedge \psi$ in $\mathscr{M}_2$, similarly for $f_{\varphi' \wedge \psi'}^{(2)}$. In other words, data matrix $\mathscr{M}_2$ has A rows and each of them behaves as the row $o_1^{(1)} \in M_1$.

row	$f_1^{(2)}$	$f_2^{(2)}$	$\ldots$	$f_K^{(2)}$	$f_{\varphi \wedge \psi}^{(2)}$	$f_{\varphi' \wedge \psi'}^{(2)}$
$o_1^{(2)}$	$f_1^{(1)}(o_1^{(1)})$	$f_2^{(1)}(o_1^{(1)})$	$\ldots$	$f_K^{(1)}(o_1^{(1)})$	1	0
$o_2^{(2)}$	$f_1^{(1)}(o_1^{(1)})$	$f_2^{(1)}(o_1^{(1)})$	$\ldots$	$f_K^{(1)}(o_1^{(1)})$	1	0
$\vdots$	$\vdots$	$\vdots$	$\ddots$	$\vdots$	$\vdots$	$\vdots$
$o_A^{(2)}$	$f_1^{(1)}(o_1^{(1)})$	$f_2^{(1)}(o_1^{(1)})$	$\ldots$	$f_K^{(1)}(o_1^{(1)})$	1	0

Fig. 11.3 Data matrix $\mathscr{M}_2$

This means $f_{\varphi \wedge \psi}^{(2)}(o_j^2) = 1$ and $f_{\varphi' \wedge \psi'}^{(2)}(o_j^2) = 0$ for $j = 1, \ldots, A$. Thus 4ft-tables $4ft(\varphi, \psi, \mathscr{M}_2)$ of φ and ψ in $\mathscr{M}_2$ and $4ft(\varphi', \psi', \mathscr{M}_2)$ of φ' and ψ' in $\mathscr{M}_2$ satisfy

$$4ft(\varphi, \psi, \mathscr{M}_2) = \langle A, 0, 0, 0 \rangle \quad \text{and} \quad 4ft(\varphi', \psi', \mathscr{M}_2) = \langle 0, b', c', d' \rangle$$

where $b' + c' + d' = A$.

We can conclude that $F_{\Rightarrow^*}(A,B) = 1$ and $4ft(\varphi,\psi,\mathscr{M}_2) = \langle A,0,0,0 \rangle$, thus $Val(\varphi \Rightarrow^* \psi, \mathscr{M}_2) = 1$. In addition, we can conclude that $F_{\Rightarrow^*}(0,b') = 0$ for $b' \geq 0$ and $4ft(\varphi',\psi',\mathscr{M}_2) = \langle 0,b',c',d' \rangle$ and thus $Val(\varphi' \Rightarrow^* \psi', \mathscr{M}_2) = 0$. This finishes the proof of point (III.1).

(III.2): Let us assume that neither (1.b) nor (2) are satisfied. The quantifier $\Rightarrow^*$ is interesting implicational and thus it is b-dependent. This means that there are non-negative integers A, B, G such that $F_{\Rightarrow^*}(A,B) = 1$ and $F_{\Rightarrow^*}(A,G) = 0$ (see definition 7.4). The fact that $F_{\Rightarrow^*}(A,B) = 1$ means that $F_{\Rightarrow^*}(A,0) = 1$, thus $A > 0$ because $F_{\Rightarrow^*}(0,0) = 0$. Furthermore, it must hold that $G > 0$ because $F_{\Rightarrow^*}(A,G) = 0$ and $F_{\Rightarrow^*}(A,0) = 1$.

Condition (1.b) i.e. $\varphi' \wedge \neg\psi' \vdash \varphi \wedge \neg\psi$ is not satisfied, thus there is a data matrix $\mathscr{M}_3 = \langle M_3, f_1^{(3)}, \ldots, f_K^{(3)} \rangle$ of the type $\mathscr{T}$ and a row $o_1^{(3)} \in M_3$ such that $f_{\varphi' \wedge \neg\psi'}^{(3)}(o_1^{(3)}) = 1$ and $f_{\varphi \wedge \neg\psi}^{(3)}(o_1^{(3)}) = 0$. In addition, $f_{\varphi' \wedge \neg\psi'}^{(3)}(o_1^{(3)}) = 1$ implies $f_{\varphi' \wedge \psi'}^{(3)}(o_1^{(3)}) = 0$, see Fig. 11.4. Here $f_{\varphi \wedge \neg\psi}^{(3)}$, $f_{\varphi' \wedge \psi'}^{(3)}$, and $f_{\varphi' \wedge \neg\psi'}^{(3)}$ are interpretations of $\varphi \wedge \neg\psi$, $\varphi' \wedge \neg\psi'$, and $\varphi' \wedge \neg\psi'$ in $\mathscr{M}_3$ respectively.

row	$f_1^{(3)}$	$\cdots$	$f_K^{(3)}$	$f_{\varphi \wedge \neg\psi}^{(3)}$	$f_{\varphi' \wedge \psi'}^{(3)}$	$f_{\varphi' \wedge \neg\psi'}^{(3)}$
$o_1^{(3)}$	$f_1^{(3)}(o_1^{(3)})$	$\cdots$	$f_K^{(3)}(o_1^{(3)})$	0	0	1
$o_2^{(3)}$	$f_1^{(3)}(o_2^{(3)})$	$\cdots$	$f_K^{(3)}(o_2^{(3)})$	$f_{\varphi \wedge \neg\psi}^{(3)}(o_2^{(3)})$	$f_{\varphi' \wedge \psi'}^{(3)}(o_2^{(3)})$	$f_{\varphi' \wedge \neg\psi'}^{(3)}(o_2^{(3)})$
$\vdots$	$\vdots$	$\ddots$	$\vdots$	$\vdots$	$\vdots$	$\vdots$
$o_{n_3}^{(3)}$	$f_1^{(3)}(o_{n_3}^{(3)})$	$\cdots$	$f_K^{(3)}(o_{n_3}^{(3)})$	$f_{\varphi \wedge \neg\psi}^{(3)}(o_{n_3}^{(3)})$	$f_{\varphi' \wedge \psi'}^{(3)}(o_{n_3}^{(3)})$	$f_{\varphi' \wedge \neg\psi'}^{(3)}(o_{n_3}^{(3)})$

Fig. 11.4 Data matrix $\mathscr{M}_3$

Condition (2) i.e. $\varphi \vdash \neg\psi$ is not satisfied, thus there is a data matrix $\mathscr{M}_4 = \langle M_4, f_1^{(4)}, \ldots, f_K^{(4)} \rangle$ of the type $\mathscr{T}$ and a row $o_1^{(4)} \in M_4$ such that $f_\varphi^{(4)}(o_1^{(4)}) = 1$ and $f_{\neg\psi}^{(4)}(o_1^{(4)}) = 0$. In addition, $f_{\neg\psi}^{(4)}(o_1^{(4)}) = 0$ means $f_\psi^{(4)}(o_1^{(4)}) = 1$ and we can conclude $f_{\varphi \wedge \psi}^{(4)}(o_1^{(4)}) = 1$ and $f_{\varphi \wedge \neg\psi}^{(4)}(o_1^{(4)}) = 0$, see Fig. 11.5. Here $f_\varphi^{(4)}$, $f_{\neg\psi}^{(4)}$, $f_{\varphi \wedge \psi}^{(4)}$, and $f_{\varphi \wedge \neg\psi}^{(4)}$, are interpretations of φ, $\neg\psi$, $\varphi \wedge \psi$, and $\varphi \wedge \neg\psi$ in $\mathscr{M}_4$ respectively.

row	$f_1^{(4)}$	$\cdots$	$f_K^{(4)}$	$f_\varphi^{(4)}$	$f_{\neg\psi}^{(4)}$	$f_{\varphi \wedge \psi}^{(4)}$	$f_{\varphi \wedge \neg\psi}^{(4)}$
$o_1^{(4)}$	$f_1^{(4)}(o_1^{(4)})$	$\cdots$	$f_K^{(4)}(o_1^{(4)})$	1	0	1	0
$o_2^{(4)}$	$f_1^{(4)}(o_2^{(4)})$	$\cdots$	$f_K^{(4)}(o_2^{(4)})$	$f_\varphi^{(4)}(o_2^{(4)})$	$f_{\neg\psi}^{(4)}(o_2^{(4)})$	$f_{\varphi \wedge \psi}^{(4)}(o_2^{(4)})$	$f_{\varphi \wedge \neg\psi}^{(4)}(o_2^{(4)})$
$\vdots$	$\vdots$	$\ddots$	$\vdots$	$\vdots$	$\vdots$	$\vdots$	$\vdots$
$o_{n_4}^{(4)}$	$f_1^{(4)}(o_{n_4}^{(4)})$	$\cdots$	$f_K^{(4)}(o_{n_4}^{(4)})$	$f_\varphi^{(4)}(o_{n_4}^{(4)})$	$f_{\neg\psi}^{(4)}(o_{n_4}^{(4)})$	$f_{\varphi \wedge \psi}^{(4)}(o_{n_4}^{(4)})$	$f_{\varphi \wedge \neg\psi}^{(4)}(o_{n_4}^{(4)})$

Fig. 11.5 Data matrix $\mathscr{M}_4$

Let $\mathcal{M}_5 = \langle M_5, f_1^{(5)}, \ldots, f_K^{(5)} \rangle$ be a data matrix such that

- $M_5 = \{o_1^{(5)}, \ldots, o_G^{(5)}, o_{G+1}^{(5)}, \ldots, o_{G+A}^{(5)}\}$
- $f_i^{(5)}(o_j^{(5)}) = f_i^{(3)}(o_1^{(3)})$ *for* $i = 1, \ldots K$ *and* $j = 1, \ldots, G$, *meaning that rows* $o_1^{(5)}, \ldots, o_G^{(5)}$ *behave like the row* $o_1^{(3)} \in M_3$ *concerning Boolean attributes* $\varphi \wedge \neg \psi,\ \varphi' \wedge \psi'$, *and* $\varphi' \wedge \neg \psi'$
- $f_i^{(5)}(o_j^{(5)}) = f_i^{(4)}(o_1^{(4)})$ *for* $i = 1, \ldots K$ *and* $j = G+1, \ldots, G+A$, *meaning that rows* $o_{G+1}^{(5)}, \ldots, o_{G+A}^{(5)}$ *behave like the row* $o_1^{(4)} \in M_4$ *concerning Boolean attributes* $\varphi \wedge \psi$ *and* $\varphi \wedge \neg \psi$,

see Fig. 11.6 where $f_{\varphi \wedge \psi}^{(5)}$, $f_{\varphi \wedge \neg \psi}^{(5)}$, $f_{\varphi' \wedge \psi'}^{(5)}$, *and* $f_{\varphi' \wedge \neg \psi'}^{(5)}$ *are interpretations of* $\varphi \wedge \psi$, $\varphi \wedge \neg \psi$, $\varphi' \wedge \psi'$, *and* $\varphi' \wedge \neg \psi'$ *in* $\mathcal{M}_5$ *respectively.*

row	$f_1^{(5)}$	$\cdots$	$f_K^{(5)}$	$f_{\varphi \wedge \psi}^{(5)}$	$f_{\varphi \wedge \neg \psi}^{(5)}$	$f_{\varphi' \wedge \psi'}^{(5)}$	$f_{\varphi' \wedge \neg \psi'}^{(5)}$
$o_1^{(5)}$	$f_1^{(3)}(o_1^{(3)})$	$\cdots$	$f_K^{(3)}(o_1^{(3)})$	$f_{\varphi \wedge \psi}^{(5)}(o_1^{(5)})$	0	0	1
$\vdots$	$\vdots$	$\ddots$	$\vdots$	$\vdots$	$\vdots$	$\vdots$	$\vdots$
$o_G^{(5)}$	$f_1^{(3)}(o_1^{(3)})$	$\cdots$	$f_K^{(3)}(o_1^{(3)})$	$f_{\varphi \wedge \psi}^{(5)}(o_G^{(5)})$	0	0	1
$o_{G+1}^{(5)}$	$f_1^{(4)}(o_1^{(4)})$	$\cdots$	$f_K^{(4)}(o_1^{(4)})$	1	0	$f_{\varphi' \wedge \psi'}^{(5)}(o_{G+1}^{(5)})$	$f_{\varphi' \wedge \neg \psi'}^{(5)}(o_{G+1}^{(5)})$
$\vdots$	$\vdots$	$\ddots$	$\vdots$	$\vdots$	$\vdots$	$\vdots$	$\vdots$
$o_{G+A}^{(5)}$	$f_1^{(4)}(o_1^{(4)})$	$\cdots$	$f_K^{(4)}(o_1^{(4)})$	1	0	$f_{\varphi' \wedge \psi'}^{(5)}(o_{G+A}^{(5)})$	$f_{\varphi' \wedge \neg \psi'}^{(5)}(o_{G+A}^{(5)})$

Fig. 11.6 Data matrix $\mathcal{M}_5$

This implies that 4ft-table $4ft(\varphi, \psi, \mathcal{M}_5)$ *of* φ *and* ψ *in* $\mathcal{M}_5$ *satisfies*

$$4ft(\varphi, \psi, \mathcal{M}_5) = \langle a, 0, c, d \rangle \ \text{where}\ a \geq A \ . \tag{11.1}$$

We have $F_{\Rightarrow^*}(A, 0, c, d) = 1$, $a \geq A$ *which implies* $F_{\Rightarrow^*}(a, 0, c, d) = 1$, *thus*

$$Val(\varphi \Rightarrow^* \psi, \mathcal{M}_5) = 1 \ . \tag{11.2}$$

Let us assume that $\langle a', b', c', d' \rangle$ *is a 4ft-table* $4ft(\varphi', \psi', \mathcal{M}_5)$ *of* φ' *and* ψ' *in* $\mathcal{M}_5$. *It holds that* $a' + b' + c' + d' = A + G$, *the rows* $o_1^{(5)}, \ldots, o_G^{(5)}$ *do not satisfy* $\varphi' \wedge \psi'$ *and thus* $a' \leq A$. *In addition, the rows* $o_1^{(5)}, \ldots, o_G^{(5)}$ *satisfy* $\varphi' \wedge \neg \psi'$ *and thus* $b' \geq G$. *We can summarize* $F_{\Rightarrow^*}(A, G) = 0$, $a' \leq A$, *and* $b' \geq G$ *which means* $F_{\Rightarrow^*}(a', b') = 0$, *thus*

$$Val(\varphi' \Rightarrow^* \psi', \mathcal{M}_5) = 0 \ .$$

This finishes the proof. $\square$

Please note that there are 13 examples of interesting implicational quantifiers in theorems 7.5 and 7.7.

We used the first row of data matrix $\mathcal{M}_3 = \langle M_3, f_1^{(3)}, \ldots, f_K^{(3)} \rangle$ and the first row of data matrix $\mathcal{M}_4 = \langle M_4, f_1^{(4)}, \ldots, f_K^{(4)} \rangle$ to build data matrix $\mathcal{M}_5 = \langle M_5, f_1^{(5)}, \ldots, f_K^{(5)} \rangle$, see point (III.2) in the proof of theorem 11.2.

In addition, the relation (11.2) follows from the relation (11.1) since $F_{\Rightarrow^*}(A, 0, c, d) = 1$, $a \geq A$, $0 \leq 0$ and $\Rightarrow^*$ is an implicational quantifier, which together implies $F_{\Rightarrow^*}(a, 0, c, d) = 1$. This, however, does not work for weakly implicational quantifiers since for weakly implicational quantifier $\Rightarrow^*$ we would need $F_{\Rightarrow^*}(A, 0, c, d) = 1$, $a \geq A$, $0 \leq 0$, and $A + 0 + c + d = a + 0 + c + d$ to conclude that $F_{\Rightarrow^*}(a, 0, c, d) = 1$. That's why we prove only theorem 11.3 for weakly implicational quantifiers. This theorem is weaker than theorem 11.2.

Theorem 11.3. *Let $\mathscr{LC}_{\mathscr{T}}$ be a logical calculus of association rules of type $\mathscr{T} = \langle t_1, \ldots, t_K \rangle$ with language $\mathscr{L}_{\mathscr{T}}$ of association rules. In addition, let $\varphi \Rightarrow^* \psi$ and $\varphi' \Rightarrow^* \psi'$ be association rules of $\mathscr{L}_{\mathscr{T}}$ where $\Rightarrow^*$ is a weakly implicational quantifier satisfying $F_{\Rightarrow^*}(0, b, c, d) = 0$ for $b + c + d > 0$ where $F_{\Rightarrow^*}$ is an associated function of $\Rightarrow^*$.*

If at least one of the conditions (1) or (2) are satisfied:

(1)   both conditions (1.a) and (1.b) are satisfied

    (1.a) $\varphi \wedge \psi \vdash \varphi' \wedge \psi'$
    (1.b) $\varphi' \wedge \neg \psi' \vdash \varphi \wedge \neg \psi$

(2) $\varphi \vdash \neg \psi$

then the deduction rule

$$\frac{\varphi \Rightarrow^* \psi}{\varphi' \Rightarrow^* \psi'}$$

is sound.

Proof. Let $\mathsf{M}_{\mathscr{T}}$ be a set of all data matrices $\mathcal{M} = \langle M, f_1, \ldots, f_K \rangle$ of type $\mathscr{T}$, see definition 3.9. We have to prove:

I)   If condition (1) is satisfied, $\mathcal{M} \in \mathsf{M}_{\mathscr{T}}$ and if $Val(\varphi \Rightarrow^* \psi, \mathcal{M}) = 1$ then also $Val(\varphi' \Rightarrow^* \psi', \mathcal{M}) = 1$.

II)   If condition (2) is satisfied, $\mathcal{M} \in \mathsf{M}_{\mathscr{T}}$ and if $Val(\varphi \Rightarrow^* \psi, \mathcal{M}) = 1$ then also $Val(\varphi' \Rightarrow^* \psi', \mathcal{M}) = 1$.

The proof is analogous to the proofs of points I) and II) in theorem 11.2. We assume that $F_{\Rightarrow^*}$ is an associated function of $\Rightarrow^*$, $\langle a, b, c, d \rangle = 4ft(\varphi, \psi, \mathcal{M})$ is the 4ft-table of φ and ψ in $\mathcal{M}$ and $\langle a', b', c', d' \rangle = 4ft(\varphi', \psi', \mathcal{M})$ is the 4ft-table of φ' and ψ' in $\mathcal{M}$, see Fig. 11.1.

If $\Rightarrow^*$ is a weakly implicational quantifier then $F_{\Rightarrow^*}(a, b, c, d) = 1$, $a' \geq a$, $b' \leq b$, and $a + b + c + d = a' + b' + c' + d'$ implies $F_{\Rightarrow^*}(a', b', c', d') = 1$, see definition 7.2. Both 4ft-tables $\langle a, b, c, d \rangle = 4ft(\varphi, \psi, \mathcal{M})$ and $\langle a', b', c', d' \rangle = 4ft(\varphi', \psi', \mathcal{M})$ concern data matrix $\mathcal{M} = \langle M, f_1, \ldots, f_K \rangle$ and thus $a + b + c + d = a' + b' + c' + d'$. The proofs for particular points follow.

I): Let $Val(\varphi \Rightarrow^* \psi, \mathcal{M}) = 1$ i.e. $F_{\Rightarrow^*}(a, b, c, d) = 1$. Condition (1.a) implies $a' \geq a$ and condition (1.b) implies $b' \leq b$. We assume $F_{\Rightarrow^*}(a, b, c, d) = 1$, we have

$a' \geq a$, $b' \leq b$ and $a+b+c+d = a'+b'+c'+d'$, thus $F_{\Rightarrow^*}(a',b',c',d') = 1$ i.e. $Val(\varphi' \Rightarrow^* \psi', \mathcal{M}) = 1$. This finishes the proof of point I).

II): Condition (2) i.e. $\varphi \vdash \neg\psi$ means that there is no row $o \in M$ satisfying $\varphi \wedge \psi$ and thus $a = 0$. We assume $F_{\Rightarrow^*}(0,b,c,d) = 0$ which means $Val(\varphi \Rightarrow^* \psi, \mathcal{M}) = 0$ for each $\mathcal{M} \in M_{\mathcal{T}}$. We can conclude that if it holds $Val(\varphi \Rightarrow^* \psi, \mathcal{M}) = 1$ then also $Val(\varphi' \Rightarrow^* \psi', \mathcal{M}) = 1$.

This finishes the proof. $\square$

Please note that the quantifier $\odot_s$ of support, see row 19 in Table 4.4, is weakly implicational according to lemma 7.3. It holds that $F_{\odot_s}(a,b,c,d) = 1$ if and only if $\frac{a}{a+b+c+d} \geq s$ and $0 < s \leq 1$ which implies $F_{\odot_s}(0,b,c,d) = 0$. This means that the just proved theorem 11.3 can also be applied to for all seven 4ft-quantifiers theorem 7.4 deals with.

11.4 Deduction Rules for Σ-double Implicational Quantifiers

There are also important results concerning interesting Σ-double implicational quantifiers, introduced in definition 8.9. Theorem 11.4 shows that a question of whether a deduction rule $\frac{\varphi \Leftrightarrow^* \psi}{\varphi' \Leftrightarrow^* \psi'}$, where $\Leftrightarrow^*$ is an interesting Σ-double implicational quantifier, is a sound deduction rule is equivalent to a question of whether some relations such as $\tau \vdash \omega$ are true where ω and τ are Boolean attributes created from φ, ψ, φ', and ψ'. Recall that the question of whether $\tau \vdash \omega$ (i.e. if ω logically follows from τ) is true can be converted into a question of whether the propositional formula $\Lambda(\tau, \omega) \wedge \pi(\tau) \rightarrow \pi(\omega)$ is a tautology or not, see theorem 11.1.

Theorem 11.5 presents a weaker result concerning deduction rules $\frac{\varphi \Leftrightarrow^* \psi}{\varphi' \Leftrightarrow^* \psi'}$ where $\Leftrightarrow^*$ is a weakly Σ-double implicational quantifier. It is proved that if the same relations $\tau \vdash \omega$ as in theorem 11.4 are true, then this deduction rule is sound.

Theorem 11.4. *Let $\mathcal{LC}_{\mathcal{T}}$ be a logical calculus of association rules of type $\mathcal{T} = \langle t_1, \ldots, t_K \rangle$ with a language $\mathcal{L}_{\mathcal{T}}$ of association rules. In addition, let $\varphi \Leftrightarrow^* \psi$ and $\varphi' \Leftrightarrow^* \psi'$ be association rules of $\mathcal{L}_{\mathcal{T}}$. If $\Leftrightarrow^*$ is an interesting Σ-double implicational quantifier then deduction rule*

$$\frac{\varphi \Leftrightarrow^* \psi}{\varphi' \Leftrightarrow^* \psi'}$$

is sound if and only if at least one of conditions (1) or (2) are satisfied:

(1)   both conditions (1.a) and (1.b) are satisfied

   (1.a) $(\varphi \wedge \psi) \vdash (\varphi' \wedge \psi')$
   (1.b) $(\varphi' \wedge \neg\psi' \vee \neg\varphi' \wedge \psi') \vdash (\varphi \wedge \neg\psi \vee \neg\varphi \wedge \psi)$

(2)   at least one of conditions (2.a) and (2.b) are satisfied

   (2.a) $\varphi \vdash \neg\psi$
   (2.b) $\psi \vdash \neg\varphi$.

Proof. The proof is similar to that of theorem 11.2. Let $M_{\mathscr{T}}$ be a set of all data matrices $\mathscr{M} = \langle M, f_1, \ldots, f_K \rangle$ of type $\mathscr{T}$, see definition 3.9. We have to prove:

I)   If condition (1) is satisfied, $\mathscr{M} \in M_{\mathscr{T}}$ and $Val(\varphi \Leftrightarrow^* \psi, \mathscr{M}) = 1$ then also $Val(\varphi' \Leftrightarrow^* \psi', \mathscr{M}) = 1$.

II)   If condition (2) is satisfied, $\mathscr{M} \in M_{\mathscr{T}}$ and $Val(\varphi \Leftrightarrow^* \psi, \mathscr{M}) = 1$ then also $Val(\varphi' \Leftrightarrow^* \psi', \mathscr{M}) = 1$.

III)   If neither (1) nor (2) are satisfied then the deduction rule $\frac{\varphi \Leftrightarrow^* \psi}{\varphi' \Leftrightarrow^* \psi'}$ is not sound. This means that we have to find $\mathscr{M} \in M_{\mathscr{T}}$ such that $Val(\varphi \Leftrightarrow^* \psi, \mathscr{M}) = 1$ and $Val(\varphi' \Leftrightarrow^* \psi', \mathscr{M}) = 0$.

We assume that $F_{\Leftrightarrow^*}$ is an associated function of $\Leftrightarrow^*$, $\langle a,b,c,d \rangle = 4ft(\varphi, \psi, \mathscr{M})$ is the 4ft-table of φ and ψ in $\mathscr{M}$ and $\langle a',b',c',d' \rangle = 4ft(\varphi', \psi', \mathscr{M})$ is the 4ft-table of φ' and ψ' in $\mathscr{M}$, see Fig. 11.1. Remember that $\Leftrightarrow^*$ is Σ-double implicational if $F_{\Leftrightarrow^*}(a,b,c,d) = 1$ and $a' \geq a \wedge b' + c' \leq b + c$ implies $F_{\Leftrightarrow^*}(a',b',c',d') = 1$, see definition 8.4 and Sect. 8.1. According to note 6.2 we write only $F_{\Leftrightarrow^*}(a,b,c)$ instead of $F_{\Leftrightarrow^*}(a,b,c,d)$. Proofs for particular points follow.

I): Let $Val(\varphi \Leftrightarrow^* \psi, \mathscr{M}) = 1$ i.e. $F_{\Leftrightarrow^*}(a,b,c) = 1$. Condition (1.a) implies $a' \geq a$ and condition (1.b) implies $b' + c' \leq b + c$. We assume $F_{\Leftrightarrow^*}(a,b,c,) = 1$, we have $a' \geq a$ and $b' + c' \leq b + c$, thus $F_{\Leftrightarrow^*}(a',b',c') = 1$ i.e. $Val(\varphi' \Leftrightarrow^* \psi', \mathscr{M}) = 1$. This finishes the proof of point I).

II): Condition (2) means that there is no row $o \in M$ satisfying $\varphi \wedge \psi$ and thus $a = 0$. We assume $\Leftrightarrow^*$ is an interesting Σ-double implicational quantifier, which means $F_{\Leftrightarrow^*}(0,0,0) = 0$, see definition 8.9. 4ft-quantifier $\Leftrightarrow^*$ is Σ-double implicational and $F_{\Leftrightarrow^*}(0,0,0) = 0$ implies $F_{\Leftrightarrow^*}(0,b,c) = 0$ for each $b \geq 0$ and $c \geq 0$. Condition (2) implies $a = 0$ and this means $Val(\varphi \Leftrightarrow^* \psi, \mathscr{M}) = 0$ for each $\mathscr{M} \in M_{\mathscr{T}}$. We can conclude that if $Val(\varphi \Leftrightarrow^* \psi, \mathscr{M}) = 1$ then also $Val(\varphi' \Leftrightarrow^* \psi', \mathscr{M}) = 1$. This finishes the proof of point II).

III): We assume that neither (1) nor (2) are satisfied. This means that at least one of the following conditions are satisfied:

(III.1)   neither (1.a) nor (2) are satisfied
(III.2)   neither (1.b) nor (2) are satisfied.

We find a data matrix $\mathscr{M} \in M_{\mathscr{T}}$ both for (III.1) and (III.2) which will satisfy $Val(\varphi \Leftrightarrow^* \psi, \mathscr{M}) = 1$ and $Val(\varphi' \Leftrightarrow^* \psi', \mathscr{M}) = 0$.

(III.1): Let us assume that neither (1.a) nor (2) are satisfied. The Σ-double implicational quantifier $\Leftrightarrow^*$ is a-dependent (see definition 7.4), thus there are non-negative integers A, B, C such that $F_{\Leftrightarrow^*}(A,B,C) = 1$. This means that $A > 0$ because $F_{\Leftrightarrow^*}(0,b,c) = 0$ for each $b \geq 0$ and $c \geq 0$, see the proof for point II).

Condition (1.a) is not satisfied thus there is a data matrix $\mathscr{M}_1 = \langle M_1, f_1^{(1)}, \ldots, f_K^{(1)} \rangle$ of the type $\mathscr{T}$ where $M_1 = \{o_1^{(1)}, \ldots, o_n^{(1)}\}$ and the row $o_1^{(1)} \in M_1$ such that $f_{\varphi \wedge \psi}^{(1)}(o_1^{(1)}) = 1$ and $f_{\varphi' \wedge \psi'}^{(1)}(o_1^{(1)}) = 0$, see Fig. 11.2. Here $f_{\varphi \wedge \psi}^{(1)}$ is the interpretation of $\varphi \wedge \psi$ in $\mathscr{M}_1$, similarly for $f_{\varphi' \wedge \psi'}^{(1)}$.

Let $\mathcal{M}_2 = \langle M_2, f_1^{(2)}, \ldots, f_K^{(2)} \rangle$ be the data matrix introduced in point (III.1) of the proof of theorem 11.2, see Fig. 11.3. Thus 4ft-tables $4ft(\varphi, \psi, \mathcal{M}_2)$ of φ and ψ in $\mathcal{M}_2$ and $4ft(\varphi', \psi', \mathcal{M}_2)$ of φ' and ψ' in $\mathcal{M}_2$ satisfy

$$4ft(\varphi, \psi, \mathcal{M}_2) = \langle A, 0, 0, 0 \rangle \quad and \quad 4ft(\varphi', \psi', \mathcal{M}_2) = \langle 0, b', c', d' \rangle$$

where $b' + c' + d' = A$.

We can conclude that $F_{\Leftrightarrow^*}(A, B, C) = 1$ and $4ft(\varphi, \psi, \mathcal{M}_2) = \langle A, 0, 0, 0 \rangle$, thus $Val(\varphi \Leftrightarrow^* \psi, \mathcal{M}_2) = 1$. In addition, $F_{\Leftrightarrow^*}(0, b', c') = 0$ for $b' \geq 0$, $c' \geq 0$ and $4ft(\varphi', \psi', \mathcal{M}_2) = \langle 0, b, c,' d' \rangle'$ and thus $Val(\varphi' \Leftrightarrow^* \psi', \mathcal{M}_2) = 0$. This finishes the proof of point (III.1).

(III.2): Let us assume that neither (1.b) nor (2) are satisfied. The quantifier $\Leftrightarrow^*$ is interesting Σ-double implicational and thus it is $b + c$-dependent, see definition 8.9. This means that there are non-negative integers A, B, C, B', C' such that $F_{\Leftrightarrow^*}(A, B, C) = 1$ and $F_{\Rightarrow^*}(A, B', C') = 0$ (see definition 8.8). The fact that $F_{\Leftrightarrow^*}(A, B, C) = 1$ means $F_{\Leftrightarrow^*}(A, 0, 0) = 1$, thus $A > 0$ because $F_{\Leftrightarrow^*}(0, 0, 0) = 0$. Furthermore, it must hold that $B' + C' > 0$ because $F_{\Leftrightarrow^*}(A, B', C') = 0$ and $F_{\Leftrightarrow^*}(A, 0, 0) = 1$.

Let us denote $\chi = (\varphi \wedge \neg\psi \vee \neg\varphi \wedge \psi)$ and $\chi' = (\varphi' \wedge \neg\psi' \vee \neg\varphi' \wedge \psi')$. Condition (1.b) can be then written as $\chi' \vdash \chi$. We assume it is not satisfied, thus there is a data matrix $\mathcal{M}_6 = \langle M_6, f_1^{(6)}, \ldots, f_K^{(6)} \rangle$ of the type $\mathcal{T}$ and a row $o_1^{(6)} \in M_6$ such that $f_{\chi'}^{(6)}(o_1^{(6)}) = 1$ and $f_\chi^{(6)}(o_1^{(6)}) = 0$. In addition, $f_{\chi'}^{(6)}(o_1^{(6)}) = 1$ implies $f_{\varphi' \wedge \psi'}^{(6)}(o_1^{(6)}) = 0$, see Fig. 11.7. Here $f_\chi^{(6)}$, $f_{\chi'}^{(6)}$, and $f_{\varphi' \wedge \psi'}^{(6)}$ are interpretations of χ, χ', and $\varphi' \wedge \psi'$ in $\mathcal{M}_6$ respectively.

row	$f_1^{(6)}$	$\cdots$	$f_K^{(6)}$	$f_\chi^{(6)}$	$f_{\chi'}^{(6)}$	$f_{\varphi' \wedge \psi'}^{(6)}$
$o_1^{(6)}$	$f_1^{(6)}(o_1^{(6)})$	$\cdots$	$f_K^{(6)}(o_1^{(6)})$	0	1	0
$o_2^{(6)}$	$f_1^{(6)}(o_2^{(6)})$	$\cdots$	$f_K^{(6)}(o_2^{(6)})$	$f_\chi^{(6)}(o_2^{(6)})$	$f_{\chi'}^{(6)}(o_2^{(6)})$	$f_{\varphi' \wedge \psi'}^{(6)}(o_2^{(6)})$
$\vdots$	$\vdots$	$\ddots$	$\vdots$	$\vdots$	$\vdots$	$\vdots$
$o_{n_6}^{(6)}$	$f_1^{(6)}(o_{n_6}^{(6)})$	$\cdots$	$f_K^{(6)}(o_{n_6}^{(6)})$	$f_\chi^{(6)}(o_{n_6}^{(6)})$	$f_{\chi'}^{(6)}(o_{n_6}^{(6)})$	$f_{\varphi' \wedge \psi'}^{(6)}(o_{n_6}^{(6)})$

Fig. 11.7 Data matrix $\mathcal{M}_6$

Condition (2) is not satisfied i.e. both $\varphi \vdash \neg\psi$ is not satisfied and $\psi \vdash \neg\varphi$ is not satisfied. Thus there is a data matrix $\mathcal{M}_7 = \langle M_7, f_1^{(7)}, \ldots, f_K^{(7)} \rangle$ of the type $\mathcal{T}$ and a row $o_1^{(7)} \in M_7$ such that $f_\varphi^{(7)}(o_1^{(7)}) = 1$ and $f_{\neg\psi}^{(7)}(o_1^{(7)}) = 0$. However, $f_{\neg\psi}^{(7)}(o_1^{(7)}) = 0$ means $f_\psi^{(7)}(o_1^{(7)}) = 1$ and we can conclude $f_{\varphi \wedge \psi}^{(7)}(o_1^{(7)}) = 1$. In addition, $f_{\varphi \wedge \psi}^{(7)}(o_1^{(7)}) = 1$ implies $f_\chi^{(7)}(o_1^{(7)}) = 0$ where $\chi = (\varphi \wedge \neg\psi \vee \neg\varphi \wedge \psi)$, see Fig. 11.8. Here $f_\varphi^{(7)}$, $f_{\neg\psi}^{(7)}$, $f_\psi^{(7)}$, $f_{\varphi \wedge \psi}^{(7)}$, and $f_\chi^{(7)}$ are interpretations of φ, $\neg\psi$, ψ, $\varphi \wedge \psi$, and χ in $\mathcal{M}_7$ respectively.

Fig. 11.8 Data matrix $\mathscr{M}_7$

Let $G = B' + C'$ and let $\mathscr{M}_8 = \langle M_8, f_1^{(8)}, \dots, f_K^{(8)} \rangle$ be a data matrix such that

- $M_8 = \{o_1^{(8)}, \dots, o_G^{(8)}, o_{G+1}^{(8)}, \dots, o_{G+A}^{(8)}\}$
- $f_i^{(8)}(o_j^{(8)}) = f_i^{(6)}(o_1^{(6)})$ for $i = 1, \dots K$ and $j = 1, \dots, G$, which means that rows $o_1^{(8)}, \dots, o_G^{(8)}$ behave like the row $o_1^{(6)} \in M_6$ with respect to Boolean attributes χ, $\varphi' \wedge \psi'$, and χ'
- $f_i^{(8)}(o_j^{(8)}) = f_i^{(7)}(o_1^{(7)})$ for $i = 1, \dots K$ and $j = G+1, \dots, G+A$, which means that rows $o_{G+1}^{(8)}, \dots, o_{G+A}^{(8)}$ behave like the row $o_1^{(7)} \in M_7$ with respect to Boolean atrributes $\varphi \wedge \psi$ and χ,

see Fig. 11.9 where $f_{\varphi\wedge\psi}^{(8)}$, $f_\chi^{(8)}$, $f_{\varphi'\wedge\psi'}^{(8)}$, and $f_{\chi'}^{(8)}$ are interpretations of $\varphi \wedge \psi$, χ, $\varphi' \wedge \psi'$, and χ' in $\mathscr{M}_8$ respectively.

row	$f_1^{(8)}$	$\dots$	$f_1^{(8)}K$	$f_{\varphi\wedge\psi}^{(8)}$	$f_\chi^{(8)}$	$f_{\varphi'\wedge\psi'}^{(8)}$	$f_{\chi'}^{(8)}$
$o_1^{(8)}$	$f_1^{(6)}(o_1^{(6)})$	$\dots$	$f_K^{(6)}(o_1^{(6)})$	$f_{\varphi\wedge\psi}^{(8)}(o_1^{(8)})$	0	0	1
$\vdots$		$\ddots$		$\vdots$	$\vdots$	$\vdots$	$\vdots$
$o_G^{(8)}$	$f_1^{(6)}(o_1^{(6)})$	$\dots$	$f_K^{(6)}(o_1^{(6)}))$	$f_{\varphi\wedge\psi}^{(8)}(o_G^{(8)})$	0	0	1
$o_{G+1}^{(8)}$	$f_1^{(7)}(o_1^{(7)})$	$\dots$	$f_K^{(7)}(o_1^{(7)})$	1	0	$f_{\varphi'\wedge\psi'}^{(8)}(o_{G+1}^{(8)})$	$f_{\chi'}^{(8)}(o_{G+1}^{(8)})$
$\vdots$		$\ddots$		$\vdots$	$\vdots$	$\vdots$	$\vdots$
$o_{G+A}^{(8)}$	$f_1^{(7)}(o_1^{(7)})$	$\dots$	$f_K^{(7)}(o_1^{(7)})$	1	0	$f_{\varphi'\wedge\psi'}^{(8)}(o_{G+A}^{(8)})$	$f_{\chi'}^{(8)}(o_{G+A}^{(8)})$

Fig. 11.9 Data matrix $\mathscr{M}_8$

It holds that $\chi = (\varphi \wedge \neg\psi \vee \neg\varphi \wedge \psi)$ and $f_\chi^{(8)}(o_j^{(8)}) = 0$ for $j = G+1, \dots, G+A$ which implies $f_{\varphi\wedge\neg\psi}^{(8)}(o_j^{(8)}) = 0$ and $f_{\neg\varphi\wedge\psi}^{(8)}(o_j^{(8)}) = 0$ where $f_{\varphi\wedge\neg\psi}^{(8)}$ and $f_{\neg\varphi\wedge\psi}^{(8)}$ are interpretations of $\varphi \wedge \neg\psi$ and $\neg\varphi \wedge \psi$ in $\mathscr{M}_8$ respectively. This means that the 4ft-table $4ft(\varphi, \psi, \mathscr{M}_8)$ of φ and ψ in $\mathscr{M}_8$ satisfies

$$4ft(\varphi, \psi, \mathscr{M}_8) = \langle a, 0, 0, d \rangle \text{ where } a \geq A. \tag{11.3}$$

We have $F_{\Leftrightarrow^*}(A,0,0,d) = 1$, $a \geq A$, and $0+0 \geq 0+0$ and thus $F_{\Leftrightarrow^*}(a,0,0,d) = 1$, which implies

$$Val(\varphi \Leftrightarrow^* \psi, \mathcal{M}_8) = 1 . \tag{11.4}$$

Let us assume that $\langle a',b',c,'d'\rangle$ is a 4ft-table $4ft(\varphi',\psi',\mathcal{M}_8)$ of φ' and ψ' in $\mathcal{M}_8$. It holds that $a' + b' + c' + d' = A + G$, the rows $o_1^{(8)},\ldots,o_G^{(8)}$ do not satisfy $\varphi' \wedge \psi'$ and thus $a' \leq A$. In addition, it holds $\chi' = (\varphi' \wedge \neg\psi' \vee \neg\varphi' \wedge \psi')$ and $f_{\chi'}^{(8)}(o_j^{(8)}) = 1$ for $j = 1,\ldots,G$ which implies $f_{\varphi' \wedge \neg\psi'}^{(8)}(o_j^{(8)}) = 1$ or $f_{\neg\varphi' \wedge \psi'}^{(8)}(o_j^{(8)}) = 1$ where $f_{\varphi' \wedge \neg\psi'}^{(8)}$ and $f_{\neg\varphi' \wedge \psi'}^{(8)}$ are interpretations of $\varphi' \wedge \neg\psi'$ and $\neg\varphi' \wedge \psi'$ in $\mathcal{M}_8$ respectively. This means $b' + c' \geq G = B' + C'$.

We can conclude that $F_{\Leftrightarrow^*}(A,B',C') = 0$, $a' \leq A$, and $b' + c' \geq B' + C'$, thus

$$Val(\varphi' \Rightarrow^* \psi', \mathcal{M}_8) = 0 .$$

This finishes the proof. □

Please note that there are three examples of interesting Σ-double implicational quantifiers in Sect. 8.4, theorem 8.10.

We used the first row of data matrix $\mathcal{M}_6 = \langle M_6, f_1^{(6)},\ldots,f_K^{(6)}\rangle$ and the first row of data matrix $\mathcal{M}_7 = \langle M_7, f_1^{(7)},\ldots,f_K^{(7)}\rangle$ to build data matrix $\mathcal{M}_8 = \langle M_8, f_1^{(8)},\ldots,f_K^{(8)}\rangle$, see point (III.2) in the proof of theorem 11.4. In addition, the relation (11.4) follows from the relation (11.3) because $F_{\Leftrightarrow^*}(A,0,0,d) = 1$, $a \geq A$, $0+0 \leq 0+0$ and $\Leftrightarrow^*$ is Σ-double implicational quantifier which together implies $F_{\Leftrightarrow^*}(a,0,0,d) = 1$. This, however, does not work for weakly Σ-double implicational quantifiers because for a weakly Σ-double implicational quantifier $\Leftrightarrow^*$ we would need $F_{\Leftrightarrow^*}(A,0,0,d) = 1$, $a \geq A$, $0+0 \leq 0+0$, and $A+0+0+d = a+0+0+d$ to conclude that $F_{\Leftrightarrow^*}(a,0,0,d) = 1$.

That's why we prove only theorem 11.5 for weakly Σ-double implicational quantifiers. This theorem is weaker than theorem 11.4.

Theorem 11.5. *Let $\mathcal{LC}_{\mathcal{T}}$ be a logical calculus of association rules of type $\mathcal{T} = \langle t_1,\ldots,t_K\rangle$ with language $\mathcal{L}_{\mathcal{T}}$ of association rules. In addition, let $\varphi \Leftrightarrow^* \psi$ and $\varphi' \Leftrightarrow^* \psi'$ be association rules of $\mathcal{L}_{\mathcal{T}}$ where $\Leftrightarrow^*$ is a weakly Σ-double implicational quantifier satisfying $F_{\Leftrightarrow^*}(0,b,c,d) = 0$ for $b+c+d > 0$ where $F_{\Leftrightarrow^*}$ is an associated function of $\Leftrightarrow^*$.*

If at least one of conditions (1) or (2) are satisfied:

(1) both conditions (1.a) and (1.b) are satisfied

    (1.a) $\varphi \wedge \psi \vdash \varphi' \wedge \psi'$

    (1.b) $(\varphi' \wedge \neg\psi' \vee \neg\varphi' \wedge \psi') \vdash (\varphi \wedge \neg\psi \vee \neg\varphi \wedge \psi)$.

(2) at least one of conditions (2.a) or (2.b) are satisfied

    (2.a) $\varphi \vdash \neg\psi$

    (2.b) $\psi \vdash \neg\varphi$

then the deduction rule

$$\frac{\varphi \Leftrightarrow^* \psi}{\varphi' \Leftrightarrow^* \psi'}$$

is sound.

Proof. Let $M_{\mathscr{T}}$ *be a set of all data matrices* $\mathscr{M} = \langle M, f_1, \ldots, f_K \rangle$ *of type* $\mathscr{T}$, *see definition 3.9. We have to prove:*

I)   *If condition (1) is satisfied,* $\mathscr{M} \in M_{\mathscr{T}}$ *and if* $Val(\varphi \Rightarrow^* \psi, \mathscr{M}) = 1$ *then also* $Val(\varphi' \Leftrightarrow^* \psi', \mathscr{M}) = 1$.

II) *If condition (2) is satisfied,* $\mathscr{M} \in M_{\mathscr{T}}$ *and if* $Val(\varphi \Leftrightarrow^* \psi, \mathscr{M}) = 1$ *then also* $Val(\varphi' \Leftrightarrow^* \psi', \mathscr{M}) = 1$.

The proof is analogous to the proofs of points I) and II) in theorem 11.4. We assume that $F_{\Leftrightarrow^*}$ *is an associated function of* $\Leftrightarrow^*$, $\langle a, b, c, d \rangle = 4ft(\varphi, \psi, \mathscr{M})$ *is the 4ft-table of* φ *and* ψ *in* $\mathscr{M}$ *and* $\langle a', b', c', d' \rangle = 4ft(\varphi', \psi', \mathscr{M})$ *is the 4ft-table of* φ' *and* ψ' *in* $\mathscr{M}$, *see Fig. 11.1.*

Remember that if $\Leftrightarrow^*$ *is a weakly* Σ-*double implicational quantifier then* $F_{\Leftrightarrow^*}(a, b, c, d) = 1$, $a' \geq a \ \wedge \ b' + c' \leq b + c$ *and* $a + b + c + d = a' + b' + c' + d'$ *implies* $F_{\Leftrightarrow^*}(a', b', c', d') = 1$, *see definition 7.2. Please note that both 4ft-tables* $\langle a, b, c, d \rangle = 4ft(\varphi, \psi, \mathscr{M})$ *and* $\langle a', b', c', d' \rangle = 4ft(\varphi', \psi', \mathscr{M})$ *concern data matrix* $\mathscr{M} = \langle M, f_1, \ldots, f_K \rangle$ *and thus* $a + b + c + d = a' + b' + c' + d'$. *The proofs for particular points follow.*

I): Let $Val(\varphi \Leftrightarrow^* \psi, \mathscr{M}) = 1$ *i.e.* $F_{\Leftrightarrow^*}(a, b, c, d) = 1$. *Condition (1.a) implies* $a' \geq a$ *and (1.b) implies* $b' + c' \leq b + c$. *We assume* $F_{\Leftrightarrow^*}(a, b, c, d) = 1$, *we have* $a' \geq a$, $b' + c' \leq b + c$ *and* $a + b + c + d = a' + b' + c' + d'$, *thus* $F_{\Leftrightarrow^*}(a', b', c', d') = 1$ *i.e.* $Val(\varphi' \Leftrightarrow^* \psi', \mathscr{M}) = 1$. *This finishes the proof of point I).*

II): Condition (2) i.e. $\varphi \vdash \neg\psi$ *or* $\psi \vdash \neg\varphi$ *means that there is no row* $o \in M$ *satisfying* $\varphi \wedge \psi$ *and thus* $a = 0$. *We assume that* $F_{\Leftrightarrow^*}(0, b, c, d) = 0$ *and this means* $Val(\varphi \Leftrightarrow^* \psi, \mathscr{M}) = 0$ *for each* $\mathscr{M} \in M_{\mathscr{T}}$. *We can conclude that if* $Val(\varphi \Leftrightarrow^* \psi, \mathscr{M}) = 1$ *then also* $Val(\varphi' \Leftrightarrow^* \psi', \mathscr{M}) = 1$.

This finishes the proof.   □

Please note that the quantifier $\odot_s$ of support, see row 19 in table 4.4 is weakly Σ-double implicational according to lemma 8.5. It holds that $F_{\odot_s}(a, b, c, d) = 1$ if and only if $\frac{a}{a+b+c+d} \geq s$ and $0 < s \leq 1$ which implies $F_{\odot_s}(0, b, c, d) = 0$. This means that the just proved theorem 11.5 can also be applied to all three 4ft-quantifiers theorem 8.7 deals with.

11.5   Deduction Rules for Σ-equivalence Quantifiers

Additional results on deduction rules concern interesting Σ-equivalence quantifiers. Theorem 11.6 shows that a question of whether a deduction rule $\frac{\varphi \equiv^* \psi}{\varphi' \equiv^* \psi'}$, where $\equiv^*$ is an interesting Σ-equivalence quantifier, is a sound deduction rule is equivalent to a question of whether a relation $\tau \vdash \omega$ is true, where ω and τ are Boolean attributes created from φ, ψ, φ', and ψ'. Let us remember that the question of whether $\tau \vdash \omega$ (i.e. if ω logically follows from τ) is true can be converted into a question of whether the propositional formula $\Lambda(\tau,\omega) \wedge \pi(\tau) \rightarrow \pi(\omega)$ is a tautology or not, see theorem 11.1.

Theorem 11.7 presents a weaker result concerning deduction rules $\frac{\varphi \equiv^* \psi}{\varphi' \equiv^* \psi'}$ where $\equiv^*$ is a weakly Σ-equivalence quantifier. It is proved that if the same relations $\tau \vdash \omega$ as in theorem 11.6 are true, then this deduction rule is sound.

Theorem 11.6. *Let $\mathscr{LC}_{\mathscr{T}}$ be a logical calculus of association rules of type $\mathscr{T} = \langle t_1, \ldots, t_K \rangle$ with a language $\mathscr{L}_{\mathscr{T}}$ of association rules. In addition, let $\varphi \equiv^* \psi$ and $\varphi' \equiv^* \psi'$ be association rules of $\mathscr{L}_{\mathscr{T}}$. If $\equiv^*$ is an interesting Σ-equivalence quantifier, then deduction rule*

$$\frac{\varphi \equiv^* \psi}{\varphi' \equiv^* \psi'}$$

is sound if and only if condition (1) is satisfied.

(1) $\quad (\varphi \wedge \psi \vee \neg\varphi \wedge \neg\psi) \vdash (\varphi' \wedge \psi' \vee \neg\varphi' \wedge \neg\psi')$.

Proof. The proof is similar to that of theorem 11.2. Let $\mathrm{M}_{\mathscr{T}}$ be a set of all data matrices $\mathscr{M} = \langle M, f_1, \ldots, f_K \rangle$ of type $\mathscr{T}$, see definition 3.9. We have to prove:

I) *If condition (1) is satisfied, $\mathscr{M} \in \mathrm{M}_{\mathscr{T}}$ and $Val(\varphi \equiv^* \psi, \mathscr{M}) = 1$ then also $Val(\varphi' \equiv^* \psi', \mathscr{M}) = 1$.*

II) *If condition (1) is not satisfied then the deduction rule $\frac{\varphi \equiv^* \psi}{\varphi' \equiv^* \psi'}$ is not sound. This means that we have to find $\mathscr{M} \in \mathrm{M}_{\mathscr{T}}$ such that $Val(\varphi \equiv^* \psi, \mathscr{M}) = 1$ and $Val(\varphi' \equiv^* \psi', \mathscr{M}) = 0$.*

We assume that $F_{\equiv^*}$ is an associated function of $\equiv^*$, $\langle a,b,c,d \rangle = 4ft(\varphi,\psi,\mathscr{M})$ is the 4ft-table of φ and ψ in $\mathscr{M}$ and $\langle a',b',c',d' \rangle = 4ft(\varphi',\psi',\mathscr{M})$ is the 4ft-table of φ' and ψ' in $\mathscr{M}$, see Fig. 11.1. Please note that $a+b+c+d = a'+b'+c'+d'$. Proofs for particular points follow.

I): Let $Val(\varphi \equiv^* \psi, \mathscr{M}) = 1$ i.e. $F_{\equiv^*}(a,b,c,d) = 1$. Condition (1) implies that $a'+d' \geq a+d$ and $a+b+c+d = a'+b'+c'+d'$ means that $b'+c' \leq b+c$. We assume $F_{\equiv^*}(a,b,c,d) = 1$, we have $a'+d' \geq a+d$ and $b'+c' \leq b+c$, thus $F_{\equiv^*}(a',b',c',d) = 1$ i.e. $Val(\varphi' \equiv^* \psi', \mathscr{M}) = 1$. This finishes the proof of point I).

II): We assume that (1) is not satisfied. We find a data matrix $\mathscr{M}$ such that $Val(\varphi \equiv^* \psi, \mathscr{M}) = 1$ and $Val(\varphi' \equiv^* \psi', \mathscr{M}) = 0$.

The quantifier $\equiv^*$ is an interesting Σ-equivalence quantifier and thus it is $a+d$-dependent, see definition 9.9. This means that there are non-negative integers A, B, C, D such that $F_{\equiv^*}(A,B,C,D) = 1$, see definition 9.8. It must hold that $A+D > 0$ because $F_{\equiv^*}(0,b,c,0) = 0$ for $b+c > 0$ for interesting Σ-equivalence quantifiers, see definition 9.9.

Condition (1) is not satisfied and this means that there is a data matrix $\mathscr{M}_9 = \langle M_9, f_9^{(9)}, \ldots, f_K^{(9)} \rangle$ of the type $\mathscr{T}$ and a row $o_1^{(9)} \in M_9$ such that $f_\chi^{(9)}(o_1^{(9)}) = 1$ and $f_{\chi'}^{(9)}(o_1^{(9)}) = 0$. Here $f_\chi^{(9)}$ is an interpretation of Boolean attribute $\chi = \varphi \wedge \psi \vee \neg\varphi \wedge \neg\psi$ and $f_{\chi'}^{(9)}$ is an interpretation of Boolean attribute $\chi' = \varphi' \wedge \psi' \vee \neg\varphi' \wedge \neg\psi'$, see Fig. 11.10.

row	$f_1^{(9)}$	$f_2^{(9)}$	$\cdots$	$f_K^{(9)}$	$f_\chi^{(9)}$	$f_{\chi'}^{(9)}$
$o_1^{(9)}$	$f_1^{(9)}(o_1^{(9)})$	$f_2^{(9)}(o_1^{(9)})$	$\cdots$	$f_K^{(9)}(o_1^{(9)})$	1	0
$o_2^{(9)}$	$f_1^{(9)}(o_2^{(9)})$	$f_2^{(9)}(o_2^{(9)})$	$\cdots$	$f_K^{(9)}(o_2^{(9)})$	$f_\chi^{(9)}(o_2^{(9)})$	$f_{\chi'}^{(9)}(o_2^{(9)})$
$\vdots$	$\vdots$	$\vdots$	$\ddots$	$\vdots$	$\vdots$	$\vdots$
$o_{n_9}^{(9)}$	$f_1^{(9)}(o_{n_9}^{(9)})$	$f_2^{(9)}(o_{n_9})$	$\cdots$	$f_K^{(9)}(o_{n_9})$	$f_\chi^{(9)}(o_{n_9}^{(9)})$	$f_{\chi'}^{(9)}(o_{n_9}^{(9)})$

Fig. 11.10 Data matrix $\mathcal{M}_9$

row	$f_1^{(10)}$	$f_2^{(10)}$	$\cdots$	$f_K^{(10)}$	$f_\chi^{(10)}$	$f_{\chi'}^{(10)}$
$o_1^{(10)}$	$f_1^{(9)}(o_1^{(9)})$	$f_2^{(9)}(o_1^{(9)})$	$\cdots$	$f_K^{(9)}(o_1^{(9)})$	1	0
$o_2^{(10)}$	$f_1^{(9)}(o_1^{(9)})$	$f_2^{(9)}(o_1^{(9)})$	$\cdots$	$f_K^{(9)}(o_1^{(9)})$	1	0
$\vdots$	$\vdots$	$\vdots$	$\ddots$	$\vdots$	$\vdots$	$\vdots$
$o_{A+D}^{(10)}$	$f_1^{(9)}(o_1^{(9)})$	$f_2^{(9)}(o_1^{(9)})$	$\cdots$	$f_K^{(9)}(o_1^{(9)})$	1	0

Fig. 11.11 Data matrix $\mathcal{M}_{10}$

Let $\mathcal{M}_{10} = \langle M_{10}, f_1^{(10)}, \ldots, f_K^{(10)} \rangle$ be a data matrix where $M_{10} = \{o_1^{(10)}, \ldots, o_{A+D}^{(10)}\}$ and $f_i^{(10)}(o_j^{(10)}) = f_i^{(9)}(o_1^{(1)})$ for $i = 1, \ldots K$ and $j = 1, \ldots, A+D$, see Fig. 11.11 where $f_\chi^{(10)}$ is the interpretation of χ in $\mathcal{M}_{10}$, similarly for $f_{\chi'}^{(10)}$. In other words, data matrix $\mathcal{M}_{10}$ has $A+D$ rows and each of them behaves as the row $o_1^{(9)} \in M_9$.

This means $f_\chi^{(10)}(o_j^{10}) = 1$ and $f_{\chi'}^{(10)}(o_j^{10}) = 0$ for $j = 1, \ldots, A+D$. Please remember that $\chi = \varphi \wedge \psi \vee \neg\varphi \wedge \neg\psi$ and $\chi' = \varphi' \wedge \psi' \vee \neg\varphi' \wedge \neg\psi'$. Thus, the 4ft-tables $4ft(\varphi, \psi, \mathcal{M}_{10})$ of φ and ψ in $\mathcal{M}_{10}$ and $4ft(\varphi', \psi', \mathcal{M}_{10})$ of φ' and ψ' in $\mathcal{M}_{10}$ satisfy

$$4ft(\varphi, \psi, \mathcal{M}_{10}) = \langle a, 0, 0, d \rangle \quad and \quad 4ft(\varphi', \psi', \mathcal{M}_{10}) = \langle 0, b', c', 0 \rangle$$

where $a + d = b' + c' = A + D$.

We can conclude that $F_{\equiv^*}(A, B, C, D) = 1$, $\equiv^*$ is an Σ-equivalence quantifier, $a + d = A + D$, $0 + 0 \leq B + C$ and thus also $F_{\equiv^*}(a, 0, 0, d) = 1$ which means $Val(\varphi \equiv^* \psi, \mathcal{M}_{10}) = 1$. In addition, $\equiv^*$ is an interesting Σ-equivalence quantifier and thus $F_{\equiv^*}(0, b', c', 0) = 0$ which means $Val(\varphi' \equiv^* \psi', \mathcal{M}_{10}) = 0$.

This finishes the proof.   □

Please note that there are 3 examples of interesting Σ-equivalence quantifiers in theorem 9.9.

We used the assumptions $F_{\equiv^*}(A, B, C, D) = 1$, $a + d = A + D$, $0 + 0 \leq B + C$ and that $\equiv^*$ is an Σ-equivalence quantifier to deduce that $F_{\equiv^*}(a, 0, 0, d) = 1$ at the end of the proof of theorem 11.6. We also used the fact that $4ft(\varphi, \psi, \mathcal{M}_{10}) = \langle a, b, c, d \rangle$ and $b = c = 0$. This, however, does not work for weakly Σ-equivalence quantifiers if $b + c \neq B + C$, see also lemma 11.3. That's why we prove only theorem 11.7 for weakly implicational quantifiers. This theorem is weaker than theorem 11.6.

Theorem 11.7. *Let $\mathcal{LC}_{\mathcal{T}}$ be a logical calculus of association rules of type $\mathcal{T} = \langle t_1, \ldots, t_K \rangle$ with language $\mathcal{L}_{\mathcal{T}}$ of association rules. In addition, let $\varphi \equiv^* \psi$ and $\varphi' \equiv^* \psi'$ be association rules of $\mathcal{L}_{\mathcal{T}}$ where $\equiv^*$ is a weakly Σ-equivalence quantifier.*

If condition (1) is satisfied

$(1) \quad (\varphi \wedge \psi \vee \neg \varphi \wedge \neg \psi) \vdash (\varphi' \wedge \psi' \vee \neg \varphi' \wedge \neg \psi')\,.$

then deduction rule

$$\frac{\varphi \equiv^* \psi}{\varphi' \equiv^* \psi'}$$

is sound.

Proof. Let $\mathsf{M}_{\mathcal{T}}$ be a set of all data matrices $\mathcal{M} = \langle M, f_1, \ldots, f_K \rangle$ of type $\mathcal{T}$, see definition 3.9. We have to prove that if condition (1) is satisfied, $\mathcal{M} \in \mathsf{M}_{\mathcal{T}}$ and $Val(\varphi \equiv^* \psi, \mathcal{M}) = 1$ then also $Val(\varphi' \equiv^* \psi', \mathcal{M}) = 1$.

If $\equiv^*$ is a weakly Σ-equivalence quantifier then $F_{\equiv^*}(a, b, c, d) = 1$, $a' + d' \geq a + d$ and $a + b + c + d = a' + b' + c' + d'$ implies $F_{\Rightarrow^*}(a', b', c', d') = 1$, see definition 9.6. We assume that $F_{\equiv^*}$ is an associated function of $\equiv^*$, $\langle a, b, c, d \rangle = 4ft(\varphi, \psi, \mathcal{M})$ is the 4ft-table of φ and ψ in $\mathcal{M}$ and $\langle a', b', c', d' \rangle = 4ft(\varphi', \psi', \mathcal{M})$ is the 4ft-table of φ' and ψ' in $\mathcal{M}$, see Fig. 11.1. Please note that $a + b + c + d = a' + b' + c' + d'$.

Let $Val(\varphi \equiv^* \psi, \mathcal{M}) = 1$ i.e. $F_{\equiv^*}(a, b, c, d) = 1$. Condition (1) implies that $a' + d' \geq a + d$. We assume $F_{\equiv^*}(a, b, c, d) = 1$, we have $a' + d' \geq a + d$ and $a + b + c + d = a' + b' + c' + d'$, thus $F_{\equiv^*}(a', b', c', d) = 1$ i.e. $Val(\varphi' \equiv^* \psi', \mathcal{M}) = 1$. This finishes the proof.   □

Please note that there are 3 examples of interesting Σ-equivalence quantifiers in theorem 9.9. Each of these quantifiers is according to lemma 9.2 also a weakly Σ-equivalence quantifier. However, among 4ft-quantifiers studied in previous sections, there is no 4ft-quantifier which is a weakly Σ-equivalence quantifier but not a Σ-equivalence quantifier. The following lemma shows that there is a weakly Σ-equivalence quantifier which is not Σ-equivalence.

Lemma 11.3. *The 4ft-quantifier $\equiv_W$ with associated function $F_{\equiv_W}$ defined such that*

$$F_{\equiv_W}(a, b, c, d) = \begin{cases} 1 \ \text{if } a + b + c + d = 10 \ \text{and } a + d \geq 5 \\ 0 \ \text{otherwise.} \end{cases}$$

is a weakly Σ-equivalence quantifier but not a Σ-equivalence quantifier.

Proof. We have to prove:

A)   If $F_{\equiv_W}(a, b, c, d) = 1$, $a' + d' \geq a + d$ and $a' + b' + c' + d' = a + b + c + d$ then $F_{\equiv_W}(a', b', c', d') = 1$, see definition 9.6.

B)   There are 4ft-tables $\langle a, b, c, d \rangle$ and $\langle a', b', c', d' \rangle$ such that $F_{\equiv_W}(a, b, c, d) = 1$, $a' + d' \geq a + d \wedge b' + c' \leq b + c$ and $F_{\equiv_W}(a', b', c', d') = 0$, see definition 9.4.

Proofs for particular points follow.

A)   *If* $F_{\equiv W}(a,b,c,d) = 1$ *then it must hold that* $a+b+c+d = 10$ *and* $a+d \geq 5$. *If* $a'+d' \geq a+d$ *then also* $a'+d' \geq 5$ *and* $a+b+c+d = a'+b'+c'+d'$ *implies* $a'+b'+c'+d' = 10$ *which means that* $F_{\equiv W}(a',b',c',d') = 1$.

B)   *Let be* $\langle a,b,c,d \rangle = \langle 10,0,0,0 \rangle$, *then* $F_{\equiv W}(a,b,c,d) = 1$. *In addition, let* $\langle a',b',c',d' \rangle = \langle 20,0,0,0 \rangle$, *then* $a'+d' \geq a+d \ \wedge \ b'+c' \leq b+c$ *and* $F_{\equiv W}(a',b',c',d') = 0$.

This finishes the proof.   □

Chapter 12
Missing Information

Dealing with missing information is a serious problem related to data analysis. A presented approach to missing information was introduced in [18]. Each attribute can get a value X in addition to its regular values. The value X is interpreted as *the value of the attribute is not known* or *there is a missing information*. This implies that also values of basic Boolean attributes derived from attributes with missing information, and consequently also values of additional derived Boolean attributes, can be unknown.

However, in various cases may be clear that the Boolean attribute in question is either true or false for a given object even if the values of some of the used attributes are not known for this object. The *principle of secured extension* formulated in [18] states that a Boolean attribute is considered true for a given object if it is sure to be true for this object in all possible completions of an analysed data matrix. Similarly, the Boolean attribute is considered false for a given object if it is false for this object in all possible completions of the analysed data matrix.

A completion of a data matrix with missing values is a data matrix in which all X values are replaced by a possible regular value. However, for some objects there can be a completion in which the value of a given Boolean attribute is true as well as another completion in which the value of the given Boolean attribute is false. Then we say that the value of the Boolean attribute for this object in the analysed data matrix is X. Thus, Boolean attributes can be assigned values from $\{0,1,X\}$. We refer to such Boolean attributes as Boolean attributes with missing values or three-valued attributes.

This means that we have to deal with association rules that concern three-valued attributes with possible values from $\{0,1,X\}$. The principle of secured extensions means that values of such association rules are assigned in the following way: If the rule is true in all possible completions of a given data matrix, then we consider this rule true in the given data matrix. If the rule is false in all possible completions of the given data matrix, then we consider this rule false in the given data matrix. If there is a completion in which the rule is false and another completion in which the rule is true then we say that the value of the rule is X.

J. Rauch: *Observational Calculi and Association Rules*, SCI 469, pp. 181–200.
DOI: 10.1007/978-3-642-11737-4_12   © Springer-Verlag Berlin Heidelberg 2013

There are nine possible combinations of values $\{0,1,X\}$ of two three-valued attributes and thus we have to deal with a nine-fold table of two three-valued attributes instead of a four-fold table. It is crucial that for important 4ft-quantifiers it is possible to get a *secured four-fold table* for each nine-fold table. It can be proved that a given association rules $\varphi \approx \psi$ is true in all completions of a given data matrix $\mathcal{M}^X$ with missing information if and only if the value of associated function $F_\approx$ of the quantifier $\approx$ is 1 for the secured completion of the nine-fold table of φ and ψ in $\mathcal{M}^X$. A way of computing the secured completions of the nine-fold table depends on the class the quantifier $\approx$ belongs to.

The goal of this chapter is to present these results. Data matrices with missing information are introduced in Sect. 12.1 together with their completions. The principles of secured extension of a logical calculus of association rules with missing information are introduced in Sect. 12.2. Evaluation of Boolean attributes according to principle of secured extension is described in Sect. 12.3. Nine-fold tables which are used to define secured extensions of associated functions of 4ft-quantifiers are introduced in Sect. 12.4. Logical calculus of association rules with missing information is defined in Sect. 12.5. Secured four-fold tables for particular important classes of 4ft-quantifiers are studied in Sect. 12.6.

The goal of this chapter is also to point out two additional approaches to dealing with missing information which are used in GUHA procedure ASSOC – optimistic completion and deleting, see Sect. 12.7.

12.1 Data Matrices with Missing Information

We define a data matrix with missing information by a modification of definition 2.2. Then we define a completion of a data matrix with missing information.

Definition 12.1. Let $\mathcal{T} = \langle t_1,\ldots,t_K \rangle$ be the type. Then a *data matrix with missing information of the type* $\mathcal{T}$ is a $K+1$-tuple $\mathcal{M}^X = \langle M, f'_1,\ldots,f'_K \rangle$, where M is a non-empty finite set and f'_i is an unary function from M to $\{1,\ldots,t_i\} \cup \{X\}$ for $i = 1,\ldots,K$. Elements of M are called rows of $\mathcal{M}$, the set M is a *set of rows* of data matrix $\mathcal{M}$, $1,\ldots,t_i$ are called *regular values of* f'_i for $i = 1,\ldots,K$ and X is an abstract value called a *missing value*, $X \notin \{1,\ldots,t_i\}$ for $i = 1,\ldots,K$.

An example of data matrix $\mathcal{M}^X = \langle M, f'_1, f'_2, f'_3, f'_4, \rangle$ with missing information of type $\mathcal{T} = \langle 2,3,6,9 \rangle$ is located in Fig. 12.1. We assume that $M = \{o_1,\ldots,o_n\}$.

Definition 12.2. Let $\mathcal{M}^X = \langle M, f'_1,\ldots,f'_K \rangle$ be a *data matrix with missing information of type* $\mathcal{T} = \langle t_1,\ldots,t_K \rangle$. Then a *completion of data matrix* $\mathcal{M}^X$ with missing information of type $\mathcal{T}$ is a data matrix $\mathcal{M} = \langle M, f_1,\ldots,f_K \rangle$ of type $\mathcal{T}$ satisfying

$$f_i(o_j) = f'_i(o_j) \quad \text{if} \quad f'_i(o_j) \neq X$$

for $i = 1,\ldots,K$ and $j = 1,\ldots,n$.

row	f_1	f_2	f_3	f_4
o_1	1	3	5	9
o_2	2	X	3	X
$\vdots$	$\vdots$	$\vdots$	$\vdots$	$\vdots$
o_{n-1}	X	1	X	9
o_n	1	2	6	X

Fig. 12.1 Data matrix $\mathscr{M}^X = \langle M, f'_1, f'_2, f'_3, f'_4, \rangle$ with missing information of type $\langle 2,3,6,9 \rangle$

An example of a completion $\mathscr{M} = \langle M, f_1, f_2, f_3, f_4, \rangle$ of data matrix $\mathscr{M}^X = \langle M, f'_1, f'_2, f'_3, f'_4, \rangle$ with missing information from Fig. 12.1 is located in Fig. 12.2.

row	f_1	f_2	f_3	f_4
o_1	1	3	5	9
o_2	2	X	3	X
$\vdots$	$\vdots$	$\vdots$	$\vdots$	$\vdots$
o_{n-1}	X	1	X	9
o_n	1	2	6	X

$$\mathscr{M}^X = \langle M, f'_1, f'_2, f'_3, f'_4, \rangle$$

row	f_1	f_2	f_3	f_4
o_1	1	3	5	9
o_2	2	3	3	8
$\vdots$	$\vdots$	$\vdots$	$\vdots$	$\vdots$
o_{n-1}	1	1	4	2
o_n	1	2	6	1

$$\mathscr{M} = \langle M, f_1, f_2, f_3, f_4, \rangle$$

Fig. 12.2 Completion $\mathscr{M} = \langle M, f_1, f_2, f_3, f_4, \rangle$ of data matrix $\mathscr{M}^X = \langle M, f'_1, f'_2, f'_3, f'_4, \rangle$

We defined the set $\mathrm{M}_{\mathscr{T}}$ of all data matrices $\mathscr{M} = \langle M, f_1, \ldots, f_K \rangle$ of type $\mathscr{T}$ in definition 3.9. We define an analogous set of all data matrices with missing values.

Definition 12.3. Let $\mathscr{T} = \langle t_1, \ldots, t_K \rangle$ be the type and let $X \notin \{1, \ldots, t_i\}$ for $i = 1, \ldots, K$. Then $\mathrm{M}_{\mathscr{T}}^X$ is a set of all data matrices $\mathscr{M}^X = \langle M, f'_1, \ldots, f'_K \rangle$ of type $\mathscr{T}$ with missing information where M is a non-empty finite set and f'_i is an unary function from M to $\{1, \ldots, t_i\} \cup \{X\}$ for $i = 1, \ldots, K$.

There is a simple but useful lemma.

Lemma 12.1. *Let $\mathrm{M}_{\mathscr{T}}^X$ be a set of all data matrices $\mathscr{M}^X = \langle M, f'_1, \ldots, f'_K \rangle$ of type $\mathscr{T}$ with missing information and let $\mathscr{M}^X \in \mathrm{M}_{\mathscr{T}}^X$. If $\mathscr{M} = \langle M, f_1, \ldots, f_K \rangle$ is a completion of $\mathscr{M}^X$ then $\mathscr{M} \in \mathrm{M}_{\mathscr{T}}$ where $\mathrm{M}_{\mathscr{T}}$ is a set of all data matrices $\mathscr{M} = \langle M, f_1, \ldots, f_K \rangle$ of type $\mathscr{T}$ introduced in definition 3.9.*

Proof. The lemma is a direct consequence of definition 12.2. □

12.2 Secured Extension of Logical Calculus of Association Rules

Please remember that logical calculus $\mathscr{LC}_{\mathscr{T}}$ of association rules is defined in Chap. 3 as a semantic system $\langle \mathrm{Sent}, \mathrm{M}, V, Val \rangle$ which consists of

- a non-empty set $\mathtt{Sent}$ of *sentences*
- a non-empty set $\mathtt{M}$ of models
- a non-empty set V of *abstract values*
- an evaluating function $Val : \mathtt{Sent} \times \mathtt{M} \to V$.

Sentences are association rules $\varphi \approx \psi$, set of models is a set $\mathtt{M}_{\mathscr{T}}$ of all data matrices $\mathscr{M} = \langle M, f_1, \ldots, f_K \rangle$ of type $\mathscr{T} = \langle t_1, \ldots, t_K \rangle$. The set V of abstract values is the set $\{0, 1\}$ where 1 means *true* and 0 means *false*. Value $Val(\varphi \approx \psi, \mathscr{M})$ of rule $\varphi \approx \psi$ in data matrix $\mathscr{M} \in \mathtt{M}_{\mathscr{T}}$ is defined as $Val(\varphi \approx \psi, \mathscr{M}) = F_{\approx}(a, b, c, d)$ where $\langle a, b, c, d \rangle$ is 4ft-table $4ft(\varphi, \psi, \mathscr{M})$ of Boolean attributes φ and ψ in data matrix $\mathscr{M}$ and $F_{\approx}$ is an associated function of 4ft-quantifiers $\approx$, see definition 3.9.

Our goal is to define a logical calculus $\mathscr{LC}_{\mathscr{T}}^{X}$ of association rules of type $\mathscr{T} = \langle t_1, \ldots, t_K \rangle$ with missing information as a secured extension of a logical calculus $\mathscr{LC}_{\mathscr{T}}$ of association rules of type $\mathscr{T} = \langle t_1, \ldots, t_K \rangle$ in the sense of definition 3.3.3 in [18]. This means that the following principles $(SE1), \ldots, (SE5)$ are applied:

$(SE1)$ The set of abstract values of $\mathscr{LC}_{\mathscr{T}}^{X}$ is the set $\{0, 1, X\}$ where 0 means *false*, 1 means *true* and X means *missing information*.

$(SE2)$ The set of models of $\mathscr{LC}_{\mathscr{T}}^{X}$ is the set $\mathtt{M}_{\mathscr{T}}^{X}$ of all data matrices $\mathscr{M}^{X} = \langle M, f_1', \ldots, f_K' \rangle$ of type $\mathscr{T}$ with missing information, see definition 12.3.

$(SE3)$ $\mathscr{LC}_{\mathscr{T}}^{X}$ and $\mathscr{LC}_{\mathscr{T}}$ have the same formulas i.e. Boolean attributes and association rules. This means that a language of $\mathscr{LC}_{\mathscr{T}}^{X}$ is the language $\mathscr{L}_{\mathscr{T}}$ of $\mathscr{LC}_{\mathscr{T}}$, see definition 12.10 in Sect. 12.5.

$(SE4)$ Interpretation of Boolean attributes in $\mathscr{LC}_{\mathscr{T}}^{X}$ is a secured extension of interpretation of Boolean attributes in $\mathscr{LC}_{\mathscr{T}}$. This means that if $\mathscr{M}^{X} = \langle M, f_1', \ldots, f_K' \rangle$ is a model of $\mathscr{LC}_{\mathscr{T}}^{X}$ and φ is a Boolean attribute then a value of φ for object o in $\mathscr{M}^{X}$ is:

- 1 if value of φ for object o is 1 in each completion $\mathscr{M}$ of $\mathscr{M}^{X}$
- 0 if value of φ for object o is 0 in each completion $\mathscr{M}$ of $\mathscr{M}^{X}$
- X otherwise,

see Sect. 12.3 for details.

$(SE5)$ Associated functions $F_{\approx_1}^{X}, \ldots, F_{\approx_Q}^{X}$ of 4ft-quantifiers $\approx_1, \ldots \approx_Q$ of language $\mathscr{L}_{\mathscr{T}}$ of $\mathscr{LC}_{\mathscr{T}}^{X}$ are secured extensions of associated functions $F_{\approx_1}, \ldots, F_{\approx_Q}$ of 4ft quantifiers $\approx_1, \ldots \approx_Q$ of language $\mathscr{L}_{\mathscr{T}}$ of $\mathscr{LC}_{\mathscr{T}}$. This means that associated functions $F_{\approx_1}^{X}, \ldots, F_{\approx_Q}^{X}$ are defined using nine-fold tables introduced in Sect. 12.4.

The definition of logical calculus $\mathscr{LC}_{\mathscr{T}}^{X}$ of association rules of type $\mathscr{T} = \langle t_1, \ldots, t_K \rangle$ with missing information is given in Sect. 12.5.

12.3   Boolean Attributes and Secured Extension

Boolean attributes of a language of association rules with missing information are evaluated in data matrices with missing information. Their values are from the set $\{0, 1, X\}$ where 0 means *false*, 1 means *true*, and X means *missing information*.

A formal definition of a value of a Boolean attribute for a given row of a data matrix with missing information is contained in definitions 12.4 – 12.6 similarly to the formal definition of a value of a Boolean attribute in Sect. 3.2. However, rules of evaluation are given by the principle of secured extension. In definitions 12.4 – 12.6, we use a language $\mathscr{L}_{\mathscr{T}}$ of association rules defined in Sect. 3.1 because the language of $\mathscr{LC}_{\mathscr{T}}^{X}$ is the language $\mathscr{L}_{\mathscr{T}}$ of $\mathscr{LC}_{\mathscr{T}}$, see principle $(SE3)$.

Definition 12.4. Let $\mathscr{L}_{\mathscr{T}}$ be a language of association rules of type $\mathscr{T} = \langle t_1, \ldots, t_K \rangle$ and let $\mathscr{M}^X = \langle M, f_1', \ldots, f_K' \rangle$ be a data matrix with missing information of type $\mathscr{T}$. Then we define:

1. The *interpretation* $\mathfrak{I}(A_i)$ *of attribute* A_i in $\mathscr{M}^X$ is the function f_i' for $i = 1, \ldots, K$.
2. If $A_i(\diamond(v_1), \ldots, \diamond(v_k))$ is a basic Boolean attribute of $\mathscr{L}_{\mathscr{T}}$ then the *interpretation* $\mathfrak{I}(\diamond(v_1), \ldots, \diamond(v_k))$ *of the coefficient* $\diamond(v_1), \ldots, \diamond(v_k)$ in $\mathscr{M}^X$ is the set $\{v_1, \ldots, v_k\}$.

Definition 12.5. Let $\mathscr{L}_{\mathscr{T}}$ be a language of association rules of type $\mathscr{T} = \langle t_1, \ldots, t_K \rangle$ with basic symbols according to definition 3.1. Let $\mathscr{M}^X = \langle M, f_1', \ldots, f_K' \rangle$ be a data matrix with missing information of the type $\mathscr{T}$. Then we define:

1. The *interpretation of a basic Boolean attribute* $A_i(\alpha)$ *in* $\mathscr{M}^X$ is a $\{0, 1, X\}$-valued function $f'_{A_i(\alpha)}$ defined for each $o \in M$ such that

$$f'_{A_i(\alpha)}(o) = \begin{cases} 1 & \text{if } f_i'(o) \in \mathfrak{I}(\alpha) \\ X & \text{if } f_i'(o) = X \\ 0 & \text{otherwise,} \end{cases}$$

where $\mathfrak{I}(\alpha)$ is the interpretation of the coefficient α in $\mathscr{M}^X$.
2. Let φ be a Boolean attribute of $\mathscr{L}_{\mathscr{T}}$ and let f_φ' be the interpretation of φ in $\mathscr{M}^X$. Then the *interpretation of Boolean attribute* $\neg\varphi$ *in* $\mathscr{M}^X$ is the function $f'_{\neg\varphi}$ defined for each $o \in M$ according to Table 12.1.

Table 12.1 Definition of function $f'_{\neg\varphi}$

$f_\varphi'(o)$	1	X	0
$f'_{\neg\varphi}(o)$	0	X	1

3. Let φ and ψ be Boolean attributes of $\mathscr{L}_{\mathscr{T}}$. In addition, let f_φ' be the interpretation of φ in $\mathscr{M}^X$ and let f_ψ' be the interpretation of ψ in $\mathscr{M}^X$. Then the *interpretation of Boolean attribute* $\varphi \wedge \psi$ *in* $\mathscr{M}^X$ is the function $f'_{\varphi \wedge \psi}$ defined for each $o \in M$ according to Table 12.2.
4. Let φ and ψ be Boolean attributes of $\mathscr{L}_{\mathscr{T}}$. In addition, let f_φ' be the interpretation of φ in $\mathscr{M}^X$ and let f_ψ' be the interpretation of ψ in $\mathscr{M}^X$. Then the *interpretation of Boolean attribute* $\varphi \vee \psi$ *in* $\mathscr{M}^X$ is the function $f'_{\varphi \vee \psi}$ defined for each $o \in M$ according to Table 12.2.

Table 12.2 Definition of functions $f'_{\varphi \wedge \psi}$ and $f'_{\varphi \vee \psi}$

$f'_{\varphi}(o)$	$f'_{\psi}(o)$	$f'_{\varphi \wedge \psi}(o)$	$f'_{\varphi \vee \psi}(o)$
1	1	1	1
1	X	X	1
1	0	0	1
X	1	X	1
X	X	X	X
X	0	0	X
0	1	0	1
0	X	0	X
0	0	0	0

In addition we define:

Definition 12.6. Let $\mathscr{L}_{\mathscr{T}}$ be a language of association rules of type $\mathscr{T} = \langle t_1, \ldots, t_K \rangle$. Let $\mathscr{M}^X = \langle M, f'_1, \ldots, f'_K \rangle$ be a data matrix with missing information of type $\mathscr{T}$. Let φ be a Boolean attributes of $\mathscr{L}_{\mathscr{T}}$ and let f'_{φ} be the interpretation of φ in $\mathscr{M}^X$. Then:

- If $o \in M$ and $f'_{\varphi}(o) = 1$ then we say that *o satisfies φ in $\mathscr{M}^X$* or that *φ is true for o in $\mathscr{M}^X$* or that a *value of φ for o in $\mathscr{M}^X$ is 1.*
- If $o \in M$ and $f'_{\varphi}(o) = 0$ then we say that *o does not satisfy φ in $\mathscr{M}^X$* or that *φ is false for o in $\mathscr{M}^X$* or that a *value of φ for o in $\mathscr{M}^X$ is 0.*
- If $o \in M$ and $f'_{\varphi}(o) = X$ then we say that *the value of φ is not known for o in $\mathscr{M}^X$.*

12.4 Nine-fold Table

Value $Val(\varphi \approx \psi, \mathscr{M})$ of association rule $\varphi \approx \psi$ of $\mathscr{L}_{\mathscr{T}}$ in data matrix $\mathscr{M} \in \mathrm{M}_{\mathscr{T}}$ without missing information is defined such that $Val(\varphi \approx \psi, \mathscr{M}) = F_{\approx}(a, b, c, d)$ where $\langle a, b, c, d \rangle = 4ft(\varphi, \psi, \mathscr{M})$ is the 4ft-table of φ and ψ in data matrix $\mathscr{M}$, see definition 3.9. The 4ft-table of φ and ψ in data matrix $\mathscr{M}$ without missing information fully describes a relation of φ and ψ in $\mathscr{M}$.

Similarly, a relation of two Boolean attributes φ and ψ in a given data matrix $\mathscr{M}$ with missing information is fully described by a 9ft-table $9ft(\varphi, \psi, \mathscr{M}^X)$, see the following definition. The 9ft-table $9ft(\varphi, \psi, \mathscr{M}^X)$ is then used to define the value of association rule $\varphi \approx \psi$ in data matrix $\mathscr{M}^X \in \mathrm{M}^X_{\mathscr{T}}$ with missing information, see Sect. 12.5.

Definition 12.7. Let $\mathscr{L}_{\mathscr{T}}$ be a language of association rules of type $\mathscr{T} = \langle t_1, \ldots, t_K \rangle$ and let $\mathscr{M}^X = \langle M, f'_1, \ldots, f'_K \rangle$ be a data matrix with missing information of type $\mathscr{T}$ and φ and ψ be Boolean attributes of $\mathscr{L}_{\mathscr{T}}$. In addition, let f'_{φ} be an interpretation of φ in $\mathscr{M}^X$ and let f'_{ψ} be an interpretation of ψ in $\mathscr{M}^X$.

Then the *nine-fold table* (i.e. *9ft-table*) $9ft(\varphi, \psi, \mathcal{M}^X)$ of φ and ψ in $\mathcal{M}^X$ is the nine-tuple of integer non-negative numbers

$$9ft(\varphi, \psi, \mathcal{M}^X) = \langle f_{1,1}, f_{1,X}, f_{1,0}, f_{X,1}, f_{X,X}, f_{X,0}, f_{0,1}, f_{0,X}, f_{0,0} \rangle$$

where

- $f_{1,1}$ is the number of rows $o \in M$ satisfying both $f'_\varphi(o) = 1$ and $f'_\psi(o) = 1$
- $f_{1,X}$ is the number of rows $o \in M$ satisfying both $f'_\varphi(o) = 1$ and $f'_\psi(o) = X$
- $f_{1,0}$ is the number of rows $o \in M$ satisfying both $f'_\varphi(o) = 1$ and $f'_\psi(o) = 0$
- $f_{X,1}$ is the number of rows $o \in M$ satisfying both $f'_\varphi(o) = X$ and $f'_\psi(o) = 1$
- $f_{X,X}$ is the number of rows $o \in M$ satisfying both $f'_\varphi(o) = X$ and $f'_\psi(o) = X$
- $f_{X,0}$ is the number of rows $o \in M$ satisfying both $f'_\varphi(o) = X$ and $f'_\psi(o) = 0$
- $f_{0,1}$ is the number of rows $o \in M$ satisfying both $f'_\varphi(o) = 0$ and $f'_\psi(o) = 1$
- $f_{0,X}$ is the number of rows $o \in M$ satisfying both $f'_\varphi(o) = 0$ and $f'_\psi(o) = X$
- $f_{0,0}$ is the number of rows $o \in M$ satisfying both $f'_\varphi(o) = 0$ and $f'_\psi(o) = 0$.

The 9ft-table $9ft(\varphi, \psi, \mathcal{M}^X)$ of φ and ψ in $\mathcal{M}^X$ can also be written in the form according to Table 12.3.

Table 12.3 9ft-table $9ft(\varphi, \psi, \mathcal{M}^X)$ of φ and ψ in $\mathcal{M}^X$

$\mathcal{M}^X$	ψ	ψ_X	$\neg\psi$
φ	$f_{1,1}$	$f_{1,X}$	$f_{1,0}$
φ_X	$f_{X,1}$	$f_{X,X}$	$f_{X,0}$
$\neg\varphi$	$f_{0,1}$	$f_{0,X}$	$f_{0,0}$

We prove a simple but useful relation between 9ft-table $9ft(\varphi, \psi, \mathcal{M}^X)$ of φ and ψ in $\mathcal{M}^X$ and 4ft-table $4ft(\varphi, \psi, \mathcal{M})$ of φ and ψ in $\mathcal{M}$ where $\mathcal{M}$ is a completion of $\mathcal{M}^X$, see lemma 12.3. Before that we prove another simple lemma.

Lemma 12.2. *Let $\mathscr{L}_\mathscr{T}$ be a language of association rules of type $\mathscr{T} = \langle t_1, \ldots, t_K \rangle$, let $\mathcal{M}^X = \langle M, f'_1, \ldots, f'_K \rangle$ be a data matrix with missing information of the type $\mathscr{T}$ and let φ be a Boolean attribute of $\mathscr{L}_\mathscr{T}$. In addition, let $o \in M$.*

If $f'_\varphi(o) = X$ where f'_φ is an interpretation of φ in $\mathcal{M}^X$ then there are completions $\mathcal{M}_1 = \langle M, f_1^{(1)}, \ldots, f_K^{(1)} \rangle$ and $\mathcal{M}_2 = \langle M, f_1^{(2)}, \ldots, f_K^{(2)} \rangle$ of $\mathcal{M}^X$ such that $f_\varphi^{(1)}(o) = 1$ where $f_\varphi^{(1)}$ is an interpretation of φ in $\mathcal{M}_1$ and $f_\varphi^{(2)}(o) = 0$ where $f_\varphi^{(2)}$ is an interpretation of φ in $\mathcal{M}_2$

Proof. We prove the lemma by induction according to the number $NCon(\varphi)$ of logical connectives $\neg, \wedge, \vee$ in Boolean attribute φ. We start with $NCon(\varphi) = 0$ which means that φ is a basic Boolean attribute $A_i(\alpha)$.

If φ is a basic Boolean attribute $A_i(\alpha)$ and $f'_{A_i(\alpha)}(o) = X$ then it must hold that $f'_i(o) = X$, see definition 12.5. Let us define $\mathcal{M}_1 = \langle M, f_1^{(1)}, \ldots, f_K^{(1)} \rangle$ such that $f_i^{(1)}(o) \in \alpha$ and $\mathcal{M}_2 = \langle M, f_1^{(2)}, \ldots, f_K^{(2)} \rangle$ such that $f_i^{(2)}(o) \notin \alpha$. This means that $f_{A_i(\alpha)}^{(1)}(o) = 1$ and $f_{A_i(\alpha)}^{(2)}(o) = 0$.

Let us assume that the lemma is true for all Boolean attributes φ such that $NCon(\varphi) \leq i$. We prove that it is true also for all Boolean attributes χ such that $NCon(\chi) = i+1$.

There are three possibilities for Boolean attribute χ satisfying $NCon(\chi) = i+1$: (i): $\chi = \neg\varphi$, (ii): $\chi = \varphi \wedge \psi$, and (iii): $\chi = \varphi \wedge \psi$, see definition 3.2. Anyway, we know that $NCon(\varphi) \leq i$ and $NCon(\psi) \leq i$ and thus the lemma is true both for φ and ψ.

(i)   *If $\chi = \neg\varphi$ and $f'_\chi(o) = X$ then it also hold $f'_\varphi(o) = X$, see Table 12.1 in definition 12.5. We know that there are completions $\mathcal{M}_1 = \langle M, f_1^{(1)}, \ldots, f_K^{(1)} \rangle$ and $\mathcal{M}_2 = \langle M, f_1^{(2)}, \ldots, f_K^{(2)} \rangle$ of $\mathcal{M}^X$ such that $f_\varphi^{(1)}(o) = 1$ where $f_\varphi^{(1)}$ is an interpretation of φ in $\mathcal{M}_1$ and $f_\varphi^{(2)}(o) = 0$ where $f_\varphi^{(2)}$ is an interpretation of φ in $\mathcal{M}_2$. This implies $f_{\neg\varphi}^{(1)}(o) = 0$ and $f_{\neg\varphi}^{(2)}(o) = 1$ and this finishes the proof of point (i).*

(ii)   *If $\chi = \varphi \wedge \psi$ and $f'_\chi(o) = X$ then one of the combinations 1, 2, 3 of values of $f'_\varphi(o)$ and $f'_\psi(o)$ introduced in Table 12.4 must happen, where f'_φ is an interpretation of φ in $\mathcal{M}^X$, f'_ψ is an interpretation of ψ in $\mathcal{M}^X$ and Table 12.4 is derived from Table 12.2 in definition 12.5.*

Combination 1 means $f'_\varphi(o) = 1$ and $f'_\psi(o) = X$. We know that $NCon(\varphi) \leq i$ and $NCon(\psi) \leq i$ and thus there are completions $\mathcal{M}_1 = \langle M, f_1^{(1)}, \ldots, f_K^{(1)} \rangle$ and $\mathcal{M}_2 = \langle M, f_1^{(2)}, \ldots, f_K^{(2)} \rangle$ of $\mathcal{M}^X$ such that $f_\varphi^{(1)}(o) = 1$, $f_\psi^{(1)}(o) = 1$, $f_\varphi^{(2)}(o) = 1$, and $f_\psi^{(2)}(o) = 0$. This however means $f_{\varphi \wedge \psi}^{(1)}(o) = 1$ and $f_{\varphi \wedge \psi}^{(2)}(o) = 0$, see row 1 in Table 12.4. This finishes the proof for combination 1. Proofs for combinations 2 and 3 are analogous, see Table 12.4. This finishes the proof for point (ii).

Table 12.4 Completions $\mathcal{M}_1$ and $\mathcal{M}_2$ of $\mathcal{M}^X$ satisfying $f_{\varphi \wedge \psi}^{(1)}(o) = 1$ and $f_{\varphi \wedge \psi}^{(2)}(o) = 0$

No.	$f'_\varphi(o)$	$f'_\psi(o)$	$f'_{\varphi \wedge \psi}(o)$	$f_\varphi^{(1)}(o)$	$f_\psi^{(1)}(o)$	$f_{\varphi \wedge \psi}^{(1)}(o)$	$f_\varphi^{(2)}(o)$	$f_\psi^{(2)}(o)$	$f_{\varphi \wedge \psi}^{(2)}(o)$
1	1	X	X	1	1	1	1	0	0
2	X	1	X	1	1	1	0	1	0
3	X	X	X	1	1	1	0	0	0

(iii)   *The proof of point (iii) is similar to that of point (ii), see Table 12.5.*

Table 12.5 Completions $\mathcal{M}_1$ and $\mathcal{M}_2$ of $\mathcal{M}^X$ satisfying $f_{\varphi \vee \psi}^{(1)}(o) = 1$ and $f_{\varphi \vee \psi}^{(2)}(o) = 0$

No.	$f'_\varphi(o)$	$f'_\psi(o)$	$f'_{\varphi \vee \psi}(o)$	$f_\varphi^{(1)}(o)$	$f_\psi^{(1)}(o)$	$f_{\varphi \vee \psi}^{(1)}(o)$	$f_\varphi^{(2)}(o)$	$f_\psi^{(2)}(o)$	$f_{\varphi \vee \psi}^{(2)}(o)$
1	X	X	X	1	1	1	0	0	0
2	X	0	X	1	0	1	0	0	0
3	0	X	X	0	1	1	0	0	0

This finishes the proof.   □

Lemma 12.3. *Let* $\langle f_{1,1}, f_{1,X}, f_{1,0}, f_{X,1}, f_{X,X}, f_{X,0}, f_{0,1}, f_{0,X}, f_{0,0}, \rangle$ *be a 9ft-table* $9ft(\varphi, \psi, \mathcal{M}^X)$ *of* φ *and* ψ *in a data matrix* $\mathcal{M}^X$ *with missing information and let* $\langle a, b, c, d \rangle$ *be a 4ft-table* $4ft(\varphi, \psi, \mathcal{M})$ *of* φ *and* ψ *in a data matrix* $\mathcal{M}$ *which is a completion of* $\mathcal{M}^X$. *Then it holds that*

- $a = f_{1,1} + f_{1,X,a} + f_{X,1,a} + f_{X,X,a}$
- $b = f_{1,0} + f_{1,X,b} + f_{X,0,b} + f_{X,X,b}$
- $c = f_{0,1} + f_{0,X,c} + f_{X,1,c} + f_{X,X,c}$
- $d = f_{0,0} + f_{0,X,d} + f_{X,0,d} + f_{X,X,d}$

where $f_{1,X,a}$, $f_{X,1,a}$, $f_{X,X,a}$, $f_{1,X,b}$, $f_{X,0,b}$, $f_{X,X,b}$, $f_{0,X,c}$, $f_{X,1,c}$, $f_{X,X,c}$, $f_{0,X,d}$, $f_{X,0,d}$, *and* $f_{X,X,d}$ *are non-negative integer numbers satisfying*

- $f_{1,X,a} + f_{1,X,b} = f_{1,X}$
- $f_{X,1,a} + f_{X,1,c} = f_{X,1}$
- $f_{X,X,a} + f_{X,X,b} + f_{X,X,c} + f_{X,X,d} = f_{X,X}$
- $f_{X,0,b} + f_{X,0,d} = f_{X,0}$
- $f_{0,X,c} + f_{0,X,d} = f_{0,X}$,

see also $9ft(\varphi, \psi, \mathcal{M}^X)$ *and* $4ft(\varphi, \psi, \mathcal{M})$ *in Fig. 12.3.*

$\mathcal{M}^X$	ψ	ψ_X	$\neg\psi$
φ	$f_{1,1}$	$f_{1,X}$	$f_{1,0}$
φ_X	$f_{X,1}$	$f_{X,X}$	$f_{X,0}$
$\neg\varphi$	$f_{0,1}$	$f_{0,X}$	$f_{0,0}$

$$9ft(\varphi, \psi, \mathcal{M}^X)$$

$\mathcal{M}$	ψ	$\neg\psi$
φ	$f_{1,1} + f_{1,X,a} + f_{X,1,a} + f_{X,X,a}$	$f_{1,0} + f_{1,X,b} + f_{X,0,b} + f_{X,X,b}$
$\neg\varphi$	$f_{0,1} + f_{0,X,c} + f_{X,1,c} + f_{X,X,c}$	$f_{0,0} + f_{0,X,d} + f_{X,0,d} + f_{X,X,d}$

$$4ft(\varphi, \psi, \mathcal{M})$$

Fig. 12.3 $9ft(\varphi, \psi, \mathcal{M}^X)$ and $4ft(\varphi, \psi, \mathcal{M})$

Proof. *A direct consequence of definition 12.5 is that for each* $o \in M$ *we have: if* $f'_\varphi(o) = 1$ *then* $f_\varphi(o) = 1$, *if* $f'_\varphi(o) = 0$ *then* $f_\varphi(o) = 0$, *and however, if* $f'_\varphi(o) = X$ *then it can be both* $f_\varphi(o) = 1$ *and* $f_\varphi(o) = 0$, *see also lemma 12.2. Here* f'_φ *is an interpretation of* φ *in* $\mathcal{M}^X$ *and* f_φ *is an interpretation of* φ *in* $\mathcal{M}$ *and similarly for* f_ψ *and* f'_ψ. *Let us denote the numbers of rows* $o \in M$ *satisfying particular combinations of values of* f'_φ, f'_ψ, f_φ, *and* f_ψ *according to Table 12.6.*

Table 12.6 The numbers of $o \in M$ for particular combinations of values of f'_φ, f'_ψ, f_φ, and f_ψ

No.	combination of values of				number of $o \in M$ satisfying combination
	f'_φ	f'_ψ	f_φ	f_ψ	
1	1	1	1	1	$f_{1,1}$
2	1	X	1	1	$f_{1,X,a}$
3	1	X	1	0	$f_{1,X,b}$
4	1	0	1	0	$f_{1,0}$
5	X	1	1	1	$f_{X,1,a}$
6	X	1	0	1	$f_{X,1,c}$
7	X	X	1	1	$f_{X,X,a}$
8	X	X	1	0	$f_{X,X,b}$
9	X	X	0	1	$f_{X,X,c}$
10	X	X	0	0	$f_{X,X,d}$
11	X	0	1	0	$f_{X,0,b}$
12	X	0	0	0	$f_{X,0,d}$
13	0	1	0	1	$f_{0,1}$
14	0	X	0	1	$f_{0,X,c}$
15	0	X	0	0	$f_{0,X,d}$
16	0	0	0	0	$f_{0,0}$

Table 12.6 shows that the numbers of rows $o \in M$ for particular combinations of values of f'_φ, f'_ψ, f_φ, and f_ψ satisfy the assertion of this lemma. This finishes the proof.   □

We will also need an additional lemma 12.4 which "inverts" the just proved lemma 12.3.

Lemma 12.4. *Let $\langle f_{1,1}, f_{1,X}, f_{1,0}, f_{X,1}, f_{X,X}, f_{X,0}, f_{0,1}, f_{0,X}, f_{0,0}, \rangle$ be a 9ft-table $9ft(\varphi, \psi, \mathcal{M}^X)$ of φ and ψ in a data matrix $\mathcal{M}^X = \langle M, f'_1, \ldots, f'_K \rangle$ with missing information and let $\langle a, b, c, d \rangle$ be a quadruple of non-negative numbers satisfying*

- $a = f_{1,1} + f_{1,X,a} + f_{X,1,a} + f_{X,X,a}$
- $b = f_{1,0} + f_{1,X,b} + f_{X,0,b} + f_{X,X,b}$
- $c = f_{0,1} + f_{0,X,c} + f_{X,1,c} + f_{X,X,c}$
- $d = f_{0,0} + f_{0,X,d} + f_{X,0,d} + f_{X,X,d}$

such that

- $f_{1,X,a} + f_{1,X,b} = f_{1,X}$
- $f_{X,1,a} + f_{X,1,c} = f_{X,1}$
- $f_{X,X,a} + f_{X,X,b} + f_{X,X,c} + f_{X,X,d} = f_{X,X}$
- $f_{X,0,b} + f_{X,0,d} = f_{X,0}$
- $f_{0,X,c} + f_{0,X,d} = f_{0,X}.$

Then there is a completion $\mathcal{M}$ of $\mathcal{M}^X$ such that $4ft(\varphi, \psi, \mathcal{M}) = \langle a, b, c, d \rangle$

Proof. According to lemma 12.2 there is a completion $\mathscr{M} = \langle M, f_1, \ldots, f_K \rangle$ of $\mathscr{M}^X$ such that the numbers of rows $o \in M$ satisfying particular combinations of values of f'_φ, f'_ψ, f_φ, and f_ψ correspond to Table 12.6. The completion $\mathscr{M} = \langle M, f_1, \ldots, f_K \rangle$ satisfies the conditions of this lemma. This finishes the proof. □

12.5 Logical Calculus of Association Rules with Missing Information

We define logical calculus of association rules with missing information according to principles $(SE1), \ldots, (SE5)$ introduced in Sect. 12.2. It remains to define associated functions of 4ft-quantifiers in calculus with missing information as secured extensions of associated functions of 4ft-quantifiers.

The value of an association rule $\varphi \approx \psi$ in data matrix $\mathscr{M}$ without missing information is defined as $F_\approx(a, b, c, d)$ where $F_\approx$ is an associated function of $\approx$ and $\langle a, b, c, d \rangle = 4ft(\varphi, \psi, \mathscr{M})$ is a 4ft-table of φ and ψ in data matrix $\mathscr{M}$. This definition is based on the fact that $4ft(\varphi, \psi, \mathscr{M})$ fully describes the relation of φ and ψ in data matrix $\mathscr{M}$.

The relation of Boolean attributes φ and ψ in data matrix $\mathscr{M}^X$ with missing information is fully described by the 9ft-table $9ft(\varphi, \psi, \mathscr{M}^X)$ of φ and ψ in data matrix $\mathscr{M}^X$ with missing information introduced by definition 12.7. This leads to the definition of associated function $F_\approx^X$ of 4ft-quantifier $\approx$ of a calculus of association rules with missing information based on completions of nine-fold tables. This is done in definitions 12.8 and 12.9.

Definition 12.8. Let $\langle f_{1,1}, f_{1,X}, f_{1,0}, f_{X,1}, f_{X,X}, f_{X,0}, f_{0,1}, f_{0,X}, f_{0,0} \rangle$ be a nine-fold table $9ft(\varphi, \psi, \mathscr{M}^X)$ of φ and ψ in data matrix $\mathscr{M}^X$ with missing information. Then a *completion* of $9ft(\varphi, \psi, \mathscr{M}^X)$ is each quadruple $\langle a, b, c, d \rangle$ satisfying the conditions

- $a = f_{1,1} + f_{1,X,a} + f_{X,1,a} + f_{X,X,a}$
- $b = f_{1,0} + f_{1,X,b} + f_{X,0,b} + f_{X,X,b}$
- $c = f_{0,1} + f_{0,X,c} + f_{X,1,c} + f_{X,X,c}$
- $d = f_{0,0} + f_{0,X,d} + f_{X,0,d} + f_{X,X,d}$

where $f_{1,X,a}$, $f_{X,1,a}$, $f_{X,X,a}$, $f_{1,X,b}$, $f_{X,0,b}$, $f_{X,X,b}$, $f_{0,X,c}$, $f_{X,1,c}$, $f_{X,X,c}$, $f_{0,X,d}$, $f_{X,0,d}$, and $f_{X,X,d}$ are non-negative integer numbers satisfying

- $f_{1,X,a} + f_{1,X,b} = f_{1,X}$
- $f_{X,1,a} + f_{X,1,c} = f_{X,1}$
- $f_{X,X,a} + f_{X,X,b} + f_{X,X,c} + f_{X,X,d} = f_{X,X}$
- $f_{X,0,b} + f_{X,0,d} = f_{X,0}$
- $f_{0,X,c} + f_{0,X,d} = f_{0,X}$,

see also $9ft(\varphi, \psi, \mathscr{M}^X)$ and $4ft(\varphi, \psi, \mathscr{M})$ in Fig. 12.3 and lemma 12.3.

Definition 12.9. Let $\approx$ be 4ft-quantifier with associated function $F_\approx$ defined according to definition 3.8. Then *secured extension of $F_\approx$* is a function $F_\approx^X$ defined for all nine tuples $\langle f_{1,1}, f_{1,X}, f_{1,0}, f_{X,1}, f_{X,X}, f_{X,0}, f_{0,1}, f_{0,X}, f_{0,0} \rangle$ of non-negative integer numbers satisfying

$$f_{1,1} + f_{1,X} + f_{1,0} + f_{X,1} + f_{X,X} + f_{X,0} + f_{0,1} + f_{0,X} + f_{0,0} > 0$$

such that

- $F_{\approx}^X(f_{1,1}, f_{1,X}, f_{1,0}, f_{X,1}, f_{X,X}, f_{X,0}, f_{0,1}, f_{0,X}, f_{0,0}) = 1$ if $F_{\approx}(a,b,c,d) = 1$ for each completion $\langle a,b,c,d \rangle$ of $\langle f_{1,1}, f_{1,X}, f_{1,0}, f_{X,1}, f_{X,X}, f_{X,0}, f_{0,1}, f_{0,X}, f_{0,0} \rangle$
- $F_{\approx}^X(f_{1,1}, f_{1,X}, f_{1,0}, f_{X,1}, f_{X,X}, f_{X,0}, f_{0,1}, f_{0,X}, f_{0,0}) = 0$ if $F_{\approx}(a,b,c,d) = 0$ for each completion $\langle a,b,c,d \rangle$ of $\langle f_{1,1}, f_{1,X}, f_{1,0}, f_{X,1}, f_{X,X}, f_{X,0}, f_{0,1}, f_{0,X}, f_{0,0} \rangle$
- $F_{\approx}^X(f_{1,1}, f_{1,X}, f_{1,0}, f_{X,1}, f_{X,X}, f_{X,0}, f_{0,1}, f_{0,X}, f_{0,0}) = X$ otherwise.

Now we define a secured extension of a logical calculus of association rules.

Definition 12.10. Let $\mathscr{LC}_{\mathscr{T}}$ be a logical calculus of association rules of type $\mathscr{T} = \langle t_1, \ldots, t_K \rangle$ according to definition 3.9. Then a *logical calculus* $\mathscr{LC}_{\mathscr{T}}^X$ *of association rules of type* $\mathscr{T} = \langle t_1, \ldots, t_K \rangle$ *with missing information which is a secured extension of* $\mathscr{LC}_{\mathscr{T}}$ is given by

1. the language $\mathscr{L}_{\mathscr{T}}$ of calculus $\mathscr{LC}_{\mathscr{T}}$
2. the set $\mathrm{M}_{\mathscr{T}}^X$ of all data matrices $\mathscr{M}^X = \langle M, f_1', \ldots, f_K' \rangle$ of type $\mathscr{T}$ with missing information according to definition 12.3.
3. the associated functions $F_{\approx_1}, \ldots, F_{\approx_Q}$ of 4ft-quantifiers $\approx_1, \ldots \approx_Q$ of language $\mathscr{L}_{\mathscr{T}}$ respectively

The evaluating function Val^X which assigns a value $Val^X(\varphi \approx \psi, \mathscr{M}^X)$ of an association rule $\varphi \approx \psi$ of $\mathscr{LC}_{\mathscr{T}}^X$ in data matrix $\mathscr{M}^X \in \mathrm{M}_{\mathscr{T}}^X$ is defined such that

$$Val^X(\varphi \approx \psi, \mathscr{M}^X) = F_{\approx}^X(f_{1,1}, f_{1,X}, f_{1,0}, f_{X,1}, f_{X,X}, f_{X,0}, f_{0,1}, f_{0,X}, f_{0,0})$$

where $\langle f_{1,1}, f_{1,X}, f_{1,0}, f_{X,1}, f_{X,X}, f_{X,0}, f_{0,1}, f_{0,X}, f_{0,0} \rangle$ is a nine-fold table $9ft(\varphi, \psi, \mathscr{M}^X)$ of φ and ψ in data matrix $\mathscr{M}^X$ and $F_{\approx}^X$ is a secured extension of $F_{\approx}$. In addition

- if $Val^X(\varphi \approx \psi, \mathscr{M}^X) = 1$ then we say that the *association rule* $\varphi \approx \psi$ *is true in data matrix* $\mathscr{M}^X$ *with missing information*
- if $Val^X(\varphi \approx \psi, \mathscr{M}^X) = 0$ then we say that the *association rule* $\varphi \approx \psi$ *is false in data matrix* $\mathscr{M}^X$ *with missing information*
- if $Val^X(\varphi \approx \psi, \mathscr{M}^X) = X$ then we say that *value of the association rule* $\varphi \approx \psi$ *in data matrix* $\mathscr{M}^X$ *with missing information is not known.*

We prove a simple theorem about the relation of a logical calculus of association rules $\mathscr{LC}_{\mathscr{T}}$ and its secured completion $\mathscr{LC}_{\mathscr{T}}^X$.

Theorem 12.1. Let $\mathscr{LC}_{\mathscr{T}}$ be a logical calculus of association rules of type $\mathscr{T} = \langle t_1, \ldots, t_K \rangle$ and let $\mathscr{LC}_{\mathscr{T}}^X$ be a secured extension of $\mathscr{LC}_{\mathscr{T}}$. Let $\varphi \approx \psi$ be an association rule of $\mathscr{LC}_{\mathscr{T}}$ and let $\mathscr{M}^X \in \mathrm{M}_{\mathscr{T}}^X$. Then

1. $Val^X(\varphi \approx \psi, \mathscr{M}^X) = 1$ if and only if $Val(\varphi \approx \psi, \mathscr{M}) = 1$ for all completions $\mathscr{M}$ of $\mathscr{M}^X$
2. $Val^X(\varphi \approx \psi, \mathscr{M}^X) = 0$ if and only if $Val(\varphi \approx \psi, \mathscr{M}) = 0$ for all completions $\mathscr{M}$ of $\mathscr{M}^X$

Proof. We prove only point 1, a proof of point 2 is similar. We have to prove:

a) *If* $Val^X(\varphi \approx \psi, \mathcal{M}^X) = 1$ *and* $\mathcal{M}$ *is a completion of* $\mathcal{M}^X$ *then* $Val(\varphi \approx \psi, \mathcal{M}) = 1$.

b) *If* $Val^X(\varphi \approx \psi, \mathcal{M}^X) \neq 1$ *then there is a completion* $\mathcal{M}$ *of* $\mathcal{M}^X$ *such that* $Val(\varphi \approx \psi, \mathcal{M}) = 0$.

Proofs of points a) and b) follow.

a) *If* $Val^X(\varphi \approx \psi, \mathcal{M}^X) = 1$ *then*

$$F_{\approx}^X(f_{1,1}, f_{1,X}, f_{1,0}, f_{X,1}, f_{X,X}, f_{X,0}, f_{0,1}, f_{0,X}, f_{0,0}) = 1$$

where $\langle f_{1,1}, f_{1,X}, f_{1,0}, f_{X,1}, f_{X,X}, f_{X,0}, f_{0,1}, f_{0,X}, f_{0,0} \rangle$ *is a nine-fold table* $9ft(\varphi, \psi, \mathcal{M}^X)$ *of* φ *and* ψ *in data matrix* $\mathcal{M}^X$ *and* $F_{\approx}^X$ *is a secured extension of associated function* $F_{\approx}$ *of 4ft-quantifier* $\approx$. *This means that* $F_{\approx}(a,b,c,d) = 1$ *for each completion* $\langle a,b,c,d \rangle$ *of* $\langle f_{1,1}, f_{1,X}, f_{1,0}, f_{X,1}, f_{X,X}, f_{X,0}, f_{0,1}, f_{0,X}, f_{0,0} \rangle$.

If $\mathcal{M}$ *is a completion of* $\mathcal{M}^X$ *then, according to lemma 12.3,* $4ft(\varphi, \psi, \mathcal{M})$ *is a completion of* $9ft(\varphi, \psi, \mathcal{M}^X)$ *and thus it holds that* $F_{\approx}(a,b,c,d) = 1$ *for* $\langle a,b,c,d \rangle = 4ft(\varphi, \psi, \mathcal{M})$ *which means* $Val(\varphi \approx \psi, \mathcal{M}) = 1$.

b) *If* $Val^X(\varphi \approx \psi, \mathcal{M}^X) \neq 1$ *then there is a completion* $\langle a,b,c,d \rangle$ *of* $\langle f_{1,1}, f_{1,X}, f_{1,0}, f_{X,1}, f_{X,X}, f_{X,0}, f_{0,1}, f_{0,X}, f_{0,0} \rangle$ *such that* $F_{\approx}(a,b,c,d) = 0$. *According to lemma 12.4, there is a completion* $\mathcal{M}$ *of* $\mathcal{M}^X$ *for which we have* $4ft(\varphi, \psi, \mathcal{M}) = \langle a,b,c,d \rangle$. *This implies that* $Val(\varphi \approx \psi, \mathcal{M} = 0$.

This finishes the proof. □

12.6 Secured 4ft-tables

Evaluation of a secured extension $F_{\approx}^X$ of an associated function $F_{\approx}$ of a 4ft-quantifier $\approx$ in a given 9ft-table $\langle f_{1,1}, f_{1,X}, f_{1,0}, f_{X,1}, f_{X,X}, f_{X,0}, f_{0,1}, f_{0,X}, f_{0,0} \rangle$ requires evaluation of the function $F_{\approx}$ in all completions $\langle a,b,c,d, \rangle$ of the given 9ft-table, see definition 12.9. We show that for important 4ft-quantifiers it is possible to find for a given 9ft-table a 4ft-table $\langle a_s, b_s, c_s, d_s \rangle$ such that $F_{\approx}(a_s, b_s, c_s, d_s) = 1$ if and only if $F_{\approx}(a,b,c,d) = 1$ for each completion $\langle a,b,c,d \rangle$ of the given 9ft-table. We call such 4ft-table a secured 4ft-table, see the following definition.

Definition 12.11. Let $Tab9 = \langle f_{1,1}, f_{1,X}, f_{1,0}, f_{X,1}, f_{X,X}, f_{X,0}, f_{0,1}, f_{0,X}, f_{0,0} \rangle$ be a 9ft-table and let $\approx$ be a 4ft-quantifier. Then 4ft-table $\langle a_s, b_s, c_s, d_s \rangle$ is a *secured 4ft-table for 4ft-quantifier* $\approx$ *and the 9ft-table Tab9 if* $F_{\approx}(a_s, b_s, c_s, d_s) = 1$ if and only if $F_{\approx}(a,b,c,d) = 1$ for each completion $\langle a,b,c,d \rangle$ of $Tab9$. We say also that $\langle a_s, b_s, c_s, d_s \rangle$ is a *secured completion for 4ft-quantifier* $\approx$ *and the 9ft-table Tab9*.

First we define an F-completion and prove a theorem concerning 4ft-quantifiers with the F-property and F-completion.

Definition 12.12. Let $Tab9 = \langle f_{1,1}, f_{1,X}, f_{1,0}, f_{X,1}, f_{X,X}, f_{X,0}, f_{0,1}, f_{0,X}, f_{0,0} \rangle$ be a 9ft-table. Let us denote $b_0 = f_{1,0} + f_{1,X} + f_{X,0}$ and $c_0 = f_{0,1} + f_{0,X} + f_{X,1}$. Then 4ft-table $\langle a_F, b_F, c_F, d_F \rangle$ is an *F-completion* of *Tab9* if it satisfies

$$\langle a_F, b_F, c_F, d_F \rangle = \langle f_{1,1}, b_0 + F_b, c_0 + F_c, f_{0,0} \rangle$$

where $b_0 + F_b$ and $c_0 + F_c$ are given by Table 12.7.

Table 12.7 Frequencies b_F and c_F in an *F-completion* of $9ft(\varphi, \psi, \mathscr{M}^X)$

No.	Condition	$b_0 + F_b$	$c_0 + F_c$
1	$b_0 \geq c_0 + f_{X,X}$ i.e. $\|b_0 - c_0\| \geq f_{X,X}$	b_0	$c_0 + f_{X,X}$
2	$c_0 \geq b_0 + f_{X,X}$ i.e. $\|b_0 - c_0\| \geq f_{X,X}$	$b_0 + f_{X,X}$	c_0
3	$\|b_0 - c_0\| < f_{X,X}$ and $b_0 + c_0 + f_{X,X}$ is even	$\frac{b_0 + c_0 + f_{X,X}}{2}$	$\frac{b_0 + c_0 + f_{X,X}}{2}$
4	$\|b_0 - c_0\| < f_{X,X}$ and $b_0 + c_0 + f_{X,X}$ is odd	$\frac{b_0 + c_0 + f_{X,X} + 1}{2}$	$\frac{b_0 + c_0 + f_{X,X} - 1}{2}$

Theorem 12.2. Let $Tab9 = \langle f_{1,1}, f_{1,X}, f_{1,0}, f_{X,1}, f_{X,X}, f_{X,0}, f_{0,1}, f_{0,X}, f_{0,0} \rangle$ be a 9ft-table, let $\approx$ be a 4ft-quantifier with the F-property and let $\langle a_F, b_F, c_F, d_F \rangle$ be an F-completion of Tab9. In addition, let b_0, F_b, c_0, and F_c be in accordance with definition 12.12.

Then F-completion $\langle a_F, b_F, c_F, d_F \rangle$ is a completion of Tab9 and if $F_{\approx}(a_F, b_0 + F_b, c_0 + F_c, d_F) = 1$ then also $F_{\approx}(a_F, b_0 + u, c_0 + f_{X,X} - u, d_F) = 1$ for $u = 0, \ldots, f_{X,X}$.

Proof. To prove that F-completion $\langle a_F, b_F, c_F, d_F \rangle$ is a completion of Tab9, it is sufficient to show that $F_b + F_c = f_{X,X}$. We prove the theorem separately for particular conditions given in Table 12.7.

1. It holds that $F_b = 0$, $F_c = f_{X,X}$ and thus $F_b + F_c = f_{X,X}$.
   We assume $b_0 \geq c_0 + f_{X,X}$ and $F_{\approx}(a_F, b_0, c_0 + f_{X,X}, d_F) = 1$. We have to prove $F_{\approx}(a_F, b_0 + u, c_0 + f_{X,X} - u, d_F) = 1$ for $u = 0, \ldots, f_{X,X}$. If $u = 0$ or $f_{X,X} = 0$ then there is nothing to prove.
   Let us assume $0 < u \leq f_{X,X}$ i.e. $f_{X,X} \geq 1$. Due to point 1 of definition 6.9 we have:
   $b_0 \geq c_0 + f_{X,X} - 1 \geq 0$ thus $F_{\approx}(a_F, b_0 + 1, c_0 + f_{X,X} - 1, d_F) = 1$,
   $b_0 + 1 \geq c_0 + (f_{X,X} - 1) - 1 \geq 0$ thus $F_{\approx}(a_F, b_0 + 2, c_0 + f_{X,X} - 2, d_F) = 1$,
   $\ldots$,
   $b_0 + f_{X,X} - 2 \geq c_0 + 2 - 1 \geq 0$ thus $F_{\approx}(a_F, b_0 + f_{X,X} - 1, c_0 + 1, d_F) = 1$,
   $b_0 + f_{X,X} - 1 \geq c_0 + 1 - 1 \geq 0$ thus $F_{\approx}(a_F, b_0 + f_{X,X}, c_0, d_F) = 1$.
   This finishes the proof for condition 1.
2. It holds that $F_b = f_{X,X}$, $F_c = 0$ and thus $F_b + F_c = f_{X,X}$.
   We assume $c_0 \geq b_0 + f_{X,X}$ and $F_{\approx}(a_F, b_0 + f_{X,X}, c_0, d_F) = 1$. We have to prove $F_{\approx}(a_F, b_0 + f_{X,X} - u, c_0 + u, d_F) = 1$ for $u = 0, \ldots, f_{X,X}$. If $u = 0$ or $f_{X,X} = 0$ then there is nothing to prove.

Let us assume $0 < u \leq f_{X,X}$ i.e. $f_{X,X} \geq 1$. Due to point 2 of definition 6.9 we have:
$c_0 \geq b_0 + f_{X,X} - 1 \geq 0$ *thus* $F_\approx(a_F, b_0 + f_{X,X} - 1, c_0 + 1, d_F) = 1$,
$c_0 + 1 \geq b_0 + (f_{X,X} - 1) - 1 \geq 0$ *thus* $F_\approx(a_F, b_0 + f_{X,X} - 2, c_0 + 2, d_F) = 1$,
...,
$c_0 + f_{X,X} - 2 \geq b_0 + 2 - 1 \geq 0$ *thus* $F_\approx(a_F, b_0 + 1, c_0 + f_{X,X} - 1, d_F) = 1$,
$c_0 + f_{X,X} - 1 \geq b_0 + 1 - 1 \geq 0$ *thus* $F_\approx(a_F, b_0, c_0 + f_{X,X}, d_F) = 1$.
This finishes the proof for condition 2.

3. *It holds that* $F_b + F_c = \frac{b_0 + c_0 + f_{X,X}}{2} - b_0 + \frac{b_0 + c_0 + f_{X,X}}{2} - c_0 = f_{X,X}$.

   We assume $|b_0 - c_0| < f_{X,X}$ *and* $F_\approx(a_F, z, z, d_F) = 1$ *where* $z = \frac{b_0 + c_0 + f_{X,X}}{2}$. *We have to prove* $F_\approx(a_F, b_0 + u, c_0 + f_{X,X} - u, d_F) = 1$ *for* $u = 0, \ldots, f_{X,X}$.

   It holds that $z - b_0 = \frac{c_0 - b_0 + f_{X,X}}{2}$ *which is positive because* $|b_0 - c_0| < f_{X,X}$ *and integer because* $b_0 + c_0 + f_{X,X}$ *is even. We distinguish three cases*

   (i) $b_0 + u > z$
   (ii) $b_0 + u = z$
   (iii) $b_0 + u < z$.

(i): Let us denote $v = b_0 + u - z$. *Then we have* $b_0 + u = z + v$ *and it also holds that* $c_0 + f_{X,X} - u = c_0 + f_{X,X} - (v - b_0 + z) = c_0 + f_{X,X} + b_0 - z - v = 2z - z - v = z - v$ *and thus we have to prove* $F_\approx(a_F, z + v, z - v, d_F) = 1$. *Please note that* $0 \leq u \leq f_{X,X}$ *and thus* $c_0 + f_{X,X} - u \geq 0$ *i.e.* $z - v \geq 0$ *which means* $z - j \geq 0$ *for* $j = 1, \ldots, v$. *In addition we have* $b_0 + u > z$ *which means* $v \geq 1$. *We assume* $F_\approx(a_F, z, z, d_F) = 1$ *and thus due to point 1 of definition 6.9 we have:*
$z \geq z - 1 \geq 0$ *thus* $F_\approx(a_F, z + 1, z - 1, d_F) = 1$,
$z + 1 \geq z - 1 - 1 \geq 0$ *thus* $F_\approx(a_F, z + 2, z - 2, d_F) = 1$,
...,
$z + (v - 2) \geq z - (v - 2) - 1 \geq 0$ *thus* $F_\approx(a_F, z + (v - 1), z - (v - 1), d_F) = 1$,
$z + (v - 1) \geq z - (v - 1) - 1 \geq 0$ *thus* $F_\approx(a_F, z + v, z - v, d_F) = 1$.
This finishes the proof for condition (i).

(ii): If $b_0 + u = z$, *then there is nothing to prove.*

(iii): Let us denote $w = z - (b_0 + u)$, *then we have* $b_0 + u = z - w$ *and it also holds* $c_0 + f_{X,X} - u = c_0 + f_{X,X} - (z - w - b_0) = z + w$ *and thus we have to prove* $F_\approx(a_F, z - w, z + w, d_F) = 1$. *Please note that* $0 \leq u \leq f_{X,X}$ *and thus* $b_0 + u = z - w \geq 0$ *which means* $z - j \geq 0$ *for* $j = 1, \ldots, w$. *In addition we have* $b_0 + u < z$ *which means* $w \geq 1$. *We assume* $F_\approx(a_F, z, z, d_F) = 1$ *and thus due to point 2 of definition 6.9 we have:*
$z \geq z - 1 \geq 0$ *thus* $F_\approx(a_F, z - 1, z + 1, d_F) = 1$,
$z + 1 \geq z - 1 - 1 \geq 0$ *thus* $F_\approx(a_F, z - 2, z + 2, d_F) = 1$,
...,
$z + (w - 2) \geq z - (w - 2) - 1 \geq 0$ *thus* $F_\approx(a_F, z - (w - 1), z + (w + 1), d_F) = 1$,
$z + (w - 1) \geq z - (w - 1) - 1 \geq 0$ *thus* $F_\approx(a_F, z - w, z + w, d_F) = 1$.
This finishes the proof for condition 3.

4. It holds that $F_b + F_c = \frac{b_0 + c_0 + f_{X,X} + 1}{2} - b_0 + \frac{b_0 + c_0 + f_{X,X} - 1}{2} - c_0 = f_{X,X}$.

We assume $|b_0 - c_0| < f_{X,X}$ and $F_{\approx}(a_F, z, z-1, d_F) = 1$ where $z = \frac{b_0 + c_0 + f_{X,X} + 1}{2}$.
We have to prove $F_{\approx}(a_F, b_0 + u, c_0 + f_{X,X} - u, d_F) = 1$ for $u = 0, \dots, f_{X,X}$.
It holds that $z - b_0 = \frac{c_0 - b_0 + f_{X,X} + 1}{2}$ which is positive because $|b_0 - c_0| < f_{X,X}$ and integer because $b_0 + c_0 + f_{X,X}$ is odd. We distinguish three cases

(i) $b_0 + u > z$
(ii) $b_0 + u = z$
(iii) $b_0 + u < z$.

(i): Let us denote $v = b_0 + u - z$. Then we have $b_0 + u = z + v$ and it also holds
$c_0 + f_{X,X} - u = c_0 + f_{X,X} - (v - b_0 + z) = c_0 + b_0 + f_{X,X} - \frac{b_0 + c_0 + f_{X,X} + 1}{2} - v =$
$= \frac{b_0 + c_0 + f_{X,X} + 1}{2} - 1 - v = z - 1 - v$ and this means that we have to prove
$F_{\approx}(a_F, z + v, z - 1 - v, d_F) = 1$. We assume $0 \le u \le f_{X,X}$ and it holds that
$c_0 + f_{X,X} - u = z - 1 - v$ which means $c_0 + f_{X,X} - u \ge 0$ i.e. $z - 1 - v \ge 0$, thus
$z - 1 - j \ge 0$ for $j = 1, \dots, v$. In addition we have $b_0 + u > z$ which means $v \ge 1$.
We assume $F_{\approx}(a_F, z, z-1, d_F) = 1$ and thus due to point 1 of definition 6.9 we
have:
$z \ge z - 1 - 1 \ge 0$ thus $F_{\approx}(a_F, z + 1, z - 1 - 1, d_F) = 1$,
$z + 1 \ge z - 1 - 1 - 1 \ge 0$ thus $F_{\approx}(a_F, z + 2, z - 1 - 2, d_F) = 1$,
$z + 2 \ge z - 1 - 2 - 1 \ge 0$ thus $F_{\approx}(a_F, z + 3, z - 1 - 3, d_F) = 1$,
$\dots$,
$z + (v - 2) \ge z - 1 - (v - 2) \ge 0$ thus $F_{\approx}(a_F, z + (v - 1), z - 1 - (v - 1), d_F) = 1$,
$z + (v - 1) \ge z - 1 - (v - 1) - 1 \ge 0$ thus $F_{\approx}(a_F, z + v, z - v, d_F) = 1$.
This finishes the proof for condition (i).

(ii): If $b_0 + u = z$, then there is nothing to prove.

(iii): Let us denote $w = z - (b_0 + u)$, then we have $b_0 + u = z - w$ and it also
holds $c_0 + f_{X,X} - u = c_0 + f_{X,X} - (z - w - b_0) = z - 1 + w$ and thus we have to
prove $F_{\approx}(a_F, z - w, z - 1 + w, d_F) = 1$.
It holds that $b_0 + u < z$, thus $w > 0$. It also holds that $u \ge 0$ which
implies $z - w \ge b_0 \ge 0$. We can conclude that for $j = 1, \dots w - 1$ it holds that
$z - 1 + j \ge z - j - 1 \ge 0$. We assume $F_{\approx}(a_F, z, z-1, d_F) = 1$ and thus due to point
2 of definition 6.9 we have:
$z - 1 \ge z - 1 \ge 0$ thus $F_{\approx}(a_F, z - 1, z - 1 + 1, d_F) = 1$,
$z - 1 + 1 \ge z - 1 - 1 \ge 0$ thus $F_{\approx}(a_F, z - 2, z - 1 + 2, d_F) = 1$,
$z - 1 + 2 \ge z - 2 - 1 \ge 0$ thus $F_{\approx}(a_F, z - 3, z - 1 + 3, d_F) = 1$,
$\dots$,
$z - 1 + (w - 2) \ge z - (w - 2) - 1 \ge 0$ thus
$$F_{\approx}(a_F, z - (w - 1), z - 1 + (w - 1), d_F) = 1,$$
$z - 1 + (w - 1) \ge z - 1 - (w - 1) - 1 \ge 0$ thus $F_{\approx}(a_F, z - w, z - 1 + w, d_F) = 1$.
This finishes the proof for condition 4.

This finishes the proof.   □

Note 12.1. Please note note that the missing values in theorem 12.2 are completed such that a_F and d_F are as small as possible and $f_{X,X}$ is distributed such that b_F and c_F are as close as possible.

The following theorem shows that for important 4ft-quantifiers there is a secured 4ft-table for a given 9ft-table.

Theorem 12.3. *Let* $Tab9 = \langle f_{1,1}, f_{1,X}, f_{1,0}, f_{X,1}, f_{X,X}, f_{X,0}, f_{0,1}, f_{0,X}, f_{0,0} \rangle$ *be a 9ft-table and let* $\approx$ *be a 4ft-quantifier. Then:*

1. If $\approx$ *is implicational then*

$$\langle f_{1,1}, f_{1,0} + f_{1,X} + f_{X,X} + f_{X,0}, 0, 0 \rangle$$

is a secured 4ft-table for 4ft-quantifier $\approx$ *and the 9ft-table Tab9.*
2. If $\approx$ *is Σ-double implicational then*

$$\langle f_{1,1}, f_{1,0} + f_{1,X} + f_{X,X} + f_{X,0}, f_{0,1} + f_{X,1} + f_{0,X}, 0 \rangle$$

is a secured 4ft-table for 4ft-quantifier $\approx$ *and the 9ft-table Tab9.*
3. If $\approx$ *is Σ-equivalence then*

$$\langle f_{1,1}, f_{1,0} + f_{1,X} + f_{X,X} + f_{X,0}, f_{0,1} + f_{X,1} + f_{0,X}, f_{0,0} \rangle$$

is a secured 4ft-table for 4ft-quantifier $\approx$ *and the 9ft-table Tab9.*
4. If $\approx$ *is an equivalence quantifier with the F-property then F-completion*

$$\langle a_F, b_F, c_F, d_F \rangle = \langle f_{1,1}, b_F, c_F, f_{0,0} \rangle$$

of Tab9 is a secured 4ft-table for 4ft-quantifier $\approx$ *and the 9ft-table Tab9.*

Proof. We have to prove for each $\langle a_s, b_s, c_s, d_s \rangle$ *of secured 4ft-tables in question:*

a)   *If it holds that* $F_{\approx}(a, b, c, d) = 1$ *for each completion* $\langle a, b, c, d \rangle$ *of Tab9, then also* $F_{\approx}(a_s, b_s, c_s, d_s) = 1$.
b)   *If it holds that* $F_{\approx}(a_s, b_s, c_s, d_s) = 1$, *then* $F_{\approx}(a, b, c, d) = 1$ *for each completion* $\langle a, b, c, d \rangle$ *of Tab9.*

Proof for point a):

1. $\approx$ *is implicational, 4ft-table*

$$\langle a_I, b_I, c_I, d_I \rangle = \langle f_{1,1}, f_{1,0} + f_{1,X} + f_{X,X} + f_{X,0}, f_{0,1} + f_{X,1} + f_{0,X}, f_{0,0} \rangle$$

is a completion of Tab9, we assume $F_{\approx}(a_I, b_I, c_I, d_I) = 1$, *it holds that* $f_{1,1} \geq a_I$ *and* $f_{1,0} + f_{1,X} + f_{X,X} + f_{X,0} \leq b_I$, *thus*

$$F_{\approx}(f_{1,1}, f_{1,0} + f_{1,X} + f_{X,X} + f_{X,0}, 0, 0) = 1 \, .$$

2. $\approx$ is Σ-double implicational, 4ft-table

$$\langle a_\Sigma, b_\Sigma, c_\Sigma, d_\Sigma \rangle = \langle f_{1,1}, f_{1,0} + f_{1,X} + f_{X,X} + f_{X,0}, f_{0,1} + f_{X,1} + f_{0,X}, f_{0,0} \rangle$$

is a completion of Tab9, we assume $F_\approx(a_\Sigma, b_\Sigma, c_\Sigma, d_\Sigma) = 1$, it holds that $f_{1,1} \geq a_\Sigma$ and $(f_{1,0} + f_{1,X} + f_{X,X} + f_{X,0}) + (f_{0,1} + f_{X,1} + f_{0,X}) \leq b_\Sigma + c_\Sigma$, thus

$$F_\approx(f_{1,1}, f_{1,0} + f_{1,X} + f_{X,X} + f_{X,0}, f_{0,1} + f_{X,1} + f_{0,X}, 0) = 1 \,.$$

3. $\approx$ is Σ-equivalence, 4ft-table $\langle a_\Sigma, b_\Sigma, c_\Sigma, d_\Sigma \rangle$ defined above is a completion of Tab9, we assume $F_\approx(a_\Sigma, b_\Sigma, c_\Sigma, d_\Sigma) = 1$, it holds that $f_{1,1} + f_{0,0} \geq a_\Sigma + d_\Sigma$ and $(f_{1,0} + f_{1,X} + f_{X,X} + f_{X,0}) + (f_{0,1} + f_{X,1} + f_{0,X}) \leq b_\Sigma + c_\Sigma$, thus

$$F_\approx(f_{1,1}, f_{1,0} + f_{1,X} + f_{X,X} + f_{X,0}, f_{0,1} + f_{X,1} + f_{0,X}, f_{0,0}) = 1 \,.$$

4. $\approx$ is an equivalence quantifier with the F-property, F-completion $\langle f_{1,1}, b_F, c_F, f_{0,0} \rangle$ given by definition 12.12 is according to theorem 12.2 a completion of Tab9. Thus, according to assumption

$$F_\approx(f_{1,1}, b_F, c_F, f_{0,0}) = 1 \,.$$

Proof for point b):
Let $\langle a, b, c, d \rangle$ *be a completion of Tab9. Then, according to definition 12.8:*

- $a = f_{1,1} + f_{1,X,a} + f_{X,1,a} + f_{X,X,a}$
- $b = f_{1,0} + f_{1,X,b} + f_{X,0,b} + f_{X,X,b}$
- $c = f_{0,1} + f_{0,X,c} + f_{X,1,c} + f_{X,X,c}$
- $d = f_{0,0} + f_{0,X,d} + f_{X,0,d} + f_{X,X,d}$

where

- $f_{1,X,a} + f_{1,X,b} = f_{1,X}$
- $f_{X,1,a} + f_{X,1,c} = f_{X,1}$
- $f_{X,X,a} + f_{X,X,b} + f_{X,X,c} + f_{X,X,d} = f_{X,X}$
- $f_{X,0,b} + f_{X,0,d} = f_{X,0}$
- $f_{0,X,c} + f_{0,X,d} = f_{0,X}.$

We have to prove that if $F_\approx(a_s, b_s, c_s, d_s) = 1$ *then it also holds that* $F_\approx(a, b, c, d) = 1$.

1. If $\approx$ is implicational and

$$F_\approx(f_{1,1}, f_{1,0} + f_{1,X} + f_{X,X} + f_{X,0}, 0, 0) = 1$$

then we use

- $a = f_{1,1} + f_{1,X,a} + f_{X,1,a} + f_{X,X,a} \geq f_{1,1}$
- $b = f_{1,0} + f_{1,X,b} + f_{X,0,b} + f_{X,X,b} \leq f_{1,0} + f_{1,X} + f_{X,0} + f_{X,X}$

which means $F_\approx(a, b, c, d) = 1$.

2. If $\approx$ is Σ-double implicational and

$$F_\approx(f_{1,1}, f_{1,0} + f_{1,X} + f_{X,X} + f_{X,0}, f_{0,1} + f_{X,1} + f_{0,X}, 0) = 1 \,,$$

then we use

- $a = f_{1,1} + f_{1,X,a} + f_{X,1,a} + f_{X,X,a} \geq f_{1,1}$
- $b + c = f_{1,0} + f_{1,X,b} + f_{X,0,b} + f_{X,X,b} + f_{0,1} + f_{0,X,c} + f_{X,1,c} + f_{X,X,c} \leq$
  $\leq f_{1,0} + f_{1,X} + f_{X,X} + f_{X,0} + f_{0,1} + f_{0,X} + f_{X,1}$

which means $F_{\approx}(a,b,c,d) = 1$.

3. *If* $\approx$ *is* Σ-*equivalence and*

$$F_{\approx}(f_{1,1}, f_{1,0} + f_{1,X} + f_{X,X} + f_{X,0}, f_{0,1} + f_{X,1} + f_{0,X}, f_{0,0}) = 1,$$

then we use

- $a + d = f_{1,1} + f_{1,X,a} + f_{X,1,a} + f_{X,X,a} + f_{0,0} + f_{0,X,d} + f_{X,0,d} + f_{X,X,d} \geq$
  $f_{1,1} + f_{0,0}$
- $b + c = f_{1,0} + f_{1,X,b} + f_{X,0,b} + f_{X,X,b} + f_{0,1} + f_{0,X,c} + f_{X,1,c} + f_{X,X,c} \leq$
  $\leq f_{1,0} + f_{1,X} + f_{X,0} + f_{X,X} + f_{0,1} + f_{0,X} + f_{X,1}$

which means $F_{\approx}(a,b,c,d) = 1$.

4. *If* $\approx$ *is an equivalence quantifier with the F-property and*

$$\langle a_F, b_F, c_F, d_F \rangle = \langle f_{1,1}, b_F, c_F, f_{0,0} \rangle$$

where $\langle f_{1,1}, b_F, c_F, f_{0,0} \rangle$ *is a F-completion and*

- $a_F = f_{1,1}$
- $b_F = b_0 + F_b$ *where* $b_0 = f_{1,0} + f_{1,X} + f_{X,0}$
- $c_F = c_0 + F_c$ *where* $c_0 = f_{0,1} + f_{0,X} + f_{X,1}$
- $d_F = f_{0,0}$,

then we use theorem 12.2 which also states that

$$F_{\approx}(f_{1,1}, b_0 + u, c_0 + f_{X,X} - u, f_{0,0}) = 1$$

for $u = 0, \ldots, f_{X,X}$. *This implies that*

$$F_{\approx}(f_{1,1}, b_0 + f_{X,X,a} + f_{X,X,b}, c_0 + f_{X,X,c} + f_{X,X,d}, f_{0,0}) = 1$$

and now we use

- $a = f_{1,1} + f_{1,X,a} + f_{X,1,a} + f_{X,X,a} \geq f_{1,1}$
- $b = f_{1,0} + f_{1,X,b} + f_{X,0,b} + f_{X,X,b} \leq b_0 + f_{X,X,a} + f_{X,X,b}$
- $c = f_{0,1} + f_{0,X,c} + f_{X,1,c} + f_{X,X,c} \leq c_0 + f_{X,X,c} + f_{X,X,d}$
- $d = f_{0,0} + f_{0,X,d} + f_{X,0,d} + f_{X,X,d} \geq f_{0,0}$

which means $F_{\approx}(a,b,c,d) = 1$.

This finishes the proof. □

Note 12.2. Please note that the secured 4ft-tables $\langle a_s, b_s, c_s, d_s \rangle$ for implicational and Σ-double implicational quantifiers need not be completions for a given 9ft-table *Tab9*. This is because the sum $a_s + b_s + c_s + d_s$ does not involve the frequency $f_{0,0}$.

12.7   Optimistic Completion and Deleting

We point out to two additional approaches to dealing with missing information used in GUHA procedure ASSOC – optimistic completion and deleting introduced in [24]. Logical calculus $\mathscr{LC}_{\mathscr{T}}^{X}$ of association rules of type $\mathscr{T} = \langle t_1, \ldots, t_K \rangle$ with missing information is defined in definition 12.10 as a secured extension of a logical calculus $\mathscr{LC}_{\mathscr{T}}$ of association rules of type $\mathscr{T} = \langle t_1, \ldots, t_K \rangle$ in the sense of definition 3.3.3 in [18]. The principles $(SE1), \ldots, (SE5)$ introduced in Sect. 12.2 are used. A crucial role in the definition is played by the secured extension $F_{\approx}^{X}$ of associated function $F_{\approx}$ of 4ft-quantifier $\approx$, see definition 12.9.

We define two additional extensions of associated function $F_{\approx}$ of 4ft-quantifier $\approx$, see next two definitions.

Definition 12.13. Let $\approx$ be 4ft-quantifier with associated function $F_{\approx}$ defined according to definition 3.8. Then an *optimistic extension of $F_{\approx}$* is a function $F_{\approx}^{O}$ defined for all nine tuples $\langle f_{1,1}, f_{1,X}, f_{1,0}, f_{X,1}, f_{X,X}, f_{X,0}, f_{0,1}, f_{0,X}, f_{0,0} \rangle$ of non-negative integer numbers satisfying

$$f_{1,1} + f_{1,X} + f_{1,0} + f_{X,1} + f_{X,X} + f_{X,0} + f_{0,1} + f_{0,X} + f_{0,0} > 0$$

such that

- $F_{\approx}^{O}(f_{1,1}, f_{1,X}, f_{1,0}, f_{X,1}, f_{X,X}, f_{X,0}, f_{0,1}, f_{0,X}, f_{0,0}) = 1$ if there is a completion $\langle a, b, c, d \rangle$ of $\langle f_{1,1}, f_{1,X}, f_{1,0}, f_{X,1}, f_{X,X}, f_{X,0}, f_{0,1}, f_{0,X}, f_{0,0} \rangle$ satisfying $F_{\approx}(a, b, c, d) = 1$.
- $F_{\approx}^{O}(f_{1,1}, f_{1,X}, f_{1,0}, f_{X,1}, f_{X,X}, f_{X,0}, f_{0,1}, f_{0,X}, f_{0,0}) = 0$ otherwise.

Definition 12.14. Let $\approx$ be a 4ft-quantifier with associated function $F_{\approx}$ defined according to definition 3.8. Then a *deleting extension of $F_{\approx}$* is a function $F_{\approx}^{D}$ defined for all nine tuples $\langle f_{1,1}, f_{1,X}, f_{1,0}, f_{X,1}, f_{X,X}, f_{X,0}, f_{0,1}, f_{0,X}, f_{0,0} \rangle$ of non-negative integer numbers such that

- $F_{\approx}^{D}(f_{1,1}, f_{1,X}, f_{1,0}, f_{X,1}, f_{X,X}, f_{X,0}, f_{0,1}, f_{0,X}, f_{0,0}) = F_{\approx}(f_{1,1}, f_{1,0}, f_{0,1}, f_{0,0})$ if $f_{1,1} + f_{1,0} + f_{0,1} + f_{0,0} > 0$
- $F_{\approx}^{D}(f_{1,1}, f_{1,X}, f_{1,0}, f_{X,1}, f_{X,X}, f_{X,0}, f_{0,1}, f_{0,X}, f_{0,0}) = X$ otherwise.

The idea of optimistic extension is to consider an association rule $\varphi \approx \psi$ true in a data matrix $\mathscr{M}^{X}$ with missing information if there is at least one completion of $\mathscr{M}^{X}$ in which the rule $\varphi \approx \psi$ is true.

The idea of deleting extension is to consider an association rule $\varphi \approx \psi$ true in a data matrix $\mathscr{M}^{X}$ with missing information if the rule is true in a date matrix which we get by deleting all rows of $\mathscr{M}^{X}$ for which there is an unknown value of φ and all rows for which there is an unknown value of ψ.

The functions $F_{\approx}^{O}$ and $F_{\approx}^{D}$ can be used to define logical calculi $\mathscr{LC}_{\mathscr{T}}^{O}$ and $\mathscr{LC}_{\mathscr{T}}^{D}$ similarly to definition 12.10 of logical calculus $\mathscr{LC}_{\mathscr{T}}^{X}$. However, we will not do this here, see also Sect. 16.9.

Chapter 13
Definability of Association Rules in Classical Predicate Calculi

Observational predicate calculi are defined and deeply studied in [18]. These can be seen as a modification of classical predicate calculi – only finite models are allowed and generalized quantifiers are added to classical quantifiers $\forall$ and $\exists$. Finite models correspond to analysed data and generalized quantifiers make it possible to express various assertion on the analysed data. All 4ft-quantifiers defined and studied in this book are generalized quantifiers.

Formulas $(\forall x)P_1(x)$ and $(\exists x)(P_2(x) \wedge P_3(y))$ are examples of formulas with classical quantifiers $\forall$ and $\exists$. The formula $(\Rightarrow_{p,Base} x)(P_1(x),P_2(x))$ is an example of a formula of an observational predicate calculus with a special generalized quantifier i.e. with 4ft-quantifier $\Rightarrow_{p,Base}$. All these formulas are formulas of an observational predicate calculus.

A natural question is if a given generalized quantifier corresponding to a 4ft-quantifier can be expressed by means of classical observational predicate calculus (i. e. only with quantifiers $\forall$ and $\exists$) with equality. A generalized quantifier which can be expressed in a such way is called classically definable. Interesting general results on the definability of generalized quantifiers were achieved in [18]. Additional new results on the definability of generalized quantifiers corresponding to 4ft-quantifiers were achieved in [63, 69, 72]. The goal of this chapter is to recapitulate the main results on definability achieved in [18] and present new results on definability in detail.

We start with a formal definition of observational predicate calculi in Sect. 13.1. We also need to specify which formulas of observational predicate calculi correspond to association rules, this is done in Sect. 13.2. Results on definability are presented in [18] as Tharp's theorem, which is introduced in Sect. 13.3 together with a definition of classical definability. We use a normal form theorem proved in [18], introduced in Sect. 13.4. A new general criterion of definability of generalized quantifiers is proved in Sect. 13.5. Sects. 13.6 and 13.7 show that this criterion can be formulated using tables of critical frequencies for equivalence quantifiers and for implicational quantifiers. Undefinability of important quantifiers is proved in Sect. 13.8.

J. Rauch: *Observational Calculi and Association Rules*, SCI 469, pp. 201–229.
DOI: 10.1007/978-3-642-11737-4_13 © Springer-Verlag Berlin Heidelberg 2013

13.1 Observational Predicate Calculi

The goal of this section is to introduce observational predicate calculi in the extent necessary to present results on definability of association rules. All definitions come from [18]. Observational predicate calculus is defined as a semantic system, as introduced also in [18]. Please note that calculi of association rules and calculi of association rules with missing information are also defined as semantic systems, see Chaps. 3 and 12.

The study of recursiveness is also an important feature of dealing with observational predicate calculi in [18]. However we will not deal with recursiveness in this chapter even if the notion is used in some of the definitions we do use. Recursiveness is not explicitly used in results concerning definability we are interested in. For more information see e.g. chapter 1 in [18].

There is a problem with the notion of *type of a logical calculus of association rules* introduced in definition 2.1. A notion of type is also defined in [18], however in a different context. To avoid possible problems we will call it the *type of OPC* where OPC means *observational predicate calculus*. We start with a definition of V-structure as defined in definition 2.1.2 in [18].

Definition 13.1

1. A *type of OPC* is a finite sequence $\langle t_1, \ldots, t_n \rangle$ of positive integer numbers.
2. A *V-structure* of the type $t = \langle t_1, \ldots, t_n \rangle$ of OPC is a n+1-tuple

$$\mathscr{M} = \langle M, f_1, \ldots, f_n \rangle,$$

   where M is a non-empty set and each f_i $(i = 1, \ldots, n)$ is a mapping from M^{t_i} into V. The set M is called the *domain* of $\mathscr{M}$.
3. Denote by $\underline{M}_t^V$ the set of all V-structures $\mathscr{M}$ of the type t of OPC such that the domain of $\mathscr{M}$ is a finite set of non-negative integer numbers.

The set $\underline{M}_t^{\{0,1\}}$ of all $\{0,1\}$-structures $\mathscr{M} = \langle M, f_1, \ldots, f_n \rangle$ of the type $t = \langle t_1, \ldots, t_n \rangle$ of OPC is intended to be a set of models of an observational predicate calculus of type $t = \langle t_1, \ldots, t_n \rangle$ of OPC. Now we define predicate language of the type $t = \langle t_1, \ldots, t_n \rangle$ of OPC. Our definition 13.2 corresponds to definition 2.2.2 in [18].

Definition 13.2. A *predicate language* $\mathscr{L}$ *of type* $t = \langle t_1, \ldots, t_n \rangle$ *of OPC* is defined in the following way.

- The *symbols* of the language $\mathscr{L}$ are:

  – *predicates* $P_1, \ldots, P_n$ of arity $t_1, \ldots, t_n$ respectively
  – an infinite sequence $x_0, x_1, x_2, \ldots$ of *variables*
  – *junctors* $\underline{0}$, $\underline{1}$ (nullary), $\neg$ (unary) and $\wedge, \vee, \rightarrow, \leftrightarrow$ (binary), called falsehood, truth, negation, conjunction, disjunction, implication and equivalence respectively

- *quantifiers* $q_0, q_1, q_2, \ldots$ of types $s_0, s_1, s_2, \ldots$ respectively. The sequence of quantifiers is either infinite or finite (non-empty). Each *quantifier type* is a sequence $\langle \underbrace{1, 1, \ldots, 1}_{s_i-times} \rangle$ of s_i consecutive symbols "1". We write $\langle 1^{s_i} \rangle$ instead of $\langle \underbrace{1, 1, \ldots, 1}_{s_i-times} \rangle$. If there are infinitely many quantifiers then the function associating the type s_i with each i is recursive.

- A *predicate language with equality* contains an additional binary predicate $=$ distinct from $P_1, \ldots, P_n$ (the equality predicate).
- *Formulas* are defined inductively as usual:

  - Each expression $P_i(u_1, \ldots, u_{t_i})$ where $u_1, \ldots, u_{t_i}$ are variables is an *atomic formula* (and $u_1 = u_2$ is an atomic formula).
  - Atomic formula is a formula, $\underline{0}$ and $\underline{1}$ are formulas. If φ and ψ are formulas, then $\neg\varphi$, $\varphi \wedge \psi$, $\varphi \vee \psi$, $\varphi \rightarrow \psi$ and $\varphi \leftrightarrow \psi$ are formulas.
  - If q_i is a quantifier of the type $\langle 1^{s_i} \rangle$, if u is a variable and $\varphi_1, \ldots, \varphi_{s_i}$ are formulas then $(q_i u)(\varphi_1, \ldots, \varphi_{s_i})$ is a formula.

- *Free and bound variables* are defined as usual. The induction step for the formula $(q_i u)(\varphi_1, \ldots, \varphi_s)$ is as follows:

  - A variable is free in $(q_i u)(\varphi_1, \ldots, \varphi_s)$ if and only if it is free in one of the formulas $\varphi_1, \ldots, \varphi_s$ and it is distinct from u.
  - A variable is bound in $(q_i u)(\varphi_1, \ldots, \varphi_s)$ if and only if it is bound in one of the formulas $\varphi_1, \ldots, \varphi_s$ or it is u.

Now we can define an observational predicate calculus, definition 13.3 corresponds to definition 2.2.5 in [18].

Definition 13.3. *Observational predicate calculus* OPC of the type $t = \langle t_1, \ldots, t_n \rangle$ of OPC is given by

- predicate language $\mathscr{L}$ of the type t of OPC
- associated function Asf_{q_i} for each quantifier q_i of the language $\mathscr{L}$. Asf_{q_i} maps the set $\underline{M}_{s_i}^{\{0,1\}}$ of all models (i.e. $\{0,1\}$-structures) of the type s_i (i.e. $\langle \underbrace{1, 1, \ldots, 1}_{s_i-times} \rangle$) whose domain is a finite subset of the set of non-negative integer numbers into $\{0,1\}$ such that the following is satisfied:

  - each Asf_{q_i} is invariant under isomorphism, i.e. if $\mathscr{M}_1, \mathscr{M}_2 \in \underline{M}_{s_i}^{\{0,1\}}$ are isomorphic, then $\mathrm{Asf}_{q_i}(\mathscr{M}_1) = \mathrm{Asf}_{q_i}(\mathscr{M}_2)$
  - $\mathrm{Asf}_{q_i}(\mathscr{M})$ is a recursive function of two variables $q_i, \mathscr{M}$.

Values of formulas are defined by definition 13.4 which corresponds to definition 2.2.6 in [18]. Associated functions of junctors are used, see section 2.2.3 in [18]. Associated functions of the junctors of a predicate calculus define values of composed formulas. The associated functions $Asf_\neg, Asf_\wedge, Asf_\vee, Asf_\rightarrow, Asf_\leftrightarrow$ of junctors

$Asf_\neg$	
0	1
1	0

$Asf_\wedge$	0	1
0	0	0
1	0	1

$Asf_\vee$	0	1
0	0	1
1	1	1

$Asf_\rightarrow$	0	1
0	1	1
1	0	1

$Asf_\leftrightarrow$	0	1
0	1	0
1	0	1

Fig. 13.1 Associated functions $Asf_\neg, Asf_\wedge, Asf_\vee, Asf_\rightarrow, Asf_\leftrightarrow$ of junctors

$\neg, \wedge, \vee, \rightarrow, \leftrightarrow$ respectively are given in tables in Fig. 13.1. The associated function of a nullary junctor is its value: the value of $\underline{0}$ is 0 and value of $\underline{1}$ is 1.

Definition 13.4. Let $\mathscr{P}$ be an observational predicate calculus of type $t = \langle t_1, \ldots, t_n \rangle$ of OPC, let $\mathscr{M} = \langle M, f_1, \ldots, f_n \rangle \in \underline{M}_t^{\{0,1\}}$ be a model of $\mathscr{P}$ and let φ be a formula; write $FV(\varphi)$ for the set of free variables of φ. An $\mathscr{M}$-*sequence for* φ is a mapping ε of $FV(\varphi)$ into M. If the domain of ε is $\{u_1, \ldots u_n\}$ and if $\varepsilon(u_i) = m_i$ then we write $\varepsilon = \frac{u_1, \ldots, u_n}{m_1, \ldots, m_n}$. We define inductively $\|\varphi\|_{\mathscr{M}}[\varepsilon]$ - the $\mathscr{M}$-value of φ for ε.

- $\|P_i(u_1, \ldots, u_k)\|_{\mathscr{M}}[\frac{u_1, \ldots, u_k}{m_1, \ldots, m_k}] = f_i(m_1, \ldots, m_k)$

- $\|u_1 = u_2)\|_{\mathscr{M}}[\frac{u_1, u_2}{m_1, m_2}] = 1$ if and only if $m_1 = m_2$
- $\|\underline{0}\|_{\mathscr{M}}[\emptyset] = 0, \|\underline{1}\|_{\mathscr{M}}[\emptyset] = 1$,
- $\|\neg\varphi\|_{\mathscr{M}}[\varepsilon] = 1 - \|\varphi\|_{\mathscr{M}}[\varepsilon]$
- If $FV(\varphi) \subseteq \text{domain}(\varepsilon)$ then we write ε/φ instead of restriction of ε to $FV(\varphi)$. Let ι be one of $\wedge, \vee, \rightarrow, \leftrightarrow$ and let Asf_ι be its associated function given by the usual truth table, see Fig. 13.1. Then

$$\|\varphi \iota \psi\|_{\mathscr{M}}[\varepsilon] = Asf_\iota (\|\varphi\|_{\mathscr{M}}[\varepsilon/\varphi], \|\psi\|_{\mathscr{M}}[\varepsilon/\psi]) .$$

- If $\text{domain}(\varepsilon) \supseteq FV(\varphi) - \{x\}$ and $x \notin \text{domain}(\varepsilon)$ then by letting x vary over M we obtain an unary function $\|\varphi\|_{\mathscr{M}}^\varepsilon$ on M such that for $m \in M$ it holds

$$\|\varphi\|_{\mathscr{M}}^\varepsilon(m) = \|\varphi\|_{\mathscr{M}}[(\varepsilon \cup \frac{x}{m})/\varphi] .$$

($\|\varphi\|_{\mathscr{M}}$ can be viewed as a k-ary function, k being the number of free variables of φ. Now all variables except x are fixed according to ε; x varies over M). We define: $\|(q_i x)(\varphi_1, \ldots, \varphi_k)\|_{\mathscr{M}}[\varepsilon] = Asf_q(\langle M, \|\varphi_1\|_{\mathscr{M}}^\varepsilon, \ldots, \|\varphi_k\|_{\mathscr{M}}^\varepsilon \rangle)$.

Note 13.1. Let $\mathscr{P}$ be an OPC of type t of OPC and let φ be its closed formula (i.e. formula without free variables). Then we write only $\|\varphi\|_{\mathscr{M}}$ instead of $\|\varphi\|_{\mathscr{M}}[\emptyset]$. If $\|\varphi\|_{\mathscr{M}} = 1$ then we say that φ is *true* in $\mathscr{M}$.

We will also use the following notions defined in [18].

Definition 13.5. An OPC is *monadic* if all its predicates are unary, i.e. if its type of OPC is $t = \langle 1, \ldots, 1 \rangle$. We write MOPC for "monadic observational predicate calculus". A MOPC whose only quantifiers are the classical quantifiers $\forall, \exists$ is called a *classical* MOPC or CMOPC. Similarly for MOPC with equality, in particular a CMOPC with equality.

13.2 Association Rules in Observational Predicate Calculi

Let $\mathscr{P}_4$ be a MOPC of the type $\langle 1,1,1,1 \rangle$ of OPC with unary predicates P_1, P_2, P_3, P_4 and with the quantifier $\Rightarrow_{p,Base}$ of the type $\langle 1,1 \rangle$. Let x be a variable of $\mathscr{P}_4$. Then the closed formula

$$(\Rightarrow_{p,Base}, x)(P_1(x) \wedge P_4(x), P_2(x) \vee P_3(x)) \tag{13.1}$$

of MOPC $\mathscr{P}_4$ corresponds to an association rule

$$A_1(1) \wedge A_4(1) \Rightarrow_{p,Base} A_2(1) \vee A_3(1) \tag{13.2}$$

where $A_1(1)$, $A_2(1)$, $A_3(1)$, $A_4(1)$ are basic Boolean attributes created from basic attributes A_1, A_2, A_3, A_4 of a language $\mathscr{L}_4$ of logical calculus $\mathscr{L}\mathscr{C}_4$ of association rules of type $\mathscr{T} = \langle 2,2,2,2 \rangle$, see definition 3.9. Basic Boolean attribute $A_1(1)$ corresponds to predicate P_1 and similarly for $A_2(1)$ and P_2, etc.

In addition, there is a one-one correspondence between data matrices from the set $\mathrm{M}_{\langle 2,2,2,2 \rangle}$ of all data matrices of type $\langle 2,2,2,2 \rangle$ in which association rules of logical calculus of association rules $\mathscr{L}\mathscr{C}_4$ are interpreted and all $\{0,1\}$-structures of the type $t = \langle 1,1,1,1 \rangle$ of OPC i.e. models of monadic predicate calculus $\mathscr{P}_4$. This means that if $\mathscr{M} = \langle M, f_1, f_2, f_3, f_4 \rangle \in \underline{M}^{\{0,1\}}_{\langle 1,1,1,1 \rangle}$ is a model of $\mathscr{P}_4$, see the upper part of Fig. 13.2, then there is a data matrix $\mathscr{M}' = \langle M', f_1', f_2', f_3', f_4' \rangle$ of type $\langle 2,2,2,2 \rangle$ which belongs to the set $\mathrm{M}_{\langle 2,2,2,2 \rangle}$, see the bottom part of Fig. 13.2, such that $\mathscr{M}$ and $\mathscr{M}'$ are isomorphic.

$o \in M$	f_1	f_2	f_3	f_4	$\|P_1(o)\|_{\mathscr{M}}$	$\|P_2(o)\|_{\mathscr{M}}$	$\|P_3(o)\|_{\mathscr{M}}$	$\|P_4(o)\|_{\mathscr{M}}$
o_1	1	0	1	0	1	0	1	0
o_2	0	0	1	1	0	0	1	1
$\vdots$	$\vdots$	$\vdots$	$\vdots$	$\vdots$	$\vdots$	$\vdots$	$\vdots$	$\vdots$
o_n	1	1	0	0	1	1	0	0

$$\mathscr{M} = \langle M, f_1, f_2, f_3, f_4 \rangle \in \underline{M}^{\{0,1\}}_{\langle 1,1,1,1 \rangle}$$

$o' \in M'$	f_1'	f_2'	f_3'	f_4'	$f_{A_1(1)}'(o')$	$f_{A_2(1)}'(o')$	$f_{A_3(1)}'(o')$	$f_{A_4(1)}(o')$
o_1'	1	2	1	2	1	0	1	0
o_2'	2	2	1	1	0	0	1	1
$\vdots$	$\vdots$	$\vdots$	$\vdots$	$\vdots$	$\vdots$	$\vdots$	$\vdots$	$\vdots$
o_n'	1	1	0	0	1	1	0	0

$$\mathscr{M}' = \langle M', f_1', f_2', f_3', f_4' \rangle \in \mathrm{M}_{\langle 2,2,2,2 \rangle}$$

Fig. 13.2 $\mathscr{M} = \langle M, f_1, f_2, f_3, f_4 \rangle \in \underline{M}^{\{0,1\}}_{\langle 1,1,1,1 \rangle}$ and $\mathscr{M}' = \langle M', f_1', f_2', f_3', f_4' \rangle \in \mathrm{M}_{\langle 2,2,2,2 \rangle}$

An isomorphism between $\{0,1\}$-structure $\mathscr{M} = \langle M, f_1, f_2, f_3, f_4 \rangle \in \underline{M}_{\langle 1,1,1,1 \rangle}^{\{0,1\}}$ and data matrix $\mathscr{M}' = \langle M', f_1', f_2', f_3', f_4' \rangle \in M_{\langle 2,2,2,2 \rangle}$ means that there is a one-one mapping ξ of M onto M' such that $f_i(o_j) = 1$ if and only if $f_i'(o_j') = 1$ for $i = 1,2,3,4$ and $j = 1,\ldots,n$ (and thus also $f_i(o_j) = 0$ if and only if $f_i'(o_j') = 2$ for $i = 1,2,3,4$ and $j = 1,\ldots,n$).

This isomorphism implies that $\|P_1(x)\|_{\mathscr{M}}[\frac{x}{o_j}] = 1$ if and only if $f_{A_1(1)}'(o_j') = 1$. Please note that $\|P_1(x)\|_{\mathscr{M}}[\frac{x}{o_j}] = 1$ means the M-value of $P_1(x)$ (where x is a free variable) for mapping e of $FV(P_1(x)) = \{x\}$ into M satisfying $e = \frac{x}{o_j}$, see definition 13.4. We write only $\|P_1(o)\|_{\mathscr{M}}$ instead of $\|P_1(x)\|_{\mathscr{M}}[\frac{x}{o}]$ in the upper part of Fig. 13.2. In addition, $f_{A_1(1)}'(o_j')$, is an interpretation of a basic Boolean attribute $A_1(1)$ in $\mathscr{M}'$, see definition 3.5.

The symbol $\Rightarrow_{p,Base}$ is considered a 4ft-quantifier of founded p-implication in association rule 13.2, see e.g. theorem 7.3. The same symbol $\Rightarrow_{p,Base}$ is used in formula 13.1 as a quantifier of type $\langle 1^2 \rangle = \langle 1,1 \rangle$ (i.e. 2 consecutive "1") of a predicate language see definition 13.2.

The formula $(\Rightarrow_{p,Base}, x)(P_1(x) \wedge P_4(x), P_2(x) \vee P_3(x))$ is closed and, according to definition 13.4 (see also note 13.1), is the $\mathscr{M}$-value

$$\|(\Rightarrow_{p,Base}, x)(P_1(x) \wedge P_4(x), P_2(x) \vee P_3(x)\|_{\mathscr{M}}$$

of $(\Rightarrow_{p,Base}, x)(P_1(x) \wedge P_4(x), P_2(x) \vee P_3(x))$ defined as

$$\|(\Rightarrow_{p,Base}, x)(P_1(x) \wedge P_4(x), P_2(x) \vee P_3(x)\|_{\mathscr{M}} =$$

$$= Asf_{\Rightarrow_{p,Base}}(\langle M, \|P_1(x) \wedge P_4(x)\|_{\mathscr{M}}^0, \|P_2(x) \vee P_3(x)\|_{\mathscr{M}}^0 \rangle)$$

where both $\|P_1(x) \wedge P_4(x)\|_{\mathscr{M}}^0$ and $P_2(x) \vee P_3(x)\|_{\mathscr{M}}^0$ are unary functions obtained such that we let x vary over M and for $o \in M$ it holds that

$$\|P_1(x) \wedge P_4(x)\|_{\mathscr{M}}^0(o) = \|P_1(x) \wedge P_4(x)\|_{\mathscr{M}}[\frac{x}{o}]$$

and

$$\|P_2(x) \vee P_3(x)\|_{\mathscr{M}}^0(o) = \|P_2(x) \vee P_3(x)\|_{\mathscr{M}}[\frac{x}{o}] .$$

The associated function $Asf_{\Rightarrow_{p,Base}}$ of the quantifier $\Rightarrow_{p,Base}$ is defined as the function mapping the set $\underline{M}_{\langle 1,1 \rangle}^{\{0,1\}}$ of all models $\mathscr{M} = \langle M, f_1, f_2 \rangle$ into $\{0,1\}$.

If we evaluate the formula $(\Rightarrow_{p,Base}, x)(P_1(x) \wedge P_4(x), P_2(x) \vee P_3(x))$ in the model $\mathscr{M} = \langle M, f_1, f_2, f_3, f_4 \rangle \in \underline{M}_{\langle 1,1,1,1 \rangle}^{\{0,1\}}$ of the above introduced MOPC $\mathscr{P}_4$ outlined in upper part of Fig. 13.2, we have to apply the function $Asf_{\Rightarrow_{p,Base}}$ to $\{0,1\}$-structure

$$\langle M, \|P_1(x) \wedge P_4(x)\|_{\mathscr{M}}^0(o), \|P_2(x) \vee P_3(x)\|_{\mathscr{M}}^0(o) \rangle$$

of the type $\langle 1^2 \rangle = \langle 1,1 \rangle$ of OPC outlined in Fig. 13.3.

$o \in M$	$\|P_1(x) \wedge P_4(x)\|_{\mathcal{M}}^{\emptyset}(o)$	$\|P_2(x) \vee P_3(x)\|_{\mathcal{M}}^{\emptyset}(o)$
o_1	0	1
o_2	0	1
$\vdots$	$\vdots$	$\vdots$
o_n	0	1

Fig. 13.3 $\{0,1\}$-structure $\langle M, \|P_1(x) \wedge P_4(x)\|_{\mathcal{M}}^{\emptyset}, \|P_2(x) \vee P_3(x)\|_{\mathcal{M}}^{\emptyset} \rangle$ of the type $\langle 1, 1 \rangle$ of OPC

The associated function $Asf_{\Rightarrow p, Base}$ is defined for a $\{0,1\}$-structure $\langle M, f_1, f_2 \rangle$ of the type $\langle 1^2 \rangle = \langle 1, 1 \rangle$ of OPC such that

$$Asf_{\Rightarrow p, Base} = \begin{cases} 1 & if \ \frac{a}{a+b} \geq p \wedge a \geq Base \\ 0 & \text{otherwise.} \end{cases}$$

where a is the number of $o \in M$ such that both $f_1(o) = 1$, $f_2(o) = 1$ and b is the number of $o \in M$ such that $f_1(o) = 1$ and $f_2(o) = 0$, see section 2.2.4 in [18].

This means that the formula

$$(\Rightarrow_{p, Base}, x)(P_1(x) \wedge P_4(x), P_2(x) \vee P_3(x))$$

is true in the model $\mathcal{M} = \langle M, f_1, f_2, f_3, f_4 \rangle \in \underline{M}_{\langle 1,1,1,1 \rangle}^{\{0,1\}}$ outlined in upper part of Fig. 13.2 if and only if the association rule

$$A_1(1) \wedge A_4(1) \Rightarrow_{p, Base} A_2(1) \vee A_3(1)$$

is true in the data matrix $\mathcal{M}' = \langle M', f_1', f_2', f_3', f_4' \rangle \in M_{\langle 2,2,2,2 \rangle}$ outlined in the bottom part of Fig. 13.2.

We define a useful notion of 4ft-table of $\{0,1\}$-structures of types $\langle 1,1 \rangle$ and $\langle 1,1,2 \rangle$ of OPC.

Definition 13.6

1. Let $\mathcal{M} = \langle M; f_1, f_2 \rangle \in \underline{M}_{\langle 1,1 \rangle}^{\{0,1\}}$ be a $\{0,1\}$-structure of the type $\langle 1, 1 \rangle$ of OPC. Then the *4ft-table* $T_{\mathcal{M}} = \langle a_{\mathcal{M}}, b_{\mathcal{M}}, c_{\mathcal{M}}, d_{\mathcal{M}} \rangle$ of $\mathcal{M}$ is defined such that

   • $a_{\mathcal{M}}$ is the number of $o \in M$ for which $f_1(o) = f_2(o) = 1$
   • $b_{\mathcal{M}}$ is the number of $o \in M$ for which $f_1(o) = 1$ and $f_2(o) = 0$
   • $c_{\mathcal{M}}$ is the number of $o \in M$ for which $f_1(o) = 0$ and $f_2(o) = 1$
   • $d_{\mathcal{M}}$ is the number of $o \in M$ for which $f_1(o) = f_2(o) = 0$.

2. Let $\mathcal{N} = \langle M; f_1, f_2, f_3 \rangle \in \underline{M}_{\langle 1,1,2 \rangle}^{\{0,1\}}$ be a $\{0,1\}$-structure of the type $\langle 1,1,2 \rangle$ of OPC. Then a $\langle 1,1 \rangle$-*shortened structure* $\mathcal{N}$ is a structure $\mathcal{N}_{\langle 1,1 \rangle} = \langle M; f_1, f_2 \rangle$ of the type $\langle 1, 1 \rangle$ of OPC.

3. Let $\mathcal{N} = \langle M; f_1, f_2, f_3 \rangle \in \underline{M}_{\langle 1,1,2 \rangle}^{\{0,1\}}$ be a $\{0,1\}$-structure of the type $\langle 1,1,2 \rangle$ of OPC. Then the *4ft-table* $T_{\mathcal{N}}$ of $\mathcal{N}$ is defined as the 4ft-table $T_{\mathcal{N}_{\langle 1,1 \rangle}}$ of $\mathcal{N}_{\langle 1,1 \rangle}$.

Note 13.2 The associated function Asf_q of a quantifier q of the type $\langle 1,1 \rangle$ is invariant under isomorphism of the structures we deal with. This means that if $\mathscr{M} = \langle M; f_1, f_2 \rangle$ is a $\{0,1\}$-structure of the type $\langle 1,1 \rangle$ then the value $\mathrm{Asf}_q(\mathscr{M}) = \mathrm{Asf}_q(\langle M; f_1, f_2 \rangle)$ is fully determined by the 4ft-table $T_{\mathscr{M}} = \langle a_{\mathscr{M}}, b_{\mathscr{M}}, c_{\mathscr{M}}, d_{\mathscr{M}} \rangle$.

This leads to the following defintion.

Definition 13.7. Let q be a quantifier of the type $\langle 1,1 \rangle$.

1. Then a *4ft-associated function Asf_q^{4ft} of quantifier q* is defined for all quadruples $\langle a,b,c,d \rangle$ of non-negative integer numbers satisfying $a+b+c+d > 0$ such that

$$\mathrm{Asf}_q^{4ft}(a,b,c,d) = \mathrm{Asf}_q(\mathscr{M})$$

where $\mathscr{M} = \langle M; f_1, f_2 \rangle$ is a $\{0,1\}$-structure of the type $\langle 1,1 \rangle$ and 4ft-table $T_{\mathscr{M}}$ of $\mathscr{M}$ satisfies $T_{\mathscr{M}} = \langle a,b,c,d \rangle$.
2. A 4ft-quantifier $\approx$ of a logical calculus of association rules is a *4ft-associated quantifier of quantifier q* if it holds for each 4ft-table $\langle a,b,c,d \rangle$

$$F_{\approx}(a,b,c,d) = \mathrm{Asf}_q^{4ft}(a,b,c,d)$$

where $F_{\approx}$ is an associated function of $\approx$.

The correspondence between formulas $(\Rightarrow_{p,Base}, x)(P_1(x) \wedge P_4(x), P_2(x) \vee P_3(x))$ and $A_1(1) \wedge A_4(1) \Rightarrow_{p,Base} A_2(1) \vee A_3(1)$ can be generalized in points 1 – 8:

1. Let $\mathscr{P}_K$ be a MOPC of the type $\langle 1^K \rangle$ of OPC (where $\langle 1^K \rangle$ abbreviates $\underbrace{\langle 1,1,\ldots,1 \rangle}_{K-times}$) with unary predicates $P_1, \ldots, P_K$ and with a quantifier q of the type $\langle 1,1 \rangle$.
2. Let $\mathscr{LC}_K^2$ be a logical calculus of association rules of type $\mathscr{T}_K^2 = \langle 2^K \rangle$ (where $\langle 2^K \rangle$ abbreviates $\underbrace{\langle 2,2,\ldots,2 \rangle}_{K-times}$) with a 4ft-quantifier $\approx$.
3. Let 4ft-quantifier $\approx$ be a 4ft-associated quantifier of quantifier q.
4. Let φ be a quantifier-free formula of $\mathscr{P}_K$ with only variable x. Then by φ^{4ft} we mean a Boolean attribute of $\mathscr{LC}_K^2$ which we get by replacing each atomic formula $P_i(x)$ in φ by a basic Boolean attribute $A_i(1)$ of $\mathscr{LC}_K^2$. Thus, e.g. we have $(P_1(x) \wedge P_4(x))^{4ft} = A_1(1) \wedge A_4(1)$.
5. Let $\mathscr{M} = \langle M, f_1, \ldots, f_n \rangle \in \underline{M}_{\langle 1K \rangle}^{\{0,1\}}$ be a model of $\mathscr{P}_K$ and let $\mathscr{M}' = \langle M', f_1', \ldots, f_n' \rangle$ be a data matrix of type $\langle 2^K \rangle$ which belongs to the set $M_{\langle 2^K \rangle}$ such that $\mathscr{M}$ and $\mathscr{M}'$ are isomorphic. The isomorphism means that $M = \{o_1, \ldots, o_n\}$ and $M' = \{o_1', \ldots, o_n'\}$ and there is a one-one mapping ξ of M onto M' such that it holds that $f_i(o_j) = 1$ if and only if $f_i'(o_j') = 1$ for $i = 1, \ldots, K$ and $j = 1, \ldots, n$.
6. Let φ and ψ be quantifier-free formulas of $\mathscr{P}_K$ with only variable x and let $q, \approx$, $\mathscr{M}$, and $\mathscr{M}'$ be as in points 3 and 5. Then the formula $(qx)(\varphi, \psi)$ is true in $\mathscr{M}$ if and only if the association rule $\varphi^{4ft} \approx \psi^{4ft}$ is true in $\mathscr{M}'$.

7. Let φ be a Boolean attribute of $\mathscr{LC}_K^2$ created from basic Boolean attributes $A_i(1)$ and $A_i(2)$ for $i = 1, \ldots, K$. Then by $\varphi^{\mathscr{P}}$ we mean a formula of $\mathscr{P}_K$ which we get by replacing each basic Boolean attribute $A_i(1)$ by an atomic formula $P_i(x)$ and each basic Boolean attribute $A_i(2)$ by a formula $\neg P_i(x)$. Here x is a variable of $\mathscr{P}_K$ which is the same for all basic Boolean attributes.

8. Let q, $\approx$, $\mathscr{M}$, and $\mathscr{M}'$ be as in points 3 and 5 and let $\varphi^{\mathscr{P}}$ and $\psi^{\mathscr{P}}$ be according to point 7. Then the association rule $\varphi \approx \psi$ is true in $\mathscr{M}'$ if and only if the formula $(qx)(\varphi^{\mathscr{P}}, \psi^{\mathscr{P}})$ is true in $\mathscr{M}$.

Note 13.3. We will not formulate the above described correspondence as a theorem even though it is possible. A correct formulation and a proof of a corresponding theorem is thus a challenge, see also Sect. 16.9.

Anyway, this correspondence can be also applied to MOPC with equality and this leads to the following reasonable definition.

Definition 13.8. Let $\mathscr{P}$ be a MOPC (with or without equality) with unary predicates $P_1, \ldots, P_n$, $n \geq 2$. Each formula

$$(\approx x)(\varphi(x), \psi(x))$$

of $\mathscr{P}$ where $\approx$ is a quantifier of the type $\langle 1, 1 \rangle$, x is a variable and $\varphi(x)$, $\psi(x)$ are open formulas built from the unary predicates, junctors and from the variable x is an *association rule in* $\mathscr{P}$. We usually write the association rule $(\approx x)(\varphi(x), \psi(x))$ in the form $\varphi \approx \psi$.

The above introduced formula $(\Rightarrow_{p,Base}, x)(P_1(x) \wedge P_4(x), P_2(x) \vee P_3(x))$ is an example of an association rule in $\mathscr{P}_4$. It is usually written in the form of $P_1(x) \wedge P_4(x) \Rightarrow_{p,Base} P_2(x) \vee P_3(x)$.

In addition, we outline how an association rule $\varphi \approx \psi$ of logical calculus $\mathscr{LC}_{\mathscr{T}}$ of association rules of type $\mathscr{T} = \langle t_1, \ldots, t_K \rangle$ can be converted into an equivalent association rule in a suitable logical calculus $\mathscr{LC}_{\mathscr{T}_2}^2$ of association rules of type $\mathscr{T}_2$. (See also calculi with qualitative values introduced in section 3.4 in [18].) We assume that $A_1, \ldots, A_K$ are basic attributes of $\mathscr{LC}_{\mathscr{T}}$. Then the conversion can be done in steps A) – F).

A) Type $\mathscr{T}_2$ of $\mathscr{LC}_{\mathscr{T}_2}^2$ satisfies $\mathscr{T}_2 = \langle 2^{t_1 + \cdots + t_K} \rangle$ i.e. $\mathscr{T}_2 = \langle \underbrace{2, \ldots, 2}_{t_1-times}, \ldots, \underbrace{2, \ldots, 2}_{t_K-times} \rangle$.

B) Each data matrix $\mathscr{M} = \langle M, f_1, \ldots, f_K \rangle$ of type $\mathscr{T} = \langle t_1, \ldots, t_K \rangle$ where $M = \{o_1, \ldots, o_n\}$ is converted into a data matrix

$$\mathscr{M}^2 = \langle M, f_{1,1}, \ldots, f_{1,t_1}, \ldots, f_{K,1}, \ldots, f_{K,t_K} \rangle$$

of type $\mathscr{T}_2 = \langle 2^{t_1 + \cdots + t_K} \rangle$ such that it holds

$$f_i(o_u) = j \quad \text{if and only if} \quad f_{i,j}(o_u) = 1$$

for $i = 1, \ldots, K$, $j = 1, \ldots, t_i$ and $u = 1, \ldots, n$. An example is located in Fig. 13.4 where data matrix $\mathscr{M} = \langle M, f_1, \ldots, f_K \rangle$ is of the type $\langle 3, \ldots, 4 \rangle$.

$o \in M$	f_1	$\cdots$	f_K
o_1	1	$\cdots$	4
o_2	3	$\cdots$	2
$\vdots$	$\vdots$	$\ddots$	$\vdots$
o_n	2	$\cdots$	1

$o \in M$	$f_{1,1}$	$f_{1,2}$	$f_{1,3}$	$\cdots$	$f_{K,1}$	$f_{K,2}$	$f_{K,3}$	$f_{K,4}$
o_1	1	0	0	$\cdots$	0	0	0	1
o_2	0	0	1	$\cdots$	0	1	0	0
$\vdots$	$\vdots$	$\vdots$	$\vdots$	$\ddots$	$\vdots$	$\vdots$	$\vdots$	$\vdots$
o_n	0	1	0	$\cdots$	1	0	0	0

$$\mathscr{M} = \langle M, f_1, \ldots, f_K \rangle \qquad\qquad \mathscr{M} = \langle M, f_{1,1}, \ldots, f_{1,3}, \ldots, f_{K,1}, \ldots, f_{K,4} \rangle$$

Fig. 13.4 $\mathscr{M} = \langle M, f_1, \ldots, f_K \rangle$ of type $\langle 3, \ldots, 4 \rangle$ and $\mathscr{M} = \langle M, f_{1,1}, \ldots, f_{1,3}, \ldots, f_{K,1}, \ldots, f_{K,4} \rangle$

C) $A_{1,1}, \ldots, A_{1,t_1}, \ldots, A_{K,1}, \ldots, A_{K,t_k}$ are basic attributes of calculus $\mathscr{LC}^2_{\mathscr{T}_2}$.

D) Let $A_i(\alpha) = A_i(m_1, m_2, \ldots, m_k)$ be a basic Boolean attribute of $\mathscr{LC}_{\mathscr{T}}$. Then by $A_i^{(2)}(\alpha)$ we mean a disjunction $A_{i,m_1}(1) \vee A_{i,m_2}(1) \vee \ldots \vee A_{i,m_k}(1)$ i.e. a Boolean attribute of calculus $\mathscr{LC}^2_{\mathscr{T}_2}$. For example, if $A_1(2,3)$ is a basic Boolean attribute of logical calculus $\mathscr{LC}_{\langle 3, \ldots, 4 \rangle}$ of association rules of type $\langle 3, \ldots, 4 \rangle$, then

$$A_1^{(2)}(2,3) = A_{1,2}(1) \vee A_{1,3}(1) .$$

E) Let φ be a Boolean attribute of $\mathscr{LC}_{\mathscr{T}}$. Then by $\varphi^{(2)}$ we mean a formula of $\mathscr{LC}^2_{\mathscr{T}_2}$ which we get by replacing each basic Boolean attribute $A_i(\alpha)$ by a disjunction $A_i^{(2)}(\alpha)$.

F) If $\mathscr{M} = \langle M, f_1, \ldots, f_K \rangle$ is a data matrix of type $\mathscr{T} = \langle t_1, \ldots, t_K \rangle$ which is converted into a data matrix $\mathscr{M}^2 = \langle M, f_{1,1}, \ldots, f_{1,t_1}, \ldots, f_{K,1}, \ldots, f_{K,t_K} \rangle$ of type $\mathscr{T}_2 = \langle 2^{t_1 + \ldots + t_K} \rangle$ in a way introduced above, then

- $A_i(\alpha) = A_i(m_1, m_2, \ldots, m_k)$ is true in $\mathscr{M}$ if and only if $A_i^{(2)}(\alpha)$ is true in $\mathscr{M}^2$
- φ is true for o in $\mathscr{M}$ if and only if $\varphi^{(2)}$ is true for o in $\mathscr{M}^2$
- association rule $\varphi \approx \psi$ is true in $\mathscr{M}$ if and only if association rule $\varphi^{(2)} \approx \psi^{(2)}$ is true in $\mathscr{M}^2$.

Note 13.4. We will not formulate the above described conversion as a theorem even though it is possible. A correct formulation and a proof of a corresponding theorem is thus a challenge, see also Sect. 16.9.

However, it is important that association rules of the calculus $\mathscr{LC}^2_{\mathscr{T}_2}$ correspond in a way outlined in the points 1 – 8 above to association rules in a suitable MOPC of type $\langle 1^{t_1 + \ldots + t_K} \rangle$.

We have shown that each association rule $\varphi \approx \psi$ can be reasonably converted into a formula of a monadic observational calculus of the form $(\approx x)(\varphi'(x), \psi'(x))$ which is usually written as $\varphi' \approx \psi'$, where $\approx$ is a 4ft-quantifier. Thus, it is an interesting question which association rules can be expressed by predicates, variables, and classical quantifiers $\forall$, $\exists$, Boolean connectives and by the predicate of equality. The remainder of the chapter focuses on answering this question.

13.3 Tharp's Theorem

The following definition comes from section 2.2.9 in [18].

Definition 13.9. Let $\mathscr{P}$ be an OPC. Suppose that φ and ψ are formulas such that $FV(\varphi) = FV(\psi)$ (φ and ψ have the same free variables). Then φ and ψ are said to be *logically equivalent* if $\|\varphi\|_{\mathscr{M}} = \|\psi\|_{\mathscr{M}}$ for each model $\mathscr{M}$.

Please note that the equality $\|\varphi\|_{\mathscr{M}} = \|\psi\|_{\mathscr{M}}$ in the previous definition is in general an equality of functions. We will use it for the closed formulas. In this case it is the equality of two values.

We are going to deal with MOPC's with equality. A MOPC with equality is a result of adding an additional binary predicate = to MOPC. If MOPC $\mathscr{P}$ without equality has n predicates then $\langle 1^n \rangle = \underbrace{\langle 1, 1, \ldots, 1 \rangle}_{n-times}$ is a type of OPC of $\mathscr{P}$. If we add the predicate = to MOPC then a type of OPC of resulting MOPC with equality is $\underbrace{\langle 1, 1, \ldots, 1, 2 \rangle}_{n-times}$. In a following definition we identify a type of MOPC without equality with a type of MOPC with equality.

Definition 13.10. Let $\mathscr{P}$ be a MOPC of the type $\langle 1^n \rangle$ of OPC and let $\mathscr{P}^=$ be a MOPC with equality which results by adding a binary predicate = of equality to $\mathscr{P}$. Then

- $\mathscr{P}^=$ is a *MOPC with equality of the type $\langle 1^n \rangle$ of OPC*
- type $\langle 1^n \rangle$ of OPC is a *type of MOPC $\mathscr{P}^=$ with equality*.

The following definition comes from section 3.1.5 in [18].

Definition 13.11. Let $\mathscr{P}$ be a MOPC (including MOPC with equality) of type $\langle 1^n \rangle$ and let q be a quantifier of type $\langle 1^k \rangle$, $k \leq n$. Then q is *definable in* $\mathscr{P}$ if there is a sentence Φ of $\mathscr{P}$ not containing q such that the sentence

$$(qx)(P_1(x), \ldots, P_k(x))$$

is logically equivalent to Φ.

The following theorem concerns the problem of definability of quantifiers. It is proved in section 3.1.26 in [18]. Here, if $\mathscr{M} = \langle M, f_1, \ldots, f_k \rangle$ and $\mathscr{N} = \langle N, g_1, \ldots, g_k \rangle$ are models of a CMOPC of type $\langle 1^k \rangle$ of OPC with equality, then $\mathscr{M} \subseteq \mathscr{N}$ means that $M \subseteq N$ and f_i is a restriction of g_i to M for $i = 1, \ldots, k$.

Theorem 13.1. *(Tharp) Let $\mathscr{P}^=$ be a CMOPC with equality and unary predicates $P_1, \ldots, P_n$ and let $\mathscr{P}'$ be its extension by adding a quantifier q of type $\langle 1^k \rangle$ ($k \leq n$). Then q is definable in $\mathscr{P}'$ if and only if there is an non-negative integer number m such that the following holds for $\varepsilon \in \{0, 1\}$ and each model $\mathscr{M}$ of type $\langle 1^k \rangle$:*

$$Asf_q(\mathscr{M}) = \varepsilon \text{ if and only if } (\exists \mathscr{M}_0 \subseteq \mathscr{M})$$

$$(\mathscr{M}_0 \text{ has } \leq m \text{ elements and } (\forall \mathscr{M}_1)(\mathscr{M}_0 \subseteq \mathscr{M}_1 \subseteq \mathscr{M} \text{ implies } Asf_q(\mathscr{M}_1) = \varepsilon)).$$

Proof. See section 3.1.26 in [18].   □

We are interested in the question of which association rules can be expressed by predicates, variables, and classical quantifiers $\forall$, $\exists$, Boolean connectives and by the predicate of equality. We call such association rules classically definable, see the following definition.

Definition 13.12. Let $\mathscr{P}^{=}$ be a CMOPC with equality and unary predicates $P_1, \ldots, P_n$ ($n \geq 2$), let $\mathscr{P}^{\approx}$ be its extension by adding a quantifier $\approx$ of type $\langle 1, 1 \rangle$. Then the quantifier $\approx$ is *classically definable* if it is definable in $\mathscr{P}^{\approx}$ and the association rule $\varphi \approx \psi$ is *classically definable* if the quantifier $\approx$ is definable in $\mathscr{P}^{\approx}$.

Tharp's theorem 13.1 is a criterion of classical definability of association rules $\varphi \approx \psi$. We show that there is an additional criterion of classical definability of association rules $\varphi \approx \psi$ closely related to the associated function $F_{\approx}$ of the 4ft-quantifier $\approx$. This criterion can be simplified for some classes of 4ft-quantifiers. This additional criterion is based on the normal form theorem introduced in Sect. 13.4.

13.4 Normal Form Theorem

The normal form theorem is proved in section 3.1.25 in [18] and presented as theorem 13.2 below. It uses several notions and a lemma introduced in sections 3.1.2, 3.1.16, 3.1.17, and 3.1.24 in [18]. These notions and lemma are also presented here, see definitions 13.13, 13.14 and lemma 13.1.

Definition 13.13

1. (see section 3.1.2 in [18]) Let $\mathscr{P}$ be a MOPC with predicates $P_1, \ldots, P_n$.

   - The first variable x_0 is called the *designated variable*. Open (= quantifier free) formulas containing no variable distinct from the designated variable x are called *designated open formulas*.
   - An *n-ary card* is a sequence $\langle u_i; 1 \leq i \leq n \rangle$ of n zeros and ones. If $\mathscr{M} = \langle M, p_1, \ldots, p_n \rangle$ is a model of $\mathscr{P}$ and if $o \in M$ then the $\mathscr{M}$-*card of* o is the tuple $C_{\mathscr{M}}(o) = \langle p_1(o), \ldots, p_n(o) \rangle$; it is evidently an n-ary card.

2. (see section 3.1.16 in [18]) For each integer number $k > 0$, $\exists^k$ is a quantifier of the type $\langle 1 \rangle$ whose associated function is defined as follows: For each finite model $\mathscr{M} = \langle M, f_1 \rangle$ it holds that $Asf_{\exists^k}(\mathscr{M}) = 1$ if and only if there are at least k elements $o \in M$ such that $f_1(o) = 1$.

Lemma 13.1. *Let $k > 0$ be an integer number and let $\mathscr{P}^k$ be the extension of CMOPC with equality $\mathscr{P}^{=}$ by adding $\exists^k$. Then $\exists^k$ is definable by the formula*

$$\phi : (\exists x_1, \ldots, x_k)(\bigwedge_{i \neq j, 1 \leq i, j \leq k} x_i \neq x_j \wedge \bigwedge_{1 \leq i \leq k} P_1(x_i))$$

Proof. See section 3.1.17 in [18].   □

Definition 13.14. (see section 3.1.24 in [18])

1. Let $u = \langle u_1, \ldots, u_n \rangle$ be an n-card. Then an *elementary conjunction given by* u is the conjunction

$$\kappa_u = \bigwedge_{i=1}^{n} \lambda_i$$

where x is the designated variable and for $i = 1, \ldots, n$ it holds:

- λ_i is $P_i(x)$ if $u_i = 1$
- λ_i is $\neg P_i(x)$ if $u_i = 0$.

2. Each formula of the form $(\exists^k x)\kappa_u$ where u is a card and x is the designated variable is called a *canonical sentence for CMOPC with equality.*

The further results concerning classical definability of association rules are based on the following normal form theorem proved in [18].

Theorem 13.2. *Let $\mathscr{P}^=$ be a CMOPC with equality and let $\mathscr{P}^*$ be the extension of $\mathscr{P}^=$ by adding the quantifiers $\exists^k$ ($k > 0$ is an integer number). Let Φ be a sentence from $\mathscr{P}^=$. Then there is a sentence Φ^* from $\mathscr{P}^*$ logically equivalent to Φ (in $\mathscr{P}^*$) and such that Φ^* is a Boolean combination of canonical sentences. (In particular, Φ^* contains neither the equality predicate nor any variable distinct from the canonical variable).*

Proof. See section 3.1.25 in [18].   □

13.5   Classical Definability of Association Rules

Tharp's theorem 13.1 proved in [18] can be used as a criterion of classical definability of association rules. The main result of this section is theorem 13.4 which is an alternative criterion of classical definability of association rules. This theorem shows that there is a relatively simple condition concerning given 4ft-quantifier $\approx$ which is equivalent to the classical definability of the rule $\varphi \approx \psi$. First we introduce some notions and prove several lemmas and one theorem.

Definition 13.15. Let $\mathscr{P}^=$ be a CMOPC with equality of type $\langle 1, 1 \rangle$ of OPC with predicates P_1 and P_2 and the designated variable x. Then we denote:

- $\kappa_a = P_1(x) \wedge P_2(x)$
- $\kappa_b = P_1(x) \wedge \neg P_2(x)$
- $\kappa_c = \neg P_1(x) \wedge P_2(x)$
- $\kappa_d = \neg P_1(x) \wedge \neg P_2(x)$.

Lemma 13.2. *Let $\mathscr{P}^=$ be a CMOPC with equality of type $\langle 1, 1 \rangle$ of OPC with predicates P_1 and P_2. Let $\mathscr{P}^*$ be an extension of $\mathscr{P}^=$ by adding all quantifiers $\exists^k$ (k* *is an integer positive number). Let* Φ *be a Boolean combination of canonical sentences of the calculus* $\mathscr{P}^*$ *(see definition 13.14). Then* Φ *is logically equivalent to the formula*

$$\bigvee_{i=1}^{K} \varphi_a^{(i)} \wedge \varphi_b^{(i)} \wedge \varphi_c^{(i)} \wedge \varphi_d^{(i)}$$

where $K \geq 0$ *and* $\varphi_a^{(i)}$ *is in one of the following formulas for each* $i = 1,\dots,K$:

- $(\exists^k x)\kappa_a$ *where* k *is a positive integer number*
- $\neg(\exists^k x)\kappa_a$ *where* k *is a positive integer number*
- $(\exists^k x)\kappa_a \wedge \neg(\exists^l x)\kappa_a$ *where* $0 < k < l$ *are positive integer numbers.*
- $\underline{1}$,

similarly for $\varphi_b^{(i)}$, $\varphi_c^{(i)}$, $\varphi_d^{(i)}$. *Let us note that the value of an empty disjunction is* $\underline{0}$.

Proof. Φ *is a Boolean combination of canonical sentences of the calculus* $\mathscr{P}^*$, *and thus we can write it in the form*

$$\bigvee_{i=1}^{K'} \bigwedge_{j=1}^{L_i} \psi_{i,j}$$

where $K' \geq 0$ *and each* $\psi_{i,j}$ *is the canonical sentence of the calculus* $\mathscr{P}^*$ *or a negation of such a canonical sentence. We create from each formula* $\bigwedge_{j=1}^{L_i} \psi_{i,j}$ $(i = 1,\dots,K')$ *a new formula*

$$\varphi_a^{(i)} \wedge \varphi_b^{(i)} \wedge \varphi_c^{(i)} \wedge \varphi_d^{(i)}$$

where all the $\varphi_a^{(i)}$, $\varphi_b^{(i)}$, $\varphi_c^{(i)}$ *and* $\varphi_d^{(i)}$ *are in the required form or equal to* $\underline{0}$.
  We can suppose without loss of generality that

$$\bigwedge_{j=1}^{L_i} \psi_{i,j} = \bigwedge_{j=1}^{A} \psi_{a,j} \wedge \bigwedge_{j=1}^{B} \psi_{b,j} \wedge \bigwedge_{j=1}^{C} \psi_{c,j} \wedge \bigwedge_{j=1}^{D} \psi_{d,j}$$

where each formula $\psi_{a,j}$ $(j = 1,\dots,A)$ *is equal to* $(\exists^k x)\kappa_a$ *or to* $\neg(\exists^k x)\kappa_a$ *where* k *is a positive integer number, analogously for* $\psi_{b,j}$, $\psi_{c,j}$ *and* $\psi_{d,j}$.
  If $A = 0$ *then we define* $\varphi_a^{(i)} = \underline{1}$ *and* $\varphi_a^{(i)}$ *is logically equivalent to* $\bigwedge_{j=1}^{A} \psi_{a,j}$ *because the value of an empty conjunction is* $\underline{1}$.
  If $A = 1$ *then we define* $\varphi_a^{(i)} = \psi_{a,1}$ *and* $\varphi_a^{(i)}$ *is again logically equivalent to* $\bigwedge_{j=1}^{A} \psi_{a,j}$.
  If $A \geq 2$ *then we can suppose without loss of generality that*

$$\bigwedge_{j=1}^{A} \psi_{a,j} = \bigwedge_{j=1}^{A_1} (\exists^{k_j} x)\kappa_a \wedge \bigwedge_{j=1}^{A_2} \neg(\exists^{l_j} x)\kappa_a .$$

If $A_1 > 0$ then we define $k = max\{k_j | j = 1,\dots,A_1\}$ and thus $\bigwedge_{j=1}^{A_1}(\exists^{k_j}x)\kappa_a$ is logically equivalent to $(\exists^k x)\kappa_a$.

If $A_2 > 0$ then we define $l = min\{l_j | j = 1,\dots,A_2\}$ and thus $\bigwedge_{j=1}^{A_2}\neg(\exists^{l_j}x)\kappa_a$ is logically equivalent to $\neg(\exists^l x)\kappa_a$.

In the case $A \geq 2$ we define the formula $\varphi_a^{(i)}$ this way:

- *if $A_1 = 0$ then $\varphi_a^{(i)} = \neg(\exists^l x)\kappa_a$*
- *if $A_2 = 0$ then $\varphi_a^{(i)} = (\exists^k x)\kappa_a$*
- *if $A_1 > 0 \wedge A_2 > 0 \wedge k < l$ then $\varphi_a^{(i)} = (\exists^k x)\kappa_a \wedge \neg(\exists^l x)\kappa_a$*
- *if $A_1 > 0 \wedge A_2 > 0 \wedge k \geq l$ then $\varphi_a^{(i)} = \underline{0}$.*

We defined the formula $\varphi_a^{(i)}$ for all the possible cases (i.e. $A = 0$, $A = 1$ and $A \geq 2$) and in all cases $\bigwedge_{j=1}^A \psi_{a,j}$ is logically equivalent to $\varphi_a^{(i)}$ and $\varphi_a^{(i)}$ is in the required form or equal to $\underline{0}$.

We analogously create $\varphi_b^{(i)}$, $\varphi_c^{(i)}$ and $\varphi_d^{(i)}$ thus they are equivalent to $\bigwedge_{j=1}^B \psi_{b,j}$, $\bigwedge_{j=1}^C \psi_{c,j}$ and $\bigwedge_{j=1}^D \psi_{d,j}$ respectively $(i = 1,\dots,K')$. Thus also the formulas

$$\bigwedge_{j=1}^{L_i} \psi_{i,j} \quad and \quad \varphi_a^{(i)} \bigwedge \varphi_b^{(i)} \bigwedge \varphi_c^{(i)} \bigwedge \varphi_d^{(i)}$$

are logically equivalent. This means that also the formulas

$$\bigvee_{i=1}^{K'} \bigwedge_{j=1}^{L_i} \psi_{i,j} \quad and \quad \bigvee_{i=1}^{K'} \varphi_a^{(i)} \wedge \varphi_b^{(i)} \wedge \varphi_c^{(i)} \wedge \varphi_d^{(i)}$$

are logically equivalent. Furthermore, all the $\varphi_a^{(i)}$, $\varphi_b^{(i)}$, $\varphi_c^{(i)}$ and $\varphi_d^{(i)}$ are in the required form or equal to $\underline{0}$.

Finally we omit all conjunctions $\varphi_a^{(i)} \wedge \varphi_b^{(i)} \wedge \varphi_c^{(i)} \wedge \varphi_d^{(i)}$ with at least one member equal to $\underline{0}$ and we arrive at the required formula

$$\bigvee_{i=1}^K \varphi_a^{(i)} \wedge \varphi_b^{(i)} \wedge \varphi_c^{(i)} \wedge \varphi_d^{(i)}.$$

This finishes the proof. □

Definition 13.16. Let $\mathscr{N}$ be the set of all non-negative integer numbers. Then we define:

- An *interval* in $\mathscr{N}^4$ is a set

$$I = I_1 \times I_2 \times I_3 \times I_4$$

such that $I_1 = \langle k,l \rangle$ or $I_1 = \langle k,\infty)$ where $0 \leq k < l$ are non-negative integer numbers, similarly for I_2, I_3, I_4. The empty set $\emptyset$ is also an interval in $\mathscr{N}^4$.

- Let $T = \langle a, b, c, d \rangle$ be a 4ft-table (see definition 3.3) and let $I = I_1 \times I_2 \times I_3 \times I_4$ be an interval in $\mathcal{N}^4$. Then

$$T \in I \quad \text{if and only if} \quad a \in I_1 \wedge b \in I_2 \wedge c \in I_3 \wedge d \in I_4.$$

Theorem 13.3. *Let $\mathscr{P}^=$ be CMOPC with equality of the type $\langle 1, 1 \rangle$ of OPC and let Φ be a sentence of $\mathscr{P}^=$. Then there are K intervals $I_1, \ldots, I_K$ in $\mathcal{N}^4$ where $K \geq 0$ such that for each model $\mathcal{M}$ of the calculus $\mathscr{P}^=$ it holds that*

$$||\Phi||_{\mathcal{M}} = 1 \quad \text{if and only if} \quad T_{\mathcal{M}} \in \bigcup_{j=1}^{K} I_j$$

where $T_{\mathcal{M}} = \langle a_{\mathcal{M}}, b_{\mathcal{M}}, c_{\mathcal{M}}, d_{\mathcal{M}} \rangle$ is a 4ft-table of the model $\mathcal{M}$, see definition 13.6. (It holds $\bigcup_{j=1}^{0} I_j = \emptyset$.)

Proof. Let $\mathscr{P}^*$ be the extension of the calculus $\mathscr{P}^=$ by adding the quantifiers $\exists^k$ $(k > 0)$. Then according to theorem 13.2 there is a sentence Φ^* of the calculus $\mathscr{P}^*$ such that Φ^* is a Boolean combination of canonical sentences and Φ^* is equivalent to Φ.

Furthermore, according to lemma 13.2 the formula Φ^* is equivalent to the formula

$$\bigvee_{i=1}^{K} \varphi_a^{(i)} \wedge \varphi_b^{(i)} \wedge \varphi_c^{(i)} \wedge \varphi_d^{(i)}$$

where $K \geq 0$ and $\varphi_a^{(i)}$ is one of the following formulas for each $i = 1, \ldots, K$:

- $(\exists^k x) \kappa_a$ *where k is a positive integer number*
- $\neg(\exists^k x) \kappa_a$ *where k is a positive integer number*
- $(\exists^k x) \kappa_a \wedge \neg(\exists^l x) \kappa_a$ *where $0 < k < l$ are positive integer numbers.*
- $\underline{1}$,

similarly for $\varphi_b^{(i)}$, $\varphi_c^{(i)}$, $\varphi_d^{(i)}$.

If $K = 0$ then $||\Phi||_{\mathcal{M}} = \underline{0}$ because the value of empty disjunction is $\underline{0}$. If $K = 0$ then also $\bigcup_{j=1}^{0} I_j = \emptyset$ and thus $T \notin \bigcup_{j=1}^{K} I_j$. This means that for $K = 0$ we have $||\Phi||_{\mathcal{M}} = 1$ if and only if $T_{\mathcal{M}} \in \bigcup_{j=1}^{K} I_j$.

If $K > 0$ then we create interval I_j for each $j = 1, \ldots, K$ in the following steps:

- *If $\varphi_a^{(j)} = (\exists^k x) \kappa_a$ then we define $I_a^{(j)} = \langle k, \infty)$.*
- *If $\varphi_a^{(j)} = \neg(\exists^k x) \kappa_a$ then we define $I_a^{(j)} = \langle 0, k)$.*
- *If $\varphi_a^{(j)} = (\exists^k x) \kappa_a \wedge \neg(\exists^l x) \kappa_a$ then we define $I_a^{(j)} = \langle k, l)$.*
- *If $\varphi_a^{(j)} = \underline{1}$ then we define $I_a^{(j)} = \langle 0, \infty)$.*
- *We analogously define $I_b^{(j)}$, $I_c^{(j)}$ and $I_d^{(j)}$ using $\varphi_b^{(j)}$, $\varphi_c^{(j)}$ and $\varphi_d^{(j)}$ respectively.*
- *We finally define $I_j = I_a^{(j)} \times I_b^{(j)} \times I_c^{(j)} \times I_d^{(j)}$*

We prove that for each model $\mathcal{M}$ of the calculus $\mathscr{P}^=$ it holds that

$$\|\Phi\|_{\mathcal{M}} = 1 \quad \text{if and only if} \quad T_{\mathcal{M}} \in \bigcup_{j=1}^{K} I_j \,.$$

We use the fact that the formula Φ is equivalent to $\bigvee_{i=1}^{K} \varphi_a^{(i)} \wedge \varphi_b^{(i)} \wedge \varphi_c^{(i)} \wedge \varphi_d^{(i)}$. We assume that $T_{\mathcal{M}} = \langle a_{\mathcal{M}}, b_{\mathcal{M}}, c_{\mathcal{M}}, d_{\mathcal{M}} \rangle$.

Let $\|\Phi\|_{\mathcal{M}} = 1$ i.e. $\|\bigvee_{i=1}^{K} \varphi_a^{(i)} \wedge \varphi_b^{(i)} \wedge \varphi_c^{(i)} \wedge \varphi_d^{(i)}\|_{\mathcal{M}} = 1$. Thus there is $j \in \{1,\dots,K\}$ such that $\|\varphi_a^{(j)} \wedge \varphi_b^{(j)} \wedge \varphi_c^{(j)} \wedge \varphi_d^{(j)}\|_{\mathcal{M}} = 1$ and this means that also $\|\varphi_a^{(j)}\|_{\mathcal{M}} = 1$. The interval $I_a^{(j)}$ is constructed such that the fact $\|\varphi_a^{(j)}\|_{\mathcal{M}} = 1$ implies $a_{\mathcal{M}} \in I_a^{(j)}$. Analogously we get $b_{\mathcal{M}} \in I_b^{(j)}$, $c_{\mathcal{M}} \in I_c^{(j)}$ and $d_{\mathcal{M}} \in I_d^{(j)}$. This means that $T_{\mathcal{M}} \in I_j$ and also $T_{\mathcal{M}} \in \bigcup_{j=1}^{K} I_j$.

Let $\|\Phi\|_{\mathcal{M}} = 0$. Then it also holds that $\|\varphi_a^{(j)} \wedge \varphi_b^{(j)} \wedge \varphi_c^{(j)} \wedge \varphi_d^{(j)}\|_{\mathcal{M}} = 0$ for each $j = 1,\dots,K$. Thus also for each such j it holds $\|\varphi_a^{(j)}\|_{\mathcal{M}} = 0$ or $\|\varphi_b^{(j)}\|_{\mathcal{M}} = 0$ or $\|\varphi_c^{(j)}\|_{\mathcal{M}} = 0$ or $\|\varphi_d^{(j)}\|_{\mathcal{M}} = 0$. If $\|\varphi_a^{(j)}\|_{\mathcal{M}} = 0$ then it must hold $a_{\mathcal{M}} \notin I_a^{(j)}$ and this means $T_{\mathcal{M}} \notin I_j$. Analogously we get $T_{\mathcal{M}} \notin I_j$ for $\|\varphi_b^{(j)}\|_{\mathcal{M}} = 0$, $\|\varphi_c^{(j)}\|_{\mathcal{M}} = 0$ and $\|\varphi_d^{(j)}\|_{\mathcal{M}} = 0$. Thus $T_{\mathcal{M}} \notin \bigcup_{j=1}^{K} I_j$. This finishes the proof. $\qquad\square$

Lemma 13.3. *Let $\langle a,b,c,d \rangle$ be a 4ft-table. Then there is a $\{0,1\}$ structure $\mathcal{M} = \langle M, f, g \rangle$ of the type $\langle 1,1 \rangle$ of OPC such that*

$$T_{\mathcal{M}} = \langle a_{\mathcal{M}}, b_{\mathcal{M}}, c_{\mathcal{M}}, d_{\mathcal{M}} \rangle = \langle a,b,c,d \rangle$$

where $T_{\mathcal{M}}$ is the 4ft-table of $\mathcal{M}$, see definition 13.6.

Proof. We construct $\mathcal{M} = \langle M, f, g \rangle$ such that M has $a+b+c+d$ elements and the functions f, g are defined according to Fig. 13.5.

element of M	f	g
$o_1,\dots,o_a$	1	1
$o_{a+1},\dots,o_{a+b}$	1	0
$o_{a+b+1},\dots,o_{a+b+c}$	0	1
$o_{a+b+c+1},\dots,o_{a+b+c+d}$	0	0

Fig. 13.5 Structure $\mathcal{M}$ for which $T_{\mathcal{M}} = \langle a,b,c,d \rangle$

This finishes the proof. $\qquad\square$

Theorem 13.4. *Let $\mathscr{P}^=$ be a CMOPC with equality of the type $\langle 1,1 \rangle$ of OPC, let $\approx$ be a quantifier of the type $\langle 1,1 \rangle$ and let $\mathscr{P}'$ be an extension of $\mathscr{P}^=$ by adding the quantifier $\approx$. Then $\approx$ is definable in $\mathscr{P}'$ if and only if there are K intervals $I_1,\dots,I_K$ in $\mathcal{N}^4$, $K \geq 0$ such that for each 4ft-table $\langle a,b,c,d \rangle$ it holds that*

$$Asf_{\approx}^{Aft}(a,b,c,d) = 1 \ \ if \ and \ only \ if \ \ \langle a,b,c,d \rangle \in \bigcup_{j=1}^{K} I_j$$

where $Asf_{\approx}^{Aft}$ is the 4ft-associated function of $\approx$, see definition 13.7.

Proof. We have to prove:

1. If $\approx$ is definable in $\mathscr{P}'$ then there are intervals $I_1, \ldots, I_K$ in $\mathscr{N}^4$, $K \geq 0$ and for each 4ft-table $\langle a,b,c,d \rangle$ $Asf_{\approx}^{Aft}(a,b,c,d) = 1$ if and only if $\langle a,b,c,d \rangle \in \bigcup_{j=1}^{K} I_j$.
2. If there are intervals $I_1, \ldots, I_K$ in $\mathscr{N}^4$, $K \geq 0$ such that for each 4ft table $\langle a,b,c,d \rangle$ $Asf_{\approx}^{Aft}(a,b,c,d) = 1$ if and only if $\langle a,b,c,d \rangle \in \bigcup_{j=1}^{K} I_j$ then $\approx$ is definable in $\mathscr{P}'$.

ad 1.: If $\approx$ is definable in $\mathscr{P}'$ then there is a sentence Φ of the calculus $\mathscr{P}^{=}$ that is logically equivalent to the sentence $(\approx x)(P_1(x), P_2(x))$. This means that

$$||\Phi||_{\mathscr{M}} = ||(\approx x)(P_1(x), P_2(x))||_{\mathscr{M}}$$

for each model $\mathscr{M}$ of $\mathscr{P}'$.

If $\mathscr{M} = \langle M, f, g, h \rangle$ is a model of $\mathscr{P}'$ (function h interprets the predicate of equality, type of OPC of $\mathscr{M}$ is $\langle 1,1,2 \rangle$), then

$$||(\approx x)(P_1(x), P_2(x))||_{\mathscr{M}} = Asf_{\approx}(\langle M, ||P_1(x)||_{\mathscr{M}}^{0}(o), ||P_2(x)||_{\mathscr{M}}^{0}(o) \rangle)$$

where both $||P_1(x)||_{\mathscr{M}}^{0}$ and $||P_2(x)||_{\mathscr{M}}^{0}$ are unary functions obtained such that we let x vary over M and for $o \in M$ it holds

$$||P_1(x)||_{\mathscr{M}}^{0}(o) = ||P_1(x)||_{\mathscr{M}}[\frac{x}{o}] \quad and \quad ||P_2(x)||_{\mathscr{M}}^{0}(o) = ||P_2(x)||_{\mathscr{M}}[\frac{x}{o}],$$

see definition 13.4. This means $\langle M, ||P_1(x)||_{\mathscr{M}}^{0}(o), ||P_2(x)||_{\mathscr{M}}^{0}(o) \rangle = \langle M, f, g \rangle$. In addition, it holds that

$$Asf_{\approx}(\langle M, ||P_1(x)||_{\mathscr{M}}^{0}(o), ||P_2(x)||_{\mathscr{M}}^{0}(o) \rangle) = Asf_{\approx}^{Aft}(a_{\mathscr{M}}, b_{\mathscr{M}}, c_{\mathscr{M}}, d_{\mathscr{M}})$$

where $T_{\mathscr{M}} = \langle a_{\mathscr{M}}, b_{\mathscr{M}}, c_{\mathscr{M}}, d_{\mathscr{M}} \rangle$ is the 4ft-table of the model $\langle M, f, g, h \rangle$, see definitions 13.6 and 13.7. We can conclude that

$$||\Phi||_{\mathscr{M}} = Asf_{\approx}^{Aft}(a_{\mathscr{M}}, b_{\mathscr{M}}, c_{\mathscr{M}}, d_{\mathscr{M}})$$

for each model $\mathscr{M} = \langle M, f, g, h \rangle$ of $\mathscr{P}'$. In addition, $\mathscr{M} = \langle M, f, g, h \rangle$ is also a model of $\mathscr{P}^{=}$ and from the same reason as above the same is true for each model $\mathscr{M} = \langle M, f, g, h \rangle$ of $\mathscr{P}^{=}$.

According to theorem 13.3 there are K intervals $I_1, \ldots, I_K$ in $\mathscr{N}^4$ where $K \geq 0$ such that for each model $\mathscr{M}$ of the calculus $\mathscr{P}^{=}$ it holds that

$$||\Phi||_{\mathscr{M}} = 1 \ \ if \ and \ only \ if \ \ T_{\mathscr{M}} \in \bigcup_{j=1}^{K} I_j .$$

We show that for each 4ft-table $\langle a,b,c,d \rangle$ it holds that

$$Asf_{\approx}^{Aft}(a,b,c,d) = 1 \quad \text{if and only if} \quad \langle a,b,c,d \rangle \in \bigcup_{j=1}^{K} I_j.$$

Let $\langle a,b,c,d \rangle$ be a 4ft-table. Then, according to lemma 13.3, there is a model $\mathcal{M}_0 = \langle M,f,g \rangle$ of the type $\langle 1,1 \rangle$ of OPC such that $T_{\mathcal{M}_0} = \langle a,b,c,d \rangle$. Then the model $\mathcal{M} = \langle M,f,g,h \rangle$ of the type $\langle 1,1,2 \rangle$ of OPC where h interprets predicate of equality is a model of $\mathcal{P}^=$ and also $T_{\mathcal{M}} = \langle a,b,c,d \rangle$.
Let $Asf_{\approx}^{Aft}(a,b,c,d) = 1$. This means that

$$||\Phi||_{\mathcal{M}} = Asf_{\approx}^{Aft}(a,b,c,d) = 1$$

and thus $\langle a,b,c,d \rangle \in \bigcup_{j=1}^{K} I_j$. Let $Asf_{\approx}(a,b,c,d) = 0$. This means that

$$||\Phi||_{\mathcal{M}} = Asf_{\approx}^{Aft}(a,b,c,d) = 0$$

and thus $\langle a,b,c,d \rangle \notin \bigcup_{j=1}^{K} I_j$.
This finishes the proof of point 1.
ad 2.: Let us assume that there are K intervals $I_1, \ldots, I_K$ in $\mathcal{N}^4$, $K \geq 0$ such that

$$Asf_{\approx}^{Aft}(a,b,c,d) = 1 \quad \text{if and only if} \quad \langle a,b,c,d \rangle \in \bigcup_{j=1}^{K} I_j.$$

We prove that $\approx$ is definable $\mathcal{P}'$. We use the fact that for each model $\mathcal{M}$ of $\mathcal{P}'$ it holds that $||(\approx x)(P_1(x),P_2(x))||_{\mathcal{M}} = Asf_{\approx}^{Aft}(a,b,c,d)$ where $T_{\mathcal{M}} = \langle a,b,c,d \rangle$, see the proof for point 1.
Let $K = 0$ then $\bigcup_{j=1}^{0} I_j = \emptyset$. This means that $Asf_{\approx}^{Aft}(a,b,c,d) = 0$ for each table $\langle a,b,c,d \rangle$. Thus for each model $\mathcal{M}$ it holds $||(\approx x)(P_1(x),P_2(x))||_{\mathcal{M}} = 0$ which means that $\approx$ is definable (e.g. by the formula $(\exists x)(x \neq x)$).
Let $K > 0$. Then we denote $\mathcal{P}^*$ the extension of $\mathcal{P}^=$ by adding all the quantifiers $\exists^k$ for $k > 0$. We create for each $j = 1,\ldots,K$ the formula

$$\psi_j = \varphi_a^{(j)} \wedge \varphi_b^{(j)} \wedge \varphi_c^{(j)} \wedge \varphi_d^{(j)}$$

in the following way (see also lemma 13.2 and theorem 13.3). We assume that $I_j = I_a \times I_b \times I_c \times I_d$.

- *If $I_a = \langle 0,\infty)$ then we define $\varphi_a^{(j)} = \underline{1}$.*
- *If $I_a = \langle k,\infty)$ then we define $\varphi_a^{(j)} = (\exists^k x)\kappa_a$.*
- *If $I_a = \langle 0,k)$ then we define $\varphi_a^{(j)} = \neg(\exists^k x)\kappa_a$.*
- *If $I_a = \langle k,l)$ then we define $\varphi_a^{(j)} = (\exists^k x)\kappa_a \wedge \neg(\exists^l x)\kappa_a$.*
- *We analogously define $\varphi_b^{(j)}$, $\varphi_c^{(j)}$ and $\varphi_d^{(j)}$ using I_b, I_c and I_d respectively.*

The formula ψ_j is created such that for each model $\mathcal{M}$ of the calculus $\mathscr{P}^*$ it holds that $||\psi_j||_{\mathcal{M}} = 1$ if and only if $T_{\mathcal{M}} \in I_j$, see also the proof of lemma 13.2. Let us define

$$\Phi = \bigvee_{j=1}^{K} \psi_j .$$

We show that the formula $(\approx x)(P_1(x), P_2(x))$ is logically equivalent to Φ in the calculus $\mathscr{P}^*$.

Let $||(\approx x)(P_1(x), P_2(x))||_{\mathcal{M}} = 1$ where $\mathcal{M}$ is a model of $\mathscr{P}'$. This means that $Asf_{\approx}^{Aft}(a,b,c,d) = 1$ where $T_{\mathcal{M}} = \langle a,b,c,d \rangle$. We assume $Asf_{\approx}^{Aft}(a,b,c,d) = 1$ if and only if $\langle a,b,c,d \rangle \in \bigcup_{j=1}^{K} I_j$ and thus there is $p \in \{1,\ldots,K\}$ such that $T_{\mathcal{M}} \in I_p$ and therefore $||\psi_p||_{\mathcal{M}} = 1$ and also $||\Phi||_{\mathcal{M}} = 1$.

Let $||\Phi||_{\mathcal{M}} = 1$. Thus there is $p \in \{1,\ldots,K\}$ such that $||\psi_p||_{\mathcal{M}} = 1$ which implies $T_{\mathcal{M}} \in I_p$ and also $T_{\mathcal{M}} \in \bigcup_{j=1}^{K} I_j$. According to the assumption this means $Asf_{\approx}^{Aft}(T_{\mathcal{M}}) = 1$ and thus $||(\approx x)(P_1(x), P_2(x))||_{\mathcal{M}} = 1$ in $\mathscr{P}^*$.

The quantifier $\exists^k$ is for each $k > 0$ definable in the extension of $\mathscr{P}^=$ by adding $\exists^k$ (see lemma 13.1). This means that there is a formula Φ^* of the calculus $\mathscr{P}^=$ which is logically equivalent to the formula Φ. This also means that Φ^* is logically equivalent to the formula $(\approx x)(P_1(x), P_2(x))$.

This finishes the proof. □

Note 13.5. The proved theorem concerns quantifiers of the type $\langle 1,1 \rangle$. Let us note that it can be generalized for quantifiers of the general type $\langle 1^k \rangle$. The generalized criterion uses intervals in $\mathcal{N}^{2^k}$ instead of intervals in $\mathcal{N}^4$, see also Sect. 16.9.

13.6 Definability of Equivalence Quantifiers and Critical Frequencies

We use the criterion of classical definability based on associated functions of 4ft-quantifiers to give an intuitive necessary condition of classical definability of equivalence rules. This criterion is based on a notion of a *partial table of maximal b for the quantifier* $\approx$ introduced in definition 13.17 and on a *step in the partial table of maximal b* introduced in definition 13.18

Please remember also the table $Tb_{\Rightarrow^*}$ of maximal b introduced in definition 7.7 for implicational quantifier $\Rightarrow^*$. It uses set $\mathcal{N}^+$ introduced in definition 7.6 as a union $\mathcal{N}^+ = \mathcal{N} \cup \{\infty\}$ where $\mathcal{N}$ is the set of all non-negative integer numbers and $n < \infty$, $n + \infty = \infty$, $\infty - n = \infty$ for each $n \in \mathcal{N}$, $\max(\mathcal{N}) = \infty$, $\min(\mathcal{N}^+) = \infty$.

Definition 13.17. Let $\approx$ be an equivalence quantifier and let $c \geq 0$ and $d \geq 0$ be integer numbers. Then a *partial table of maximal b for the quantifier* $\approx$ *and for the couple* $\langle c,d \rangle$ is the function $Tbp_{\approx,c,d}$ assigning to each integer non-negative a a $Tbp_{\approx,c,d}(a) \in \mathcal{N}^+$ such that

$$Tbp_{\approx,c,d}(a) = \min\{b | F_{\approx}(a,b,c,d) = 0\} .$$

Before defining a step in the partial table of maximal b we prove a simple theorem on properties of the function $Tbp_{\approx,c,d}$.

Theorem 13.5. *Let $Tbp_{\approx,c,d}$ be a partial table of maximal b for equivalence quantifier $\approx$ and for the couple $\langle c,d \rangle$. Then the function $Tbp_{\approx,c,d}$ satisfies:*

1. *For each $b \geq 0$ it holds $F_{\approx}(a,b,c,d) = 1$ if and only if $b < Tb_{\approx,c,d}(a)$.*
2. *If $a' > a$ then $Tb_{\approx,c,d}(a') \geq Tb_{\approx,c,d}(a)$ i.e. the function $Tb_{\approx,c,d}(a)$ is nondecreasing.*

Proof. Quantifier $\approx$ is equivalence, thus

$$F_{\approx}(a,b,c,d) = 1 \wedge a' \geq a \wedge b' \leq b \wedge c' \leq c \wedge d' \geq d \text{ implies } F_{\approx}(a',b',c',d') = 1.$$

This means:

I: if $F_{\approx}(a,b,c,d) = 0$ and $v \leq a$ then also $F_{\approx}(v,b,c,d) = 0$
II: if $F_{\approx}(a,b,c,d) = 0$ and $w \geq b$ then also $F_{\approx}(a,w,c,d) = 0$.

We prove that the function $Tbp_{\approx,c,d}$ has the properties 1. and 2.

1. *a: Let $b \geq 0$ and $F_{\approx}(a,b,c,d) = 1$. We prove $b < Tb_{\approx,c,d}(a)$. Let us assume $b \geq Tb_{\approx,c,d}(a)$. Then, according to point II, we have $F_{\approx}(a,b,c,d) = 0$ since $F_{\approx}(a,Tb_{\approx,c,d}(a),c,d) = 0$. Thus it must hold that $b < Tb_{\approx,c,d}(a)$.*
   b: Let $b \geq 0$ and $F_{\approx}(a,b,c,d) = 0$. Then it holds $b \geq Tb_{\approx,c,d}(a)$ according to definition.
   This finishes the proof of point 1.
2. *Let us assume $a' > a$ and also $Tb_{\approx,c,d}(a') < Tb_{\approx,c,d}(a)$. Thus $Tb_{\approx,c,d}(a) > 0$. Let us denote $Tb_{\approx,c,d}(a) = e$. This means $Tb_{\approx,c,d}(a') \leq e - 1$ and according to the definition of $Tb_{\approx,c,d}(a')$ it holds that $F_{\approx}(a',e-1,c,d) = 0$. Due to point I we also have $F_{\approx}(a,e-1,c,d) = 0$.*
   However, it also holds that $e - 1 < e = Tb_{\approx,c,d}(a)$. According to already proved point 1, this means $F_{\approx}(a,e-1,c,d) = 1$. This is a contradiction and thus it cannot happen that $a' > a$ and $Tb_{\approx}(a',c,d) < Tb_{\approx}(a,c,d)$. Thus $a' > a$ implies $Tb_{\approx}(a',c,d) \geq Tb_{\approx}(a,c,d)$.
   This finishes the proof of point 2.

This finishes the proof. □

Definition 13.18. Let $Tbp_{\approx,c,d}$ be a partial table of maximal b for equivalence quantifier $\approx$ and for the couple $\langle c,d \rangle$. Then a *step in partial table $Tbp_{\approx,c,d}$ of maximal b* is each such $a \geq 0$ for which $Tbp_{\approx,c,d}(a) < Tbp_{\approx,c,d}(a+1)$.

Now we can prove a necessary condition of classical definability for equivalence quantifiers.

Note 13.6. It is possible to prove that the quantifier $\approx$ of the type $\langle 1,1 \rangle$ is associational according to section 3.2.2 in [18] if and only if the 4ft-associated quantifier $\approx^{4ft}$ of quantifier $\approx$ (see definition 13.7) is equivalence. Similarly we can prove that $\approx$ is implicational according to section 3.2.10 in [18] if and only if the 4ft-associated quantifier $\approx^{4ft}$ of quantifier $\approx$ is implicational. However, we will not do this, see also Sect. 16.9. Such result can be used to formulate theorems 13.6 and 13.7 in an alternative way.

Theorem 13.6. *Let $\mathscr{P}^=$ be a CMOPC with equality and unary predicates $P_1, \ldots, P_n$ ($n \geq 2$), let $\mathscr{P}^{\approx}$ be its extension by adding a classically definable quantifier $\approx$ of type $\langle 1, 1 \rangle$ and let $\approx^{4ft}$ be a 4ft-associated quantifier of quantifier $\approx$. If $\approx^{4ft}$ is an equivalence quantifier then each partial table of maximal b for the quantifier $\approx^{4ft}$ has only a finite number of steps.*

Proof. We suppose that $\approx$ is a classically definable quantifier. Thus according to theorem 13.4 there are K intervals $I_1, \ldots, I_K$ in $\mathscr{N}^4$, $K \geq 0$ such that for each 4ft-table $\langle a, b, c, d \rangle$ the following holds:

$$Asf_{\approx}^{4ft}(a,b,c,d) = 1 \text{ if and only if } \langle a,b,c,d \rangle \in \bigcup_{j=1}^{K} I_j$$

where $Asf_{\approx}^{4ft}$ is the 4ft-associated function of $\approx$, see definition 13.7. Please note that according to point 2 of definition 13.7 we have

$$F_{\approx^{4ft}}(a,b,c,d) = Asf_{\approx}^{4ft}(a,b,c,d)$$

where $F_{\approx^{4ft}}$ is an associated function of $\approx^{4ft}$.

If $K = 0$ then $F_{\approx^{4ft}}(a,b,c,d) = 0$ for each 4ft-table $\langle a,b,c,d \rangle$ and $Tbp_{\approx,c,d}(a) = 0$ for each a and for each partial table $Tbp_{\approx,c,d}$ of maximal b of $\approx$. But this means that each such partial table of maximal b has no step.

Let us assume that $K > 0$ and that it holds

$$I_j = \langle a_j, A_j \rangle \times \langle b_j, B_j \rangle \times \langle c_j, C_j \rangle \times \langle d_j, D_j \rangle$$

for $j = 1, \ldots, K$. Suppose that there are u and v such that the partial table $Tbp_{\approx,u,v}$ of maximal b has infinitely many steps $\bar{a}_1, \bar{a}_2, \ldots, \bar{a}_n, \bar{a}_{n+1}, \ldots$. These mutually differ ($Tbp_{\approx,u,v}$ is nondecreasing) and thus we can assume

$$\bar{a}_1 < \bar{a}_2 < \ldots < \bar{a}_n < \bar{a}_{n+1} < \ldots$$

and according to definition 13.18 we have

$$Tbp_{\approx,u,v}(\bar{a}_1) < Tbp_{\approx,u,v}(\bar{a}_2) < \ldots < Tbp_{\approx,u,v}(\bar{a}_n) < Tbp_{\approx,u,v}(\bar{a}_{n+1}) < \ldots.$$

This means that for each integer $n > 0$ there is $\bar{a}_k > n$ such that $Tbp_{\approx,u,v}(\bar{a}_k) > n + 1$. According to point 1 of theorem 13.5 it holds that $F_{\approx}(\bar{a}_k, Tbp_{\approx,u,v}(\bar{a}_k - 1), u, v) = 1$. We can conclude that for each n there is $a > n$ and $b > n$ such that $F_{\approx}(a, b, u, v) = 1$. This means there is $m \in 1, \ldots, K$ such that

$$I_m = \langle a_m, \infty \rangle \times \langle b_m, \infty \rangle \times \langle c_m, C_m \rangle \times \langle d_m, D_m \rangle$$

and $u \in \langle c_m, C_m \rangle$ and $v \in \langle d_m, D_m \rangle$.

The partial table $Tbp_{\approx,u,v}$ of maximal b has infinitely many steps, thus there is a step $\bar{a} > a_m$ such that $Tbp_{\approx,u,v}(\bar{a}) < Tbp_{\approx,u,v}(\bar{a}+1)$. This means

$$F_{\approx}(\bar{a}, Tbp_{\approx,u,v}(\bar{a}+1), u, v) = 0 \, .$$

Let us denote $b' = \max(b_m, Tbp_{\approx,u,v}(a+1))$, thus it holds $F_{\approx}(\bar{a}, b', u, v) = 0$ because of $\approx$ is equivalence (see also point II in the proof of theorem 13.5).

However, $\langle a, b', u, v \rangle \in I_m$ which means that $F_{\approx}(a, b', u, v) = 1$. This means that the assumption on infinitely many steps in a partial table of maximal b for $K > 0$ leads to a contradiction. This finishes the proof. □

13.7 Definability of Implicational Quantifiers and Critical Frequencies

We show that the necessary condition of definability of equivalence rules proved in theorem 13.6 is also a sufficient condition of definability of implicational quantifiers. Please remember definition 7.7 of table $Tb_{\Rightarrow^*}$ of maximal b for an implicational quantifier $\Rightarrow^*$, specifically

$$Tb_{\Rightarrow^*}(a) = \min\{e | F_{\Rightarrow^*}(a, e) = 0\} \, .$$

Each implicational quantifier is also an equivalence quantifier, see theorem 6.1, but the value of $F_{\Rightarrow^*}(a, b, c, d)$ does not depend neither on c nor on d, see note 6.1. Thus $Tb_{\Rightarrow^*}(a) = Tbp_{\Rightarrow^*,c,d}(a)$ for all integers $c \geq 0$ and $d \geq 0$.

We define a step in a table $Tb_{\Rightarrow^*}$ of maximal b for an implicational quantifier $\Rightarrow^*$ in the same way as the step in partial table $Tbp_{\approx,c,d}$ of maximal b is introduced in definition 13.18.

Definition 13.19. Let $\Rightarrow^*$ be an implicational quantifier and let $Tb_{\Rightarrow^*}$ be a table of maximal b for $\Rightarrow^*$. Then a *step in the table $Tb_{\Rightarrow^*}$ of maximal b for $\Rightarrow^*$* is each $a \geq 0$ for which $Tb_{\Rightarrow^*}(a) < Tb_{\Rightarrow^*}(a+1)$.

Now we can prove an intuitive condition of definability of implicational quantifiers.

Theorem 13.7. *Let $\mathscr{P}^=$ be a CMOPC with equality and unary predicates $P_1, \ldots, P_n$ ($n \geq 2$), let $\mathscr{P}^{\approx}$ be its extension by adding a quantifier $\approx$ of type $\langle 1, 1 \rangle$ and let $\approx^{4ft}$ be a 4ft-associated quantifier of quantifier $\approx$. If $\approx^{4ft}$ is an implicational quantifier then $\approx$ is classically definable if and only if the table of maximal b of $\approx^{4ft}$ has only a finite number of steps.*

Proof. Let $Tb_{\approx^{4ft}}$ be a table of maximal b of $\approx^{4ft}$. We have to prove

(i)   If $\approx$ is classically definable then $Tb_{\approx^{4ft}}$ has only a finite number of steps.
(ii)   If $Tb_{\approx^{4ft}}$ has only a finite number of steps then $\approx$ is classically definable.

(i): If $\approx$ is classically definable then we prove that $Tb_{\approx^{4ft}}$ has only finite number of steps in a similar way as we proved that each partial table of maximal b for the equivalence quantifier $\approx^{4ft}$ has only finite number of steps in theorem 13.6.

(ii): *Let us suppose that* $Tb_{\approx 4ft}$ *has* K *steps where* $K \geq 0$ *is an integer number. We prove that* $\approx$ *is classically definable. We use the fact that*

$$F_{\approx 4ft}(a,b,c,d) = Asf_{\approx}^{4ft}(a,b,c,d)$$

where $F_{\approx 4ft}$ *is an associated function of* $\approx^{4ft}$ *see point 2 of definition 13.7.*

First let us suppose that $K = 0$. *Then* $Tb_{\approx 4ft}(a) = Tb_{\approx 4ft}(1)$ *for each* $a \geq 0$ *because there is no step. We distinguish two cases:* $Tb_{\approx 4ft}(1) = 0$ *and* $Tb_{\approx 4ft}(1) > 0$.

If $K = 0$ *and* $Tb_{\approx 4ft}(1) = 0$ *then* $F_{\approx 4ft}(a,b,c,d) = 0$ *for each 4ft-table* $\langle a,b,c,d \rangle$ *since it cannot be that* $b < 0$. *Thus* $F_{\approx 4ft}(a,b,c,d) = 1$ *i.e.* $Asf_{\approx}^{4ft}(a,b,c,d) = 1$ *if and only if* $\langle a,b,c,d \rangle \in \emptyset$. *However, the empty set* $\emptyset$ *is also the interval in* $\mathcal{N}^4$ *and the quantifier* $\approx$ *is classically definable according to theorem 13.4.*

If $K = 0$ *and* $Tb_{\approx 4ft}(1) > 0$ *then* $F_{\approx 4ft}(a,b,c,d) = 1$ *i.e.* $Asf_{\approx}^{4ft}(a,b,c,d) = 1$ *if and only if*

$$\langle a,b,c,d \rangle \in \langle 0, \infty) \times \langle 0, Tb_{\approx 4ft}(1)) \times \langle 0, \infty) \times \langle 0, \infty)$$

and thus the quantifier $\approx$ *is classically definable according to theorem 13.4.*

Let us suppose that $K > 0$ *is an integer number and that*

$$0 \leq a_1 < a_2 < \ldots < a_K$$

are all the steps in $Tb_{\approx 4ft}$. *This means that*

$$Tb_{\approx 4ft}(0) = \ldots = Tb_{\approx 4ft}(a_1) < Tb_{\approx 4ft}(a_1 + 1) = \ldots = Tb_{\approx 4ft}(a_2) <$$

$$< Tb_{\approx 4ft}(a_2 + 1) = \ldots \ldots = Tb_{\approx 4ft}(a_{K-1}) < Tb_{\approx 4ft}(a_K) = Tb_{\approx 4ft}(a_K + 1) = \ldots \quad .$$

We define intervals $I_1, I_2, \ldots, I_{K+1}$ *in the following way. If* $Tb_{\approx 4ft}(a_1) = 0$ *then* $I_1 = \emptyset$ *otherwise*

$$I_1 = \langle 0, a_1 + 1) \times \langle 0, Tb_{\approx 4ft}(a_1)) \times \langle 0, \infty) \times \langle 0, \infty) \, .$$

This means that for $a \in \langle 0, a_1 + 1)$ *it holds* $F_{\approx 4ft}(a,b,c,d) = 1$ *if and only if* $\langle a,b,c,d \rangle \in I_1$.

For $j = 2, \ldots, K$ *we define*

$$I_j = \langle a_{j-1} + 1, a_j + 1) \times \langle 0, Tb_{\approx 4ft}(a_j)) \times \langle 0, \infty) \times \langle 0, \infty) \, .$$

This means that for $j = 2, \ldots, K$ *and* $a \in \langle a_{j-1} + 1, a_j + 1)$ *it holds* $F_{\approx 4ft}(a,b,c,d) = 1$ *if and only if* $\langle a,b,c,d \rangle \in I_j$.

The interval I_{K+1} *is defined such that*

$$I_{K+1} = \langle a_K + 1, \infty) \times \langle 0, Tb_{\approx 4ft}(a_K)) \times \langle 0, \infty) \times \langle 0, \infty) \, .$$

This means that for $a \geq a_K + 1$ *it holds* $F_{\approx 4ft}(a,b,c,d) = 1$ *if and only if* $\langle a,b,c,d \rangle \in I_{K+1}$. *We can conclude that*

$$F_{\approx 4ft}(a,b,c,d) = 1 \ \ if \ and \ only \ if \ \langle a,b,c,d \rangle \in \bigcup_{j=1}^{K+1} I_j$$

which means

$$Asf_{\approx}^{Aft}(a,b,c,d) = 1 \ \ if \ and \ only \ if \ \langle a,b,c,d \rangle \in \bigcup_{j=1}^{K+1} I_j$$

and according to theorem 13.4 the quantifier $\approx$ is classically definable. This finishes the proof. □

13.8 Undefinability of Particular 4ft-quantifiers

The goal of this section is to show undefinability for some of the important quantifiers of the type $\langle 1,1 \rangle$. We start with four lemmas.

Lemma 13.4. *Let $\Rightarrow^*$ be an implicational quantifier that satisfies the following conditions a) and b).*

a)   *There is $A \geq 0$ such that for each $a \geq A$ there is an integer b satisfying $F_{\Rightarrow^*}(a,b) = 0$.*

b)   *For each $a \geq 0$ and $b \geq 0$ such that $F_{\Rightarrow^*}(a,b) = 0$ there is an integer a' satisfying $a' \geq a$ and $F_{\Rightarrow^*}(a',b) = 1$.*

Then the table $Tb_{\Rightarrow^*}$ of maximal b of $\Rightarrow^*$ has infinitely many steps.

Proof. If the quantifier $\Rightarrow^*$ satisfies the condition a) then for each $a \geq A$ there is b such that $F_{\Rightarrow^*}(a,b) = 0$ and this means $Tb_{\Rightarrow^*}(a) = \min\{e|F_{\Rightarrow^*}(a,e) = 0\} \leq b < \infty$. Thus, it holds that $F_{\Rightarrow^*}(a,Tb_{\Rightarrow^*}(a)) = 0$.

If the quantifier $\Rightarrow^*$ satisfies the condition b) then there is for each $a > A$ an integer a' such that $a' > a$ and $F_{\Rightarrow^*}(a',Tb_{\Rightarrow^*}(a)) = 1$. We can conclude that $F_{\Rightarrow^*}(a,v) = 0$, $F_{\Rightarrow^*}(a',v) = 1$, and $a' > a$ where $v = Tb_{\Rightarrow^*}(a)$.

This means that between a and a' there must be an integer u satisfying

$$a \leq u < u+1 \leq a', \ \ F_{\Rightarrow^*}(u,v) = 0 \ \ and \ F_{\Rightarrow^*}(u+1,v) = 1 \,,$$

which implies

$$Tb_{\Rightarrow^*}(u) = \min\{e|F_{\Rightarrow^*}(u,e) = 0\} \leq v \,.$$

According to point a) there is an integer w such that $F_{\Rightarrow^*}(u+1,w) = 0$, thus

$$Tb_{\Rightarrow^*}(u+1) = \min\{e|F_{\Rightarrow^*}(u+1,e) = 0\} \leq w \,.$$

It must hold that $Tb_{\Rightarrow^*}(u+1) > v$ because $\Rightarrow^*$ is implicational, $Tb_{\Rightarrow^*}(u+1) \leq v$ means $F_{\Rightarrow^*}(u+1,v) = 0$ and we have $F_{\Rightarrow^*}(u+1,v) = 1$.

We can conclude that $Tb_{\Rightarrow^*}(u) \leq v$ and $Tb_{\Rightarrow^*}(u+1) > v$ which means $Tb_{\Rightarrow^*}(u) < Tb_{\Rightarrow^*}(u+1)$ and thus u is a step in the table $Tb_{\Rightarrow^*}$. We have proved *that for each $a > A$ there is a step $u \geq a$ of the table $Tb_{\Rightarrow^*}$. However this means that the table $Tb_{\Rightarrow^*}$ has infinitely many steps. This finishes the proof.*   □

Lemma 13.5. *Let $\approx$ be an equivalence quantifier and let c_0 and d_0 be integer numbers such that the conditions a) and b) are satisfied.*

a)   *There is $A \geq 0$ such that for each $a \geq A$ there is an integer b for which $F_{\approx}(a,b,c_0,d_0) = 0$.*
b)   *For each $a \geq 0$ and $b \geq 0$ such that $F_{\approx}(a,b,c_0,d_0) = 0$ there is an integer a' satisfying $a' \geq a$ and $F_{\approx}(a',b,c,d) = 1$.*

Then the partial table $Tbp_{\approx}(a,c_0,d_0)$ of maximal b of $\approx$ has infinitely many steps.

Proof. The proof is similar to the proof of lemma 13.4.   □

Lemma 13.6

1. *4ft-quantifier $\Rightarrow_{p,Base}$ of founded implication satisfies condition a) from lemma 13.4 for each $0 < p \leq 1$ and $Base > 0$.*
2. *4ft-quantifier $\Rightarrow^{!}_{p,\alpha,Base}$ of lower critical implication satisfies condition a) from lemma 13.4 for each $0 < p < 1$, $0 < \alpha < 0.5$ and $Base > 0$.*

Proof

1. *We have to prove that there is $A \geq 0$ such that for each $a \geq A$ there is an integer b for which $F_{\Rightarrow_{p,Base}}(a,b) = 0$ for each $0 < p \leq 1$ and $Base > 0$. Let us remember that the 4ft-quantifier $\Rightarrow_{p,Base}$ is defined by the condition $\frac{a}{a+b} \geq p \wedge a \geq Base$. Let $A > Base$ and $a \geq A$. Let as choose an integer b satisfying $b > \frac{a-p*a}{p}$. Then it holds that $\frac{a}{a+b} < \frac{a}{a+\frac{a-p*a}{p}} = a \frac{p}{a*p+a-p*a} = p$ and thus $F_{\Rightarrow_{p,Base}}(a,b) = 0$.*
2. *We have to prove that there is $A \geq 0$ such that for each $a \geq A$ there is an integer b for which $F_{\Rightarrow^{!}_{p,\alpha,Base}}(a,b) = 0$ for each $0 < p < 1$, $0 < \alpha < 0.5$ and $Base > 0$. Let us remember that the 4ft-quantifier $\Rightarrow^{!}_{p,\alpha,Base}$ is defined by the condition*

$$\sum_{i=a}^{a+b} \binom{a+b}{i} p^i (1-p)^{a+b-i} \leq \alpha \wedge a \geq Base \,,$$

see Table 4.4 and theorem 7.7. Let $A > Base$ and $a \geq A$. We show that there is an integer $b > 0$ such that

$$\sum_{i=a}^{a+b} \binom{a+b}{i} p^i (1-p)^{a+b-i} > \alpha \,.$$

It holds that $\sum_{i=a}^{a+b} \binom{a+b}{i} p^i (1-p)^{a+b-i} > \alpha$ if and only if

$$\sum_{i=0}^{a-1} \binom{a+b}{i} p^i (1-p)^{a+b-i} < 1-\alpha$$

because $\sum_{i=0}^{a+b} \binom{a+b}{i} p^i (1-p)^{a+b-i} = 1.$
According to lemma 5.14 there is an integer $V > a$ such that it holds

$$\binom{V}{i} p^i (1-p)^{V-i} < \frac{1-\alpha}{a}$$

for $i = 0, \ldots, a-1$. Thus it holds holds that

$$\sum_{i=0}^{a-1} \binom{V}{i} p^i (1-p)^{V-i} < 1 - \alpha$$

Let us choose $b = V - a$. This means

$$\sum_{i=0}^{a-1} \binom{a+b}{i} p^i (1-p)^{a+b-i} < 1 - \alpha$$

and this finishes the proof. □

Lemma 13.7

1. *4ft-quantifier $\Rightarrow_{p,Base}$ of founded implication satisfies the condition b) from lemma 13.4 for each $0 < p < 1$ and Base > 0.*
2. *4ft-quantifier $\Rightarrow^{!}_{p,\alpha,Base}$ of lower critical implication satisfies condition b) from lemma 13.4 for each $0 < p < 1$, $0 < \alpha < 0.5$ and Base > 0.*

Proof

1. *We have to prove that for each $a \geq 0$, $b \geq 0$ satisfying $F_{\Rightarrow_{p,Base}}(a,b) = 0$ there is an integer $a' \geq a$ for which it holds that $F_{\Rightarrow_{p,Base}}(a',b) = 1$. The proof is trivial, we use the fact that $\lim_{a \to \infty} \frac{a}{a+b} = 1$.*
2. *We have to prove that for each $a \geq 0$ and $b \geq 0$ satisfying $F_{\Rightarrow^{!}_{p,\alpha,Base}}(a,b) = 0$ there is an integer $a' \geq a$ for which it holds that $F_{\Rightarrow^{!}_{p,\alpha,Base}}(a,b) = 1$. Let us assume $F_{\Rightarrow^{!}_{p,\alpha,Base}}(a,b) = 0$. This means that $a < Base$ or $\sum_{i=a}^{a+b} \binom{a+b}{i} p^i (1-p)^{a+b-i} > \alpha$. According to lemma 5.17 there is an integer n such that for each e, $e > n$ and $k = 0, \ldots, b$ it holds that*

$$\binom{e+b}{e+k} p^{e+k} (1-p)^{b-k} < \frac{\alpha}{b+1}.$$

Let us choose $a' = \max\{a, n, Base\}$. Then it holds that $a' \geq Base$ and also

$$\sum_{i=a'}^{a'+b} \binom{a'+b}{i} p^i (1-p)^{a'+b-i} = \sum_{k=0}^{b} \binom{a'+b}{a'+k} p^{a'+k} (1-p)^{b-k} < (b+1)\frac{\alpha}{b+1} < \alpha.$$

Thus we have $\Rightarrow^{!}_{p,\alpha,Base}(a',b) = 1$.

This finishes the proof. □

Theorem 13.8. *Let $\mathscr{P}^=$ be a CMOPC with equality and unary predicates $P_1, \ldots, P_n$ ($n \geq 2$), let $\mathscr{P}^\approx$ be its extension by adding a quantifier $\approx$ of type $\langle 1, 1 \rangle$ and let $\approx^{4ft}$ be a 4ft-associated quantifier of quantifier $\approx$.*

1. *If $\approx^{4ft}$ is the 4ft-quantifier $\Rightarrow_{p,Base}$ of founded implication where $0 < p < 1$ and $Base > 0$ then $\approx$ is not classically definable.*
2. *If $\approx^{4ft}$ is the 4ft-quantifier $\Rightarrow^!_{p,\alpha,Base}$ of lower critical implication where $0 < p < 1$, $0 < \alpha < 0.5$ and $Base > 0$ then $\approx$ is not classically definable.*

Proof

1. *The table of maximal b of the 4ft-quantifier $\Rightarrow_{p,Base}$ of founded implication has infinitely many steps according to lemmas 13.4, 13.6 and 13.7. Thus $\approx^{4ft}$ is not classically definable according to theorem 13.7.*
2. *The proof for the quantifier $\Rightarrow^!_{p,\alpha,Base}$ is analogous.*

This finishes the proof.   □

Now we prove that the Fisher's quantifier $\sim^1_{\alpha,Base}$ is not classically definable.

Theorem 13.9. *Let $\mathscr{P}^=$ be a CMOPC with equality and unary predicates $P_1, \ldots, P_n$ ($n \geq 2$), let $\mathscr{P}^\approx$ be its extension by adding a quantifier $\approx$ of type $\langle 1, 1 \rangle$ and let $\approx^{4ft}$ be a 4ft-associated quantifier of quantifier $\approx$. If $\approx^{4ft}$ is the Fisher's quantifier $\sim^1_{\alpha,Base}$ where $0 < \alpha < 0.5$ and $Base > 0$ then $\approx$ is not classically definable.*

Proof. Fisher's quantifier $\sim^1_{\alpha,Base}$ is defined by the condition

$$\sum_{i=a}^{\min(r,k)} \frac{\binom{k}{i}\binom{n-k}{r-i}}{\binom{n}{r}} \leq \alpha \wedge ad > bc \wedge a \geq Base,$$

see Table 4.4 and theorem 9.6, it holds that $k = a+c$, $r = a+b$ and $n = a+b+c+d$.

According to theorem 13.6 it is sufficient to prove that a partial table $Tbp_{\sim^1_{\alpha,Base}}(a, 1, 1)$ of maximal b of $\sim^1_{\alpha,Base}$ has infinitely many steps. Lemma 13.5 means that it is sufficient to prove a) and b):

a) *There is $A \geq 0$ such that for each $a \geq A$ there is an integer b for which it holds $F_{\sim^1_{\alpha,Base}}(a, b, 1, 1) = 0$.*

b) *For each $a \geq 0$ and $b \geq 0$ such that $F_{\sim^1_{\alpha,Base}}(a, b, 1, 1) = 0$ there is an integer a' for which $a' \geq a$ and $F_{\sim^1_{\alpha,Base}}(a', b, 1, 1) = 1$.*

ad a): *Let us choose $b = a+1$ for each $a \geq Base$. Then $a*1 < (a+1)*1$ i.e. $ad \not< cd$ and thus it holds $F_{\sim^1_{\alpha,Base}}(a, a+1, 1, 1) = 0$. This means that the condition a) is satisfied.*

ad b): *Fisher's quantifier is equivalence and thus if $F_{\sim^1_{\alpha,Base}}(a', b, 1, 1) = 1$ for $b > 1$ then also $F_{\sim^1_{\alpha,Base}}(a', 1, 1, 1) = 1$ and $F_{\sim^1_{\alpha,Base}}(a', 0, 1, 1) = 1$. This means that it is sufficient to prove that for each $a \geq 0$ and $b > 1$ such that $F_{\sim^1_{\alpha,Base}}(a, b, 1, 1) = 0$ there is $a' \geq a$ for which $F_{\sim^1_{\alpha,Base}}(a', b, 1, 1) = 1$.*

If $b > 1$ then $F_{\sim_{\alpha,Base}^{1}}(a',b,1,1) = 1$ if and only if

$$\sum_{i=a'}^{a'+1} \frac{\binom{a'+1}{i}\binom{a'+b+1+1-(a'+1)}{a'+b-i}}{\binom{a'+b}{a'+b+1+1}} \leq \alpha \;\wedge\; a' > b \;\wedge\; a' \geq Base .$$

According to definition 5.1 and the proof of lemma 5.9 we have

$$\sum_{i=a}^{\min(r,k)} \frac{\binom{k}{i}\binom{n-k}{r-i}}{\binom{n}{r}} = \sum_{i=a}^{\min(r,k)} \frac{k!l!r!s!}{i!(k-i)!(r-i)!(n-k-r+i)!n!} .$$

In our case this means

$$\sum_{i=a'}^{a'+1} \frac{\binom{a'+1}{i}\binom{a'+b+1+1-(a'+1)}{a'+b-i}}{\binom{a'+b+1+1}{a'+b}} =$$

$$= \frac{(a'+1)!(b+1)!(a'+b)!2!}{a'!1!b!1!(a'+b+2)!} + \frac{(a'+1)!(b+1)!(a'+b)!2!}{(a'+1)!0!(b-1)!2!(a'+b+2)!} =$$

$$= \frac{(a'+1)(b+1)*2}{(a'+b+1)(a'+b+2)} + \frac{b(b+1)}{(a'+b+1)(a'+b+2)} .$$

It holds that

$$\lim_{u \to \infty} \frac{(u+1)(b+1)*2}{(u+b+1)(u+b+2)} = 0 \quad and \quad \lim_{u \to \infty} \frac{b(b+1)}{(u+b+1)(u+b+2)} = 0 ,$$

thus there is an integer N such that for $u \geq N$ it holds $\sum_{i=u}^{u+1} \frac{\binom{u+1}{i}\binom{u+b+1+1-(u+1)}{u+b-i}}{\binom{u+b+1+1}{u+b}} \leq \alpha$.
This means that if $a' \geq \max\{a,N,b,Base\}$ then $F_{\sim_{\alpha,Base}^{1}}(a',b,1,1) = 1$.
 This finishes the proof. $\square$

Note 13.7. It is of course possible to prove undefinability for additional 4ft-quantifiers, see also Sect. 16.9.

Part IV
Applications and Research Challenges

Observational calculi were introduced as a tool of logic of discovery. An additional tool of logic of discovery is the GUHA method, the goal of which is to offer interesting hypotheses i.e. formulas of observational calculi. The GUHA method is realized by GUHA procedures. The input of a GUHA procedure consists of analysed data and of several parameters defining a large set of formulas of observational calculi i.e. relevant patterns. The output is the set of all prime patterns. A pattern is prime if it is relevant, true in the analysed data and if it does not logically follow from another simpler true pattern which is already a part of the output. The GUHA procedure used most frequently is the procedure ASSOC producing association rules studied in previous parts of this book.

The probably most contemporarily used implementation of the ASSOC procedure is the procedure 4ft-Miner, which is a part of the LISp-Miner system. The LISp-Miner system involves six additional GUHA procedures. Two of them mine for interesting couples of association rules, additional four GUHA procedures mine for various types of patterns. All implementations of the ASSOC procedure including the 4ft-Miner procedure use results presented in previous parts. Important features of both the GUHA method and the LISp-Miner system are introduced in Chap. 14.

The performance of procedures implemented in the LISp-Miner is good enough to solve a lot of practically important tasks in reasonable times. The EverMiner project was launched with the aim of studying the possibilities of automating the data mining process using formalized domain knowledge. The EverMiner project is closely related to the SEWEBAR project. The goal of the SEWEBAR project is to study the possibilities of presenting results of the LISp-Miner applications in the form of analytical reports, which are textual documents intended for domain experts with zero or little knowledge of data mining methods. The reports are presented through a web-based content management system. All these projects use results achieved in previous parts. The research projects related to the LISp-Miner system are introduced in Chap. 15.

There are important and interesting results on observational calculi tackling problems not closely related to association rules. There are results which can be seen as initial steps opening new research issues. Experience with the GUHA procedures involved in the LISp-Miner system as well as with related research project lead to new research challenges concerning both logical calculi of association rules and additional observational calculi. A short overview of results not presented in previous chapters and an inventory of research challenges and partly solved problems related to data mining and observational calculi are presented in Chap. 16.

Chapter 14
GUHA Method and the LISp-Miner System

Both observational calculi and the GUHA method were developed as tools of logic of discovery [18]. The goal of the GUHA method is to offer interesting hypotheses i.e. formulas of observational calculi. The GUHA method is realized by GUHA procedures. One of the GUHA procedures is the ASSOC procedure which deals with association rules studied in previous parts of this book. The ASSOC procedure uses results of logic of association rules achieved in [18] and also in the previous parts of this book. The relation of the GUHA procedure ASSOC to observational calculi and logic of association rules is clarified in Sect. 14.1.

The LISp-Miner system is an academic project to support research and teaching of knowledge discovery in databases [80, 89]. It focuses on the GUHA method. It has been in development since 1996. The core of the LISp-Miner system is the 4ft-Miner procedure which is one of advanced implementations of the ASSOC procedure. The main features of the 4ft-Miner procedure are outlined in Sect. 14.2, examples of its applications are located in Sects. 14.4 and 14.5. A real data set used in examples is introduced in Sect. 14.3.

The 4ft-Miner procedure as well as additional ASSOC procedures do not use the well known a-priori algorithm. Their implementation is based on suitable strings of bits, see Sect. 14.6. The bitstring approach enabled implementation of six additional GUHA procedures which are involved together with the 4ft-Miner procedure in the LISp-Miner system.

The procedures SD4ft-Miner and Ac4ft-Miner mine for interesting couples of association rules. SD4ft-Miner searches for two sets of objects which differ regarding measures of an association rule. The Ac4ft-Miner procedure is inspired by action rules. Principles of both procedures are outlined in Sects. 14.7 and 14.8. The additional four GUHA procedures of the LISp-Miner system are shortly introduced in Sect. 14.9.

Let us note that experience with the LISp-Miner system led both to the launch of several research projects outlined in Chap. 15 and to the formulation of various research problems concerning observational calculi, see Chap. 16.

Note 14.1. Please note also that the whole LISp-Miner system can be freely downloaded from the address http://lispminer.vse.cz/. The goal of this

J. Rauch: *Observational Calculi and Association Rules*, SCI 469, pp. 233–260.
DOI: 10.1007/978-3-642-11737-4_14 © Springer-Verlag Berlin Heidelberg 2013

chapter is only to point out experience with research projects that uses the results presented in the previous parts of this book and to use this experience to formulate additional research problems related to observational calculi. A systematic description of the LISp-Miner system is not a goal of this chapter.

Note 14.2. There are also two additional procedures in the LISp-Miner system. The first one is the procedure KEX [4, 90], the second one is the procedure ETree-Miner [3]. Both procedures concern machine learning and are related to the 4ft-Miner procedure.

Note 14.3. There also exists a data mining software system FERDA [56, 57] which implements some of the GUHA procedures implemented in the LISp-Miner system.

14.1 GUHA Procedure ASSOC and Observational Calculi

Logic of discovery developed in [18] is based on observational and theoretical calculi. Formulas of observational calculi talk about observational data. Logical calculi of associational rules studied in the first three parts of this book are examples of observational calculi. Theoretical calculi developed and studied in [18] are calculi formulas which correspond to assertions on the whole world as described by analysed data - a result of observation. One of the crucial questions of logic of discovery is *What are the conditions for a theoretical statement or a set of theoretical statements to be of interest (importance) with respect to the task of scientific cognition?*

Theoretical calculi are based on statistical approaches, the correspondence between observational and theoretical calculi is given by statistical inductive inference rules. This means that in many cases there is a one-one correspondence between observational and theoretical statements. Thus, the above introduced question can be converted into a similar question about observational statements.

The conceptual frame of scientific research is given by the observational language, and theoretical language and by collected data - an observational model. Observational sentences can then be classified as relevant or irrelevant questions. An observational sentence Φ is a *relevant question* if a decision whether Φ is true for the given data or not is valuable because

- we do not know whether Φ is true
- if Φ is true then it leads, via the inference rule, to an interesting theoretical hypothesis which is justified since Φ is true.

We call Φ a *relevant observational truth* if it is a relevant observational question and if it is true for the given data. This means we are interested in a set RT of all relevant observational truths to get a set of theoretical statements interesting with regard to the given task of scientific cognition.

However, the set RT could be really very large and it is reasonable to search for its concise representation. The notion of immediate conclusion is introduced and we are actually interested in a set X of truths such that each relevant truth is an immediate conclusion from X. The immediate conclusion is represented by a transparent deduction rule I. The *observational research problem* $\mathscr{P}$ is given by a set RQ of relevant questions and by inference rule I. The solution of a given observational research problem $\mathscr{P}$ is each set X of observational sentences where each relevant truth $\Phi \in RT$ can be inferred by I from X. Some of the deduction rules studied in Chap. 11 can serve as transparent inference rules.

The GUHA method is defined in [18] as a method for solving observational research problems. Several general aspects of the GUHA method are emphasized, i.e. the possibility of choosing an appropriate type of relevant question and satisfactory variable syntactical descriptions of sets of relevant questions. Please note that this is only a very brief and informal sketch of many both formal and informal considerations given in [18].

The GUHA method is carried out using GUHA procedures. A GUHA procedure is a computer program the input of which consists of analysed data and a set of parameters defining the large set of relevant questions. The output of the GUHA procedure is the set X representing the set of all relevant truths. The GUHA procedure ASSOC mines for association rules studied in this book. It has been implemented several times and applied many times, see e.g. [10, 11, 24, 25, 56]. The procedure 4ft-Miner, which is probably the most used contemporary implementation of the ASSOC procedure, is described in the next section.

14.2 4ft-Miner – Main Features

The goals of presentation of the 4ft-Miner procedure are:

- To show an effective way of mining for association rules introduced in previous parts of this book.
- To emphasize that it is possible to mine association rules where both antecedents and succedents are not only conjunctions of *attribute-value* pairs but also conjunctions or disjunctions of *attribute-set of values pairs*.
- To show that it is possible to define syntactical restrictions on the *attribute-set of values pairs* which automatically deal with some aspects of semantics.
- To point out a possibility of dealing with conditional association rules.
- To prepare a starting point for the presentation of research projects presented in Chap. 15.

The 4ft-Miner procedure mines for association rules introduced in previous parts of this book. It also mines for conditional association rules. The important general features of the 4ft-Miner procedure are summarized in this section. In Sects. 14.4 and 14.5, there are examples showing that the 4ft-Miner procedure deals both with conjunctions and disjunctions of *attribute-set of values pairs* which have some aspects of semantics. We use a real data set STULONG introduced in Sect. 14.3.

The 4ft-Miner procedure mines for association rules $\varphi \approx \psi$ and for conditional association rules $\varphi \approx \psi/\chi$ where φ, ψ, and χ are Boolean attributes. The conditional association rule $\varphi \approx \psi/\chi$ is verified in a given data matrix $\mathscr{M}$ such that we consider the rule $\varphi \approx \psi/\chi$ true in $\mathscr{M}$ if the rule $\varphi \approx \psi$ is true in a data matrix $\mathscr{M}/\chi$. The data matrix $\mathscr{M}/\chi$ consists of all rows of $\mathscr{M}/\chi$ satisfying χ. There could be a problem if no row of $\mathcal{M}$ satisfies χ. For more information on conditional association rules see Sect. 16.5.

The input of 4ft-Miner consists of

- an analysed data matrix
- specification of a set of relevant questions, i.e a set of association rules to be automatically generated and tested.

The *analysed data matrix* is created from a database table. Any database accessible by the ODBC can be used. The columns of the database table are transformed into attributes (i.e. columns) of the analysed data matrix. There is a special module called *DataSource* in the LISP-Miner system intended for these transformations. It is, for example, possible to use the original values from the given column of the database table as the categories of the defined attribute. It is also possible to define new categories as intervals of a given length. Moreover the *DataSource* module can generate the given number of equifrequency intervals as new categories.

The output of the 4ft-Miner procedure consists of all *prime association rules*. An association rule is prime if it is both true in the analysed data matrix and if it does not immediately follow from other simpler output association rules. The definition of a prime association rule depends on properties of the used 4ft-quantifier, namely on available transparent deduction rules, see also Sect. 16.4. There are also great possibilities of filtering and sorting the output set of association rules.

The *specification of a set of association rules to be automatically generated and tested* consists of

- a specification of a *set $\mathcal{A}$ of relevant antecedents*
- a specification of a *set $\mathcal{S}$ of relevant succedents*
- a specification of a *set $\mathcal{C}$ of relevant conditions*
- a specification of a 4ft-quantifier $\approx$
- a specification of dealing with missing information
- some additional details we will not describe here, as they are irrelevant to the goals of this chapter.

First we outline the main features of the *specification of the set $\mathcal{A}$ of relevant antecedents*. The antecedent is a conjunction of partial antecedents $\varphi_1, \ldots \varphi_k$. There exists definitions of sets $\mathcal{A}_1, \ldots, \mathcal{A}_k$ of partial antecedents. Each antecedent φ is a conjunction $\varphi = \varphi_1 \wedge \ldots \wedge \varphi_k$ where $\varphi_i \in \mathcal{A}_i$ for $i = 1, \ldots, k$. Particular sets of relevant partial antecedents can contain an empty conjunction that is identically true. The minimum and maximum number of attributes in the antecedent φ can also be set.

Each particular set $\mathcal{A}_i$ of relevant partial antecedents is a set of conjunctions or a set of disjunctions of literals. The literal is a basic Boolean attribute $A(\alpha)$ or its negation $\neg A(\alpha)$. Each set of partial antecedents is given in the following manner:

- the choice of conjunctions/disjunctions
- the minimum and maximum number of literals in the partial antecedent is defined
- a set of attributes $A'_1, \ldots, A'_q$ from which literals will be generated is given, it is a subset of the set $A_1, \ldots, A_k$ of all attributes given by the analysed data matrix

- a specification of sets $\mathscr{B}(A'_1), \ldots, \mathscr{B}(A'_q)$ of relevant literals that are automatically created from the attributes $A'_1, \ldots, A'_q$
- some attributes can be marked as *basic*, each partial antecedent must then contain at least one basic attribute
- *classes of equivalence* can be defined, each attribute belongs to at most one class of equivalence; no partial antecedent can contain two or more attributes from one class of equivalence.

We outline how the set $\mathscr{B}(A)$ of relevant literals is defined. Literal $A(\alpha)$ is a *positive literal*, literal $\neg A(\alpha)$ is a *negative literal*. If $\alpha = \{c_1, \ldots, c_k\}$ where $c_1, \ldots, c_k$ are categories, then k is the *length of the literals* $A(\alpha)$ and $\neg A(\alpha)$. The set $\mathscr{B}(A)$ of relevant literals to be generated for a particular attribute A is given by:

- the *type of coefficient* - there are seven types of coefficients *subsets, intervals, cyclical intervals, left cuts, right cuts, cuts, one category*, see below
- the minimum and maximum length of the literal
- one of the following possible options:
    - generate only positive literals $A(\alpha)$
    - generate only negative literals $\neg A(\alpha)$
    - generate both positive and negative literals.

We give examples of particular types of coefficients. We use an attribute A with categories $\{1, 2, 3, 4, 5\}$.

Subsets: definition of subsets of length 2-3 gives literals $A(1,2)$, $A(1,3)$, $A(1,4)$, $A(1,5)$, $A(2,3)$, ..., $A(4,5)$, $A(1,2,3)$, $A(1,2,4)$, $A(1,2,5)$, $A(2,3,4)$, ..., $A(3,4,5)$.

Intervals: definition of intervals of length 2-3 gives literals $A(1,2)$, $A(2,3)$, $A(3,4)$, $A(4,5)$, $A(1,2,3)$, $A(2,3,4)$ and $A(3,4,5)$ i.e. 2-3 consecutive categories.

Cyclical intervals: definition of cyclical intervals of length 2-3 gives literals $A(1,2)$, $A(2,3)$, $A(3,4)$, $A(4,5)$, $A(5,1)$, $A(1,2,3)$, $A(2,3,4)$, $A(3,4,5)$, $A(4,5,1)$, and $A(5,1,2)$.

Left cuts: a definition of left cuts with a maximum length of 3 defines literals $A(1)$, $A(1,2)$ and $A(1,2,3)$. Left cuts can be used to define intervals corresponding to the minimum values of the given attribute.

Right cuts: a definition of right cuts with a maximum length of 3 defines literals $A(5)$, $A(5,4)$, and $A(5,4,3)$. Right cuts can be used to define intervals corresponding to the maximum values of the given attribute.

Cuts mean both left cuts and right cuts. Cuts can be used to investigate the extreme values of attributes.

One category means one particular category.

We would like to point out that the minimum length of the partial antecedent can be 0 and this results in some conjunctions of length 0. The value of the conjunction of length 0 is identical to 1 (i.e. *truth*), thus the partial antecedent φ_i of the length 0 can be omitted from the antecedent by setting its minimal length to 0.

The *specifications of the set of relevant succedents* $\mathscr{S}$ *and the set of relevant conditions* $\mathscr{C}$ are analogous to the specification of the set of relevant antecedents $\mathscr{A}$. The procedure 4ft-Miner generates and verifies all association rules $\varphi \approx \psi/\chi$ where $\varphi \in \mathscr{A}$, $\psi \in \mathscr{S}$, and $\chi \in \mathscr{C}$ where φ, ψ, and χ have no common attributes and $\approx$ is a defined 4ft-quantifier.

The 4ft-quantifier $\approx$ *is specified* such that

- a list of q 4ft-quantifiers $\approx_1, \ldots, \approx_q$ chosen from 4ft-quantifiers specified in rows 1 - 15 of table 4.2 is given, this list can be empty
- maximally one 4ft-quantifier $\approx_B$ is chosen from quantifiers $\oplus_{Base}$ – Base, $\odot_s$ – Support, $\overline{\oplus}_{Base}$ – Ceil, and $\overline{\odot}_s$ – Ceil support specified in rows 18 - 21 of table 4.2.
- an additional 4ft-quantifier $\approx_F$ concerning particular frequencies a, b, c, d and sums $a+b, c+d, a+c$ $b+d$ can be also specified. Examples include conditions $b \geq 50, a+c \leq 30, a+c \leq 50 \wedge b+d \geq 100$. We will not specify all possibilities in details here.

The 4ft-quantifier $\approx$ is then considered a conjunction of 4ft-quantifiers $\approx_1 \wedge \ldots \wedge \approx_q \wedge \approx_B \wedge \approx_F$ where at least one 4ft-quantifiers $\approx_1, \ldots, \approx_q, \approx_B, \approx_F$ is used, see also definition 4.3.

The specification of dealing with missing information means choosing one of the possibilities introduced in Chap. 12:

- secured completion
- optimistic completion
- deleting.

14.3 STULONG Data Set

We use data matrix *ENTRY* of data set STULONG[1] [94]. Data matrix *ENTRY* involves 1 417 rows, each row corresponds to one patient. There are 64 columns corresponding to attributes of patients. There are 6 attributes concerning *social characteristics*, 18 attributes concerning *personal anamnesis* etc.

The goal of applications of the LISp-Miner system presented here is not to obtain new medical knowledge, but to outline the possibilities of GUHA procedures involved in the system. This is the reason why we use only 14 out of 64 attributes. The used 14 attributes are grouped into 6 groups of attributes, see table 14.1.

---

[1] The study (STULONG) was realized at the 2nd Department of Medicine, 1st Faculty of Medicine of Charles University and Charles University Hospital, U nemocnice 2, Prague 2 (head. Prof. M. Aschermann, MD, SDr, FESC), under the supervision of Prof. F. Boudík, MD, ScD, with collaboration of M. Tomečková, MD, PhD and Ass. Prof. J. Bultas, MD, PhD. The data were transferred to electronic form by the European Centre of Medical Informatics, Statistics and Epidemiology of Charles University and Academy of Sciences (head. Prof. RNDr. J. Zvárová, DrSc). The data resource is located on the web pages `http://euromise.vse.cz/challenge2004`.

Table 14.1 Attributes used in examples of applications of the LISp-Miner system

Group	Attribute	Categories
Measures	Height [cm]	$\langle 148;160\rangle, \langle 160;163\rangle, \dots, \langle 190;202\rangle$, see Fig. 14.1
	Weight [kg]	$\langle 50;60\rangle, \langle 60;65\rangle, \dots, \langle 110;135\rangle$, see Fig. 14.2
Social	Education	basic school, apprentice school, secondary school, university, see Fig. 14.3
	Marital_Status	married, divorced, single, widower, see Fig. 14.4
	Responsibility (i.e. responsibility in job)	managerial worker, partly independent worker, others, pensioner, see Fig. 14.5
Vices	Beer	he does not drink, up to 1 litre / day, more than 1 litre / day, see Fig. 14.6
	Coffee	he does not drink, 1-2 cups / day, 3 and more cups / day, see Fig. 14.7
Blood_pressure	Diastolic	$\langle 50;70\rangle, \langle 70;80\rangle, \dots, \langle 120;150\rangle$, see Fig. 14.8
	Systolic	$\langle 90;110\rangle, \langle 110;120\rangle, \dots, \langle 180;220\rangle$, see Fig. 14.9
Problems	Diabetes	yes – 30, no – 1378, ? i.e. not known – 9 patients
	Hyperlipidemia	yes – 54, no – 815, ? i.e. not known – 548 patients
	Ictus	yes – 2, no – 1408, ? i.e. not known – 7 patients
	Infarction	yes – 34, no – 1378, not known – 5 patients note: Infarction means myocardial infarction
Cholesterol	Cholesterol [mg%]	$\langle 100;160\rangle, \langle 160;180\rangle, \dots, \langle 320;540\rangle$, see Fig. 14.10

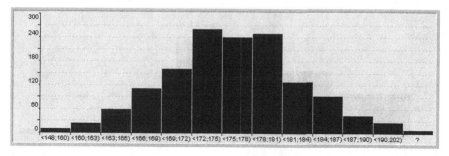

Fig. 14.1 Frequencies of particular categories of the attribute *Height*

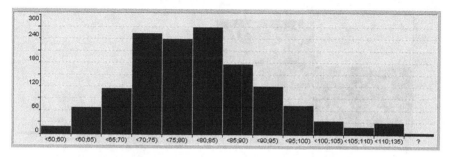

Fig. 14.2 Frequencies of particular categories of the attribute *Weight*

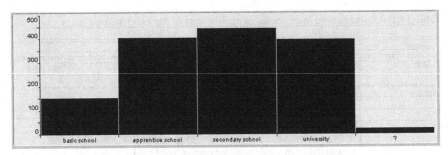

Fig. 14.3 Frequencies of particular categories of the attribute *Education*

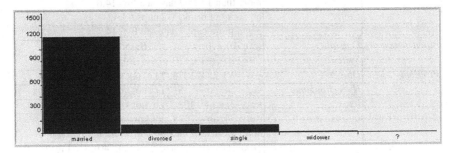

Fig. 14.4 Frequencies of particular categories of the attribute *Marital_Status*

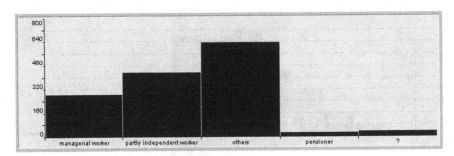

Fig. 14.5 Frequencies of particular categories of the attribute *Responsibility*

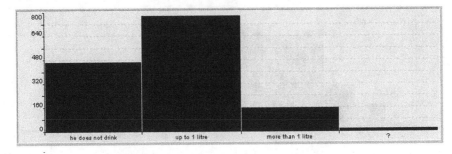

Fig. 14.6 Frequencies of particular categories of the attribute *Beer*

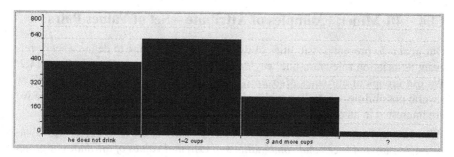

Fig. 14.7 Frequencies of particular categories of the attribute *Coffee*

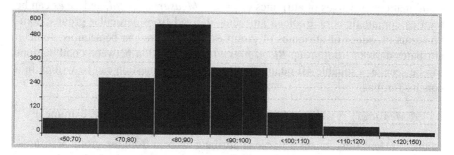

Fig. 14.8 Frequencies of particular categories of the attribute *Diastolic*

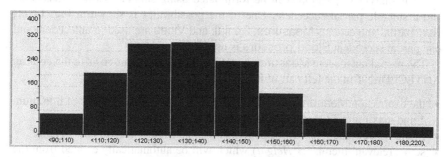

Fig. 14.9 Frequencies of particular categories of the attribute *Systolic*

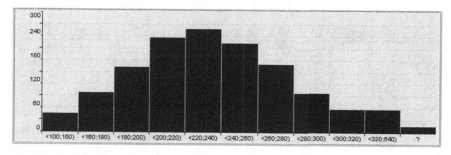

Fig. 14.10 Frequencies of particular categories of the attribute *Cholesterol*

14.4   4ft-Miner, Examples of Attribute – Set of Values Pairs

Our goal is to present possibilities of the 4ft-Miner procedure to define a set of relevant association rules in a fine way using the possibilities outlined in Sect. 14.2. We use groups of attributes *Measures, Social, Vices*, and *Blood_pressure* to show several possibilities of dealing with pairs *attribute-set of values*. Let us assume we are interested in an analytical question

What strong relations between combinations of values of attributes of groups Measures, Social, *and* Vices *and combinations of extremal values of attributes of group* Blood_pressure *are valid in the* ENTRY *data matrix?*

Combinations of values of attributes of groups *Measures, Social*, and *Vices* can be seen as conjunctions of Boolean attributes derived from particular groups, combinations of values of attributes of group *Blood_pressure* can be seen as Boolean attributes derived from group *Blood_pressure*. The relation between combinations can be seen as a suitable 4ft-quantifier. Our analytical question can be written in a concise form as

$$\mathscr{B}(\textit{Measures}) \wedge \mathscr{B}(\textit{Social}) \wedge \mathscr{B}(\textit{Vices}) \approx \mathscr{B}(\textit{Blood\_pressure}, \text{extremes})$$

where $\mathscr{B}(\textit{Measures})$ means a Boolean attribute derived from the group *Measures*, etc.

Such analytical questions can be formulated using suitable partial antecedents and a suitable partial succedent. One of the possibilities is shown in Fig. 14.11. Here partial antecedents **Measures, Social**, and **Vices** are used in antecedent and one partial succedent **Blood_pressure** is used in succedent.

The partial antecedent **Measures** is defined in the first three rows in the column ANTECEDENT in the left part of Fig. 14.11 such that:

- the expression **Measures   Conj, 1-2** means that conjunctions of minimum 1 and maximum 2 literals will be generated
- literals in conjunctions will be generated from attributes *Height* and *Weight*
- set of relevant literals $\mathscr{B}(\textit{Height})$ which will be automatically created from the attribute *Height* is given by the expression **Height(int), 1-4   B, pos**

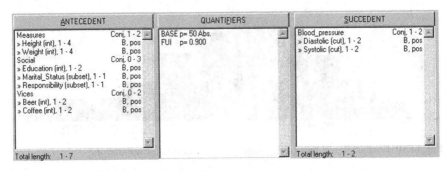

Fig. 14.11 Example of input of the 4ft-Miner procedure

- set of relevant literals $\mathscr{B}(Weight)$ which will be automatically created from the attribute *Weight* is given by the expression Weight(int), 1-4 B, pos.

The additional partial antecedents and the partial succedent are defined analogously.

The substring B, pos in the expression Height(int), 1-4 B, pos means that the attribute *Height* is marked as basic and only positive literals will be generated. The substring Height(int), 1-4 means that literals with coefficients - intervals of length 1-4 will be used. Examples of such literals follow.

Please note that we write $Height\langle 148;160\rangle$ instead of $Height(\langle 148;160\rangle)$ and $Height\langle 148;163\rangle$ instead of $Height(\langle 148;160\rangle,\langle 161;163\rangle)$ etc.

- there are 12 literals with coefficients - intervals of length 1: $Height\langle 148;160\rangle$, $Height\langle 160;163\rangle, \ldots, Height\langle 190;202\rangle$
- there are 11 literals with coefficients - intervals of length 2: $Height\langle 148;163\rangle$, $Height\langle 160;166\rangle, \ldots, Height\langle 187;202\rangle$
- there are 10 literals with coefficients - intervals of length 3: $Height\langle 148;166\rangle$, $Height\langle 160;169\rangle, \ldots, Height\langle 184;202\rangle$
- there are 9 literals with coefficients - intervals of length 4: $Height\langle 148;169\rangle$, $Height\langle 160;172\rangle, \ldots, Height\langle 181;202\rangle$.

The sets of relevant literals $\mathscr{B}(Weight)$, $\mathscr{B}(Education)$, $\mathscr{B}(Beer)$ and $\mathscr{B}(Coffee)$ are defined analogously.

The set of relevant literals $\mathscr{B}(Marital\_Status)$ is given by the expression Marital_Status(subset), 1-1 B, pos. Again, this means that the attribute *Marital_Status* is marked as basic and only positive literals will be generated. The substring Marital_Status(subset), 1-1 means that literals with coefficients - subsets of length 1 will be generated: $Marital\_Status(married)$, $Marital\_Status(divorced)$, $Marital\_Status(single)$, $Marital\_Status(widower)$. The set of relevant literals $\mathscr{B}(Responsibility)$ is given analogously.

The set of relevant literals $\mathscr{B}(Diastolic)$ is given by the expression Diastolic(Cut), 1-2 B, pos. Again, this means that the attribute *Diastolic* is marked as basic and only positive literals will be generated. The substring Diastolic(Cut), 1-2 means that literals with coefficients - cuts of length 1-2 will be generated:

- left cuts: $Diastolic\langle 50;70\rangle$, $Diastolic\langle 50;80\rangle$
- right cuts: $Diastolic\langle 120;150\rangle$, $Diastolic\langle 110;150\rangle$

Please note that the type cut of coefficients means that extremal values of the attribute are generated. The set of relevant literals $\mathscr{B}(Systolic)$ is given analogously.

The 4ft-quantifier is specified in the column QUANTIFIERS in the middle part of Fig. 14.11 by rows

Base p=50 Abs.
FUI p=0.900

which specifies two 4ft-quantifiers - quantifier Base $\oplus_{50}$ and *p*-implication $\Rightarrow_{0.9}$. Their conjunction is applied which means that the quantifier of founded *p*-implication $\Rightarrow_{0.9,50}$ is applied. Remember that its associated function $F_{\Rightarrow_{0.9,50}}$ is defined such that $F_{\Rightarrow_{0.9,50}}(a,b,c,d) = 1$ if and only if $\frac{a}{a+b} \geq 0.9 \wedge a \geq 50$. In addition, deleting of missing information is applied (not seen in Fig. 14.11).

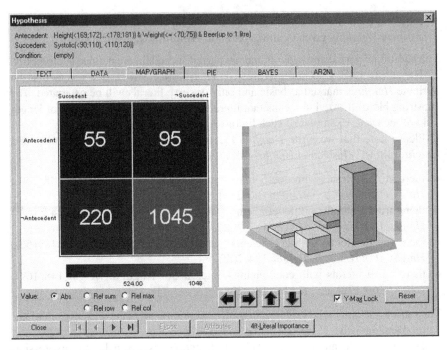

Fig. 14.12 Detail of the 4ft-Miner output

The above specified task was solved in 45 seconds (PC with 1.99 GB RAM and Intel(R) Core(TM)2 Duo CPU processor at 1.33 GHz). $967 * 10^3$ association rules were generated and tested, no true rules were found. This is, however, not surprising, we cannot expect that in a set of at least 50 patients defined by one of the specified antecedents is at least 90 per cent of patients with extremal blood pressure. An additional run of the 4ft-Miner procedure with modified 4ft-quantifier $\Rightarrow_{0.4,50}$ resulted in 20 true rules, the highest confidence was 0.42.

Such results are, however, not interesting. Probably more interesting are the results of application of 4ft-quantifier $\sim^+_{0.6,50}$ of founded above average dependence instead of $\Rightarrow_{0.4,50}$. Remember that its associated function $F_{\sim^+_{0.6,50}}$ is defined such that $F_{\sim^+_{0.6,50}}(a,b,c,d) = 1$ if and only if $\frac{a}{a+b} \geq (1+0.6)\frac{a+c}{a+b+c+d} \wedge a \geq 50$. Association rule $\varphi \sim^+_{0.6,50} \psi$ means that the relative frequency of patients satisfying ψ among patients satisfying φ is at least 60 per cent higher than the relative frequency of patients satisfying ψ among all observed patients and that there are at least 50 observed patients satisfying both φ and ψ.

The above specified task with 4ft-quantifier $\sim^+_{0.6,50}$ instead of $\Rightarrow_{0.9,50}$ was solved in the same time, again $967 * 10^3$ association rules were generated and tested and 60 true rules were found. Detail of the strongest rule (i.e. rule with the highest difference between relative frequencies) is presented in Fig. 14.12. This rule can be written as

$Height\langle169;181\rangle \wedge Weight(<75) \wedge Beer(\text{up to 1 litre}) \sim^{+}_{0.89,55} Systolic\langle90;120\rangle$.

Please note that the relative frequency of patients satisfying $Systolic\langle90;120\rangle$ among patients satisfying $Height(\langle169;181\rangle \wedge Weight(<75) \wedge Beer(\text{up to 1 litre})$ (i.e. confidence) is $\frac{55}{55+95} = 0.3667$, relative frequency of patients satisfying $Systolic\langle90;120\rangle$ among all observed patients is $\frac{55+220}{55+95+220+1045} = 0.1943$ and $\frac{55}{55+95} = (1+0.89)\frac{55+220}{55+95+220+1045}$. In addition, let us note that the sum of frequencies from the contingency table is $55+95+220+1045 = 1415$ which means that there are two patients with missing information for the attributes in question.

We can conclude that we have shown an application of 4ft-Miner dealing with association rules with *attribute-set of values pairs* and with quantifiers $\Rightarrow_{p,Base}$ and $\sim^{+}_{q,Base}$. The 4ft-quantifier $\Rightarrow_{p,Base}$ gives no interesting results and there are interesting rules with the 4ft-quantifier $\sim^{+}_{q,Base}$.

14.5 4ft-Miner, Example of Disjunctions

The goal of this section is to to give an example of application of a partial succedent defined as disjunction of literals. Let us assume we are interested in an analytical question

What strong relations between combinations of values of attributes of groups Measures, Social, *and* Vices *and combinations of problems of patients described by attributes of group* Problems *are valid in the* ENTRY *data matrix?*

This analytical question can be written in a concise form as

$$\mathscr{B}(Measures) \wedge \mathscr{B}(Social) \wedge \mathscr{B}(Vices) \approx \mathscr{B}(Problems) \ .$$

We will use the same definition of partial antecedents Measures, Social, and Vices as in Fig. 14.11. We are interested in combinations of patient problems described by attributes of group *Problems*. This means that we are interested in basic Boolean attributes *Diabetes*(yes), *Hyperlipidemia*(yes), *Ictus*(yes) and *Infarction*(yes), and not in attributes *Diabetes*(no), *Hyperlipidemia*(no), *Ictus*(no) and *Infarction*(no).

There are two options to define partial succedent Problems - as conjunctions of basic Boolean attributes, see left part of Fig. 14.13, or as disjunctions of basic Boolean attributes, see right part of Fig. 14.13. The *conjunctions* option means that the following Boolean attributes are generated:

- conjunctions of length 1:
  Diabetes(yes), *Hyperlipidemia*(yes), *Ictus*(yes), *Infarction*(yes)
- conjunctions of length 2:
  Diabetes(yes) $\wedge$ *Hyperlipidemia*(yes), *Diabetes*(yes) $\wedge$ *Ictus*(yes),
  Diabetes(yes) $\wedge$ *Infarction*(yes), *Hyperlipidemia*(yes) $\wedge$ *Ictus*(yes),
  Hyperlipidemia(yes) $\wedge$ *Infarction*(yes), *Ictus*(yes) $\wedge$ *Infarction*(yes)

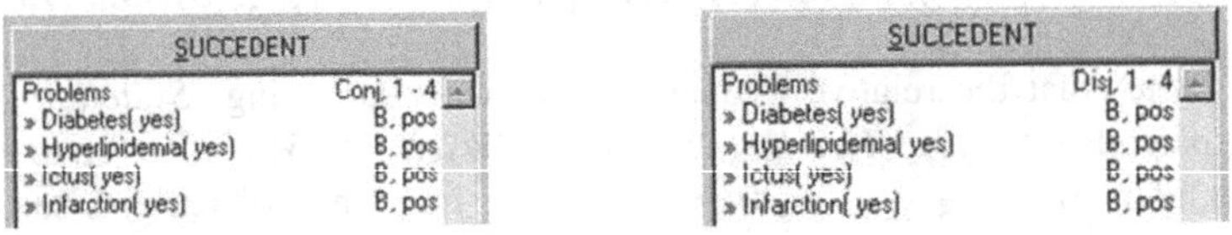

option conjunctions   option disjunctions

Fig. 14.13 Two options to define partial succedent **Problems**

- conjunctions of length 3:
  Diabetes(yes) $\wedge$ *Hyperlipidemia*(yes) $\wedge$ *Ictus*(yes),
  Diabetes(yes) $\wedge$ *Hyperlipidemia*(yes) $\wedge$ *Infarction*(yes),
  Diabetes(yes) $\wedge$ *Ictus*(yes) $\wedge$ *Infarction*(yes),
  Hyperlipidemia(yes) $\wedge$ *Ictus*(yes) $\wedge$ *Infarction*(yes),
- conjunctions of length 4:
  Diabetes(yes) $\wedge$ *Hyperlipidemia*(yes) $\wedge$ *Ictus*(yes) $\wedge$ *Infarction*(yes).

and similarly for *disjunctions*.

We use 4ft-quantifier $\sim^{+}_{0.2,50}$ of founded above average dependence. The option *conjunctions* results in verification of $107 * 10^3$ association rules in 12 seconds, no true rule is found. The option *disjunctions* results in verification of $1.6 * 10^6$ association rules in 67 seconds, 17 true rules are found. Detail of the strongest rule (i.e. rule with the highest difference between relative frequencies) is shown in Fig. 14.14.

This rule can be written as

$$Weight\langle 80; 100\rangle \wedge Beer(\text{he does not drink, up to 1 litre}) \sim^{+}_{0.35,55}$$
$$\sim^{+}_{0.35,55} Hyperlipidemia(\text{yes}) \vee Ictus(\text{yes}) \vee Infarction(\text{yes}).$$

Please note that a relative frequency of patients satisfying succedent i. e. disjunction *Hyperlipidemia*(yes) $\vee$ *Ictus*(yes) $\vee$ *Infarction*(yes) among patients satisfying *Weight*$\langle 80; 100\rangle \wedge$ *Beer*(he does not drink, up to 1 litre) (i.e. confidence) is $\frac{55}{55+298} = 0.1558$, relative frequency of patients satisfying succedent among all observed patients is $\frac{55+47}{55+298+47+483} = 0.1155$ and $\frac{55}{55+298} = (1+0.35)\frac{55+47}{55+298+47+483}$. In addition, let us note that the sum of frequencies from contingency table is $55 + 298 + 47 + 483 = 883$ which means that there are 534 patients with missing information for the attributes in question.

We can conclude that we have shown an application of 4ft-Miner dealing with association rules with *attribute-set of values pairs* and disjunctions of basic Boolean attributes in a partial succedent.

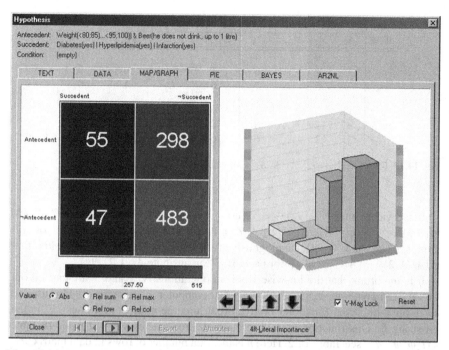

Fig. 14.14 Detail of the 4ft-Miner output - option *disjunctions* for partial succedent Problems

14.6   Applying Strings of Bits

The above outlined dealing with *attribute - set of values* pairs in partial antecedents and disjunctions is based on a representation of analysed data by strings of bits. We use the data matrix $\mathcal{M}$ from Fig. 14.15 to present the principles of the bit string approach. The basic principle is to present each attribute (i.e. a column of $\mathcal{M}$) by cards of its categories. Let $\{1,2,3,4\}$ be the set of all categories of the attribute A_1. Then the attribute A_1 is represented by cards $A_1[1]$, $A_1[2]$, $A_1[3]$, $A_1[4]$ of categories 1,2,3,4 respectively.

The card $A_1[1]$ of category 1 is the string of bits. Each row of data matrix $\mathcal{M}$ corresponds to one bit of the card $A_1[1]$. There is "1" in the bit corresponding to row o_i if and only if row o_i contains the value 1. In other words there is "1" in the i-th bit of the card $A_1[1]$ if and only if the basic Boolean attribute $A_1(1)$ is true in row o_i. The same is true for other categories and attributes. The cards $A_1[1]$, $A_1[2]$, $A_1[3]$, $A_1[4]$ are shown in Fig. 14.15.

Cards of Boolean attributes φ and ψ are used to compute frequencies from 4ft-tables. The card of the Boolean attribute φ is denoted by $\mathscr{C}(\varphi)$. The card $\mathscr{C}(\varphi)$ is a string of bits that is analogous to the card of the category. Each row of the data matrix corresponds to one bit of $\mathscr{C}(\varphi)$ and there is "1" in the i-th bit if and only if φ is true in row o_i.

row of $\mathcal{M}$	attributes of $\mathcal{M}$				cards of categories of attribute A_1			
	A_1	A_2	...	A_K	$A_1[1]$	$A_1[2]$	$A_1[3]$	$A_1[4]$
o_1	1	9	...	4	1	0	0	0
o_2	1	4	...	6	1	0	0	0
o_3	3	5	...	7	0	0	1	0
$\vdots$	$\vdots$	$\vdots$	$\ddots$	$\vdots$	$\vdots$	$\vdots$	$\vdots$	$\vdots$
o_{n-1}	4	1	...	8	0	0	0	1
o_n	2	2	...	6	0	1	0	0

Fig. 14.15 Cards $A_1[1], A_1[2], A_1[3], A_1[4]$ of categories 1,2,3,4 of attribute A_1

It is evident that $\mathcal{C}(\varphi \wedge \psi) = \mathcal{C}(\varphi) \wedge \mathcal{C}(\psi)$, $\mathcal{C}(\varphi \vee \psi) = \mathcal{C}(\varphi) \vee \mathcal{C}(\psi)$, $\mathcal{C}(\neg\varphi) = \dot{\neg}\,\mathcal{C}(\varphi)$. Here $\mathcal{C}(\varphi) \wedge \mathcal{C}(\psi)$ is a bit-wise conjunction of bit strings $\mathcal{C}(\varphi)$ and $\mathcal{C}(\psi)$, analogously for $\vee$ and $\dot{\neg}$. Moreover it holds that $\mathcal{C}(A_1(1,2)) = A_1[1] \vee A_1[2]$ for basic Boolean attribute $A_1(1,2)$ etc.

It is important that the bit-wise Boolean operations $\wedge$, $\vee$ and $\dot{\neg}$ are carried out by very fast computer instructions. Very fast computer instructions are also used to carry out a bit string function $Count(\xi)$ returning the number of values "1" in the bit string ξ. This function is used to compute frequencies a, b, c, d from 4ft-table $4ft(\varphi, \psi, \mathcal{M})$, see Table 14.2. Here n is the total number of rows in data matrix $\mathcal{M}$.

Table 14.2 4ft-table $4ft(\varphi, \psi, \mathcal{M})$

$\mathcal{M}$	ψ	$\neg\psi$
φ	$a = Count(\mathcal{C}(\varphi) \wedge \mathcal{C}(\psi))$	$b = Count(\mathcal{C}(\varphi)) - a$
$\neg\varphi$	$c = Count(\mathcal{C}(\psi)) - a$	$d = n - a - b - c$

Here we assume that there is not missing information. However, the bit-string approach can be also applied when dealing with missing information. For more details on this approach to mining association rules see [59, 61, 80]. Let us note that the bit string representation of analysed data is used also in the granular computing approach [53]. Informally speaking, a binary representation of a granule corresponds to a card of category. The algorithm used in [53] is however different from the algorithm used in the 4ft-Miner procedure [80].

14.7 Couples of Association Rules – SD4ft-Miner

The applications of the 4ft-Miner procedure such as those solved in Sects. 14.4 and 14.5 led to additional analytical questions enhancing the original analytical questions. An example of such an enhanced analytical question is the question

Are there any differences among patients with various values of attribute Marital_Status *with regards to relations between combinations of values of attributes*

of groups Measures, Social, *and* Vices *and combinations of extremal values of attributes of group* Blood_pressure *in the* ENTRY *data matrix?*

In this question *Social* means the couple of attributes *Education* and *Responsibility*. The question can be written in a concise form as

$$(Status, \bowtie) : \mathscr{B}(Measures) \wedge \mathscr{B}(Social) \wedge \mathscr{B}(Vices) \approx \mathscr{B}(Blood\_pressure, \text{extremes})$$

where we write *Status* instead of *Marital_Status*.

The procedure SD4ft-Miner [81, 85] was implemented to solve such questions. It mines for SD4ft-patterns of the form

$$\alpha \bowtie \beta : \; \varphi \approx \psi \; / \; \gamma \; .$$

Here α, β, γ, φ, and ψ are Boolean attributes derived from the columns of analysed data matrix $\mathscr{M}$. The attributes α and β define two subsets of rows (i.e. subsets of patients in our case). The attributes φ and ψ are *antecedent* and *succedent* respectively of an association rule which expresses a relation between Boolean attributes. The attribute γ defines a condition, and it may be omitted.

The SD4ft-pattern $\alpha \bowtie \beta : \varphi \approx \psi / \gamma$ means that the subsets of patients given by Boolean attributes α and β differ with regards to measures of interestingness of association rule $\varphi \approx \psi$ when the condition given by Boolean attribute γ is satisfied. A measure of difference is defined by the symbol $\approx$ which is called *SD4ft-quantifier* here.

An example of an SD4ft-pattern without condition γ is the pattern

$$\text{divorced} \bowtie \text{married} : \; Height\langle 172; 178 \rangle \; \wedge \; Weight\langle 65; 85 \rangle \Rightarrow_{0.2}^{D} Diastolic\langle 50; 80 \rangle \; .$$

This means that divorced patients differ from married patients with regards to confidences p of association rules

$$Height\langle 172; 178 \rangle \; \wedge \; Weight\langle 65; 85 \rangle \; \Rightarrow_{p} \; Diastolic\langle 50; 80 \rangle$$

computed separately for both groups of patients. The difference is given by the *SD4ft-quantifier* $\Rightarrow_{p}^{D}$ with parameter p, in our example $p = 0.2$ which means that the difference is at least 0.2.

A general *SD4ft-quantifier* $\approx$ is defined by a condition concerning two 4ft-tables $4ft(\varphi, \psi, \mathscr{M}/(\alpha \wedge \gamma))$ and $4ft(\varphi, \psi, \mathscr{M}/(\beta \wedge \gamma))$, see Fig. 14.16. The 4ft-table

$\mathscr{M}/(\alpha \wedge \gamma)$	ψ	$\neg\psi$
φ	$a_{\alpha \wedge \gamma}$	$b_{\alpha \wedge \gamma}$
$\neg\varphi$	$c_{\alpha \wedge \gamma}$	$d_{\alpha \wedge \gamma}$

$4ft(\varphi, \psi, \mathscr{M}/(\alpha \wedge \gamma))$

$\mathscr{M}/(\beta \wedge \gamma)$	ψ	$\neg\psi$
φ	$a_{\beta \wedge \gamma}$	$b_{\beta \wedge \gamma}$
$\neg\varphi$	$c_{\beta \wedge \gamma}$	$d_{\beta \wedge \gamma}$

$4ft(\varphi, \psi, \mathscr{M}/(\beta \wedge \gamma))$

Fig. 14.16 4ft-tables $4ft(\varphi, \psi, \mathscr{M}/(\alpha \wedge \gamma))$ and $4ft(\varphi, \psi, \mathscr{M}/(\beta \wedge \gamma))$

$4ft(\varphi, \psi, \mathcal{M}/(\alpha \wedge \gamma))$ of φ and ψ in $\mathcal{M}/(\alpha \wedge \gamma)$ is the contingency table of φ and ψ in $\mathcal{M}/(\alpha \wedge \gamma)$. The data matrix $\mathcal{M}/(\alpha \wedge \gamma)$ is a data sub-matrix of $\mathcal{M}$ that consists of exactly all rows of $\mathcal{M}$ satisfying $\alpha \wedge \gamma$. This means that $\mathcal{M}/(\alpha \wedge \gamma)$ corresponds to all objects (i.e. rows) from the set defined by α which satisfy condition γ.

It holds that $4ft(\varphi, \psi, \mathcal{M}/(\alpha \wedge \gamma)) = \langle a_{\alpha \wedge \gamma}, b_{\alpha \wedge \gamma}, c_{\alpha \wedge \gamma}, d_{\alpha \wedge \gamma} \rangle$ where $a_{\alpha \wedge \gamma}$ is the number of rows of data matrix $\mathcal{M}/(\alpha \wedge \gamma)$ satisfying both φ and ψ, $b_{\alpha \wedge \gamma}$ is the number of rows of $\mathcal{M}/(\alpha \wedge \gamma)$ satisfying φ and not satisfying ψ, etc. The 4ft-table $4ft(\varphi, \psi, \mathcal{M}/(\beta \wedge \gamma))$ of φ and ψ on $\mathcal{M}/(\beta \wedge \gamma)$ is defined analogously.

The SD4ft-pattern $\alpha \bowtie \beta : \varphi \approx \psi / \gamma$ is *true on data matrix* $\mathcal{M}$ if the condition related to $\approx$ is satisfied on data matrix $\mathcal{M}$. The SD4ft-pattern $\alpha \bowtie \beta : \varphi \approx \psi / \gamma$ can be seen as a couple of conditional association rules

$$\varphi \approx_{\alpha} \psi / (\gamma \wedge \alpha) \text{ and } \varphi \approx_{\beta} \psi / (\gamma \wedge \beta)$$

where $\approx_{\alpha}$ and $\approx_{\beta}$ are suitable 4ft-quantifiers.

The *SD4ft-quantifier* $\Rightarrow_p^D$ is defined by the condition

$$\frac{a_{\alpha \wedge \gamma}}{a_{\alpha \wedge \gamma} + b_{\alpha \wedge \gamma}} - \frac{a_{\beta \wedge \gamma}}{a_{\beta \wedge \gamma} + b_{\beta \wedge \gamma}} \geq p \ . \tag{14.1}$$

This condition means that the difference between the confidence of the association rule $\varphi \approx \psi$ on data matrix $\mathcal{M}/(\alpha \wedge \gamma))$ and the confidence of this association rule on data matrix $\mathcal{M}/(\beta \wedge \gamma))$ is at least p. The SD4ft-pattern $\alpha \bowtie \beta : \varphi \Rightarrow_p^D \psi / \gamma$ corresponds to a couple of association rules

$$\varphi \Rightarrow_{p_{\alpha}} \psi / (\gamma \wedge \alpha) \text{ and } \varphi \Rightarrow_{p_{\beta}} \psi / (\gamma \wedge \beta)$$

where $p_{\alpha} - p_{\beta} \geq p$.

The analytical question

$$(Status, \bowtie) : \mathcal{B}(Measures) \wedge \mathcal{B}(Social) \wedge \mathcal{B}(Vices) \approx \mathcal{B}(Blood\_pressure, \text{extremes})$$

can be solved by a run of the SD4ft-Miner procedure with input parameters defined according to Fig. 14.17.

We outline the meaning of parameters in Fig. 14.17, a more detailed description of possibilities of the SD4ft-Miner procedure is not the goal of this book. The parameters in Fig. 14.17 means that all SD4ft-patterns $\alpha \bowtie \beta : \varphi \approx \psi / \gamma$ satisfying conditions 1 – 4 will be generated and tested.

1. $\alpha = Marital\_Status(U)$, $\beta = Marital\_Status(V)$ where $U \neq V$ and both U and V belong to the set {married, divorced, single, widower} of all categories of the attribute *Marital_Status*, see boxes (1) FIRST SET and (2) SECOND SET.
2. Antecedent φ is defined by the partial antecedents Measures, Social, and Vices in a similar way as in Fig. 14.11, see box ANTECEDENT. The only exception concerns the partial antecedent Social which does not contain the attribute *Marital_Status*.

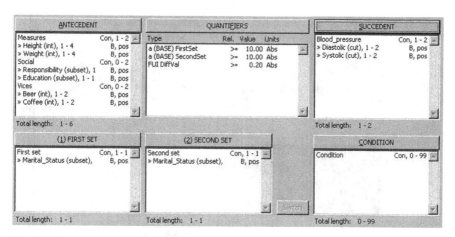

Fig. 14.17 Example of input of the SD4ft-Miner procedure

3. Succedent ψ is defined by the partial succedent Blood_pressure in the same way as in Fig. 14.11, see box <u>SUCCEDENT</u>.
4. SD4ft-quantifier $\Rightarrow^D_{p,Base_\alpha,Base_\beta}$ which is defined by the condition

$$\frac{a_{\alpha\wedge\gamma}}{a_{\alpha\wedge\gamma}+b_{\alpha\wedge\gamma}} - \frac{a_{\beta\wedge\gamma}}{a_{\beta\wedge\gamma}+b_{\beta\wedge\gamma}} \geq p \wedge a_{\alpha\wedge\gamma} \geq Base_\alpha \wedge a_{\beta\wedge\gamma} \geq Base_\beta$$

concerning 4ft-tables introduced in Fig. 14.16 is used. We have $p = 0.2$, $Base_\alpha = 10$, and $Base_\beta = 10$, see box QUANTI<u>F</u>IERS.

These parameters result in the verification of $39.8 * 10^6$ SD4ft-patterns in 27 minutes and 19 seconds, 28 true SD4ft-patterns are found. Detail of the strongest SD4ft-pattern (i.e. SD4ft-pattern with the highest difference between confidences of both rules) is shown in Fig. 14.18.

This SD4ft-pattern can be written as

$$\text{divorced} \bowtie \text{married}: \quad \varphi \Rightarrow^D_{0.27,11,62} Diastolic\langle 50; 80\rangle$$

where

$\varphi = Height\langle 169; 178\rangle \wedge Weight\langle 70; 85\rangle \wedge Beer(\text{he does not drink, up to 1 litre}) \wedge$
$\qquad \wedge Coffee(\text{1-2 cups, 3 and more cups})$

and we write "divorced" instead of "*Marital_Status*(divorced)" and "married" instead of "*Marital_Status*(married)".

This SD4ft-pattern can also be written as a couple of conditional association rules

$$\varphi \Rightarrow_{0.55,11} Diastolic\langle 50; 80\rangle \; / \; Marital\_Status(\text{divorced})$$

and

$$\varphi \Rightarrow_{0.28,62} Diastolic\langle 50; 80\rangle \; / \; Marital\_Status(\text{married}).$$

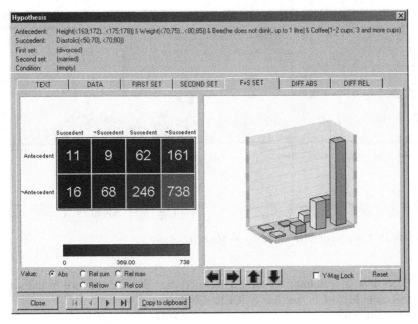

Fig. 14.18 Detail of the SD4ft-Miner output

Please note that $0.55 = \frac{11}{11+9}$, $0.28 = \frac{62}{62+161}$ and $0.27 = 0.55 - 0.28$. We can conclude that we have shown an application of SD4ft-Miner procedure mining for reasonable SD4ft-patterns which can be seen as couples of association rules.

14.8   Couples of Association Rules – Ac4ft-Miner

An additional reasonable analytical question is the question:

Is there a possibility for some groups of patients defined by combinations of values of attributes of groups Measures *and* Social *to influence the probability of having extreme level of cholesterol by changing values of attributes of group* Vices?

The question can be written in a concise form as

$$\mathcal{B}(Measures) \wedge \mathcal{B}(Social) \wedge \mathcal{B}(Vices, \text{Initial} \to \text{Final}) \approx \mathcal{B}(Cholesterol, \text{Extremes})$$

The Ac4ft-Miner procedure [84] was implemented to solve such questions. It is inspired by action rules introduced in [58]. Ac4ft-Miner mines for Ac4ft-patterns (G-action rules according to [84]). The Ac4ft-pattern is an expression

$$\varphi_{St} \wedge \Phi_{Chg} \approx^* \psi_{St} \wedge \Psi_{Chg}$$

where

- φ_{St} is a Boolean attribute called *stable antecedent*
- Φ_{Chg} is an expression called *change of antecedent*
- ψ_{St} is a Boolean attribute called *stable succedent*
- Ψ_{Chg} is an expression called *change of succedent*
- $\approx^*$ is a symbol called *Ac4ft-quantifier*.

An example of an Ac4ft-pattern is the expression

$$Weight\langle 85;95\rangle \wedge Coffee[(\text{1-2 cups}) \rightarrow (\text{not})] \Rightarrow_{0.2}^{D} Cholesterol(\geq 260),$$

here

- $Weight\langle 85;95\rangle$ is an example of stable antecedent
- $Coffee[(\text{1-2 cups}) \rightarrow (\text{not})]$ is an example of change of antecedent, it is actually a short form of $Coffee[(\text{1-2 cups}) \rightarrow (\text{he does not drink})]$
- $Cholesterol(\geq 260)$ is an example of stable succedent
- change of succedent is not used
- $\Rightarrow_{0.2}^{D}$ is an example of an Ac4ft-quantifier.

Please note that the Ac4ft-quantifier $\Rightarrow_{0.2}^{D}$ is the same symbol which is called a SD4ft-quantifier in a context of SD4ft-patterns.

This Ac4ft-pattern means that a change of the value of attribute *Coffee* from "1-2 cups" to "he does not drink" leads for patients satisfying $Weight\langle 85;95\rangle$ to a decrease of relative frequency of patients satisfying $Cholesterol(\geq 260)$ by at least 0.2. This can be seen as a couple of association rules

$$Weight\langle 85;95\rangle \wedge Coffee(\text{1-2 cups}) \Rightarrow_{p_I} Cholesterol(\geq 260)$$

and

$$Weight\langle 85;95\rangle \wedge Coffee(\text{he does not drink}) \Rightarrow_{p_F} Cholesterol(\geq 260)$$

where $p_I - p_F \geq 0.2$.

The *change of coefficient* is an expression $Z[(\kappa) \rightarrow (\kappa')]$ created from basic Boolean attributes $Z(\kappa)$ and $Z(\kappa')$ satisfying $\kappa \cap \kappa' = \emptyset$. The expression $Coffee[(\text{1-2 cups}) \rightarrow (\text{he does not drink})]$ is an example of the change of coefficient. The *change of Boolean attribute* is created from *changes of coefficient* and Boolean connectives in the same way as the Boolean attribute is created from the literals. Both the change of antecedent Φ_{Chg} and the change of succedent Ψ_{Chg} are changes of Boolean attributes.

If $\Lambda = Z[(\kappa) \rightarrow (\kappa')]$ is a change of coefficient, then an *initial state* $\mathscr{I}(\Lambda)$ of Λ is defined as $\mathscr{I}(\Lambda) = Z(\kappa)$ and a *final state* $\mathscr{F}(\Lambda)$ of Λ is defined as $\mathscr{F}(\Lambda) = Z(\kappa')$. Examples of initial and final states are are:

$$\mathscr{I}(Coffee[(\text{1-2 cups}) \rightarrow (\text{he does not drink})]) = Coffee(\text{1-2 cups})$$

$$\mathscr{F}(Coffee[(\text{1-2 cups}) \rightarrow (\text{he does not drink})]) = Coffee(\text{he does not drink}).$$

If Φ is a change of Boolean attribute, then *initial state* $\mathscr{I}(\Phi)$ of Φ is the Boolean attribute that we get by replacing all changes of coefficients Λ occurring in Φ by their initial states $\mathscr{I}(\Lambda)$. The *final state* $\mathscr{F}(\Phi)$ of Φ is the Boolean attribute that we get by replacing all changes of coefficients Λ occurring in Φ by their final states $\mathscr{F}(\Lambda)$.

The Ac4ft-pattern

$$\varphi_{St} \wedge \Phi_{Chg} \approx^* \psi_{St} \wedge \Psi_{Chg}$$

describes what happens when we change values of attributes occurring in Φ_{Chg} from the values corresponding to $\mathscr{I}(\Phi)$ to values corresponding to $\mathscr{F}(\Phi)$. Of course, it is necessary to distinguish *flexible attributes* (i.e. attributes - the values of which can be changed, e.g. *Coffee*) and stable attributes (i.e. attributes - the values of which cannot be changed, e.g. *Year of birth*). The effect of the change is described by two association rules $\mathscr{R}_I$ and $\mathscr{R}_F$:

$$\mathscr{R}_I: \quad \varphi_{St} \wedge \mathscr{I}(\Phi_{Chg}) \approx_I \psi_{St} \wedge \mathscr{I}(\Psi_{Chg})$$

$$\mathscr{R}_F: \quad \varphi_{St} \wedge \mathscr{F}(\Phi_{Chg}) \approx_F \psi_{St} \wedge \mathscr{F}(\Psi_{Chg}).$$

The first rule $\mathscr{R}_I$ characterizes the initial state. The second rule $\mathscr{R}_F$ describes the final state induced by the change of the initial state $\mathscr{I}(\Phi_{Chg})$ to the final state $\mathscr{F}(\Phi_{Chg})$. If we denote $\varphi_{St} \wedge \mathscr{I}(\Phi_{Chg})$ as φ_I, $\psi_{St} \wedge \mathscr{I}(\Psi_{Chg})$ as ψ_I, $\varphi_{St} \wedge \mathscr{F}(\Phi_{Chg})$ as φ_F, and $\psi_{St} \wedge \mathscr{F}(\Psi_{Chg})$ as ψ_F then the rules $\mathscr{R}_I$ and $\mathscr{R}_F$ can be written as

$$\mathscr{R}_I: \quad \varphi_I \approx_I \psi_I \qquad\qquad \mathscr{R}_F: \quad \varphi_F \approx_F \psi_F.$$

A general *Ac4ft-quantifier* $\approx^*$ is defined by a condition concerning two 4ft-tables $4ft(\varphi_I, \psi_I, \mathscr{M})$ and $4ft(\varphi_F, \psi_F, \mathscr{M})$ concerning the analysed data matrix, see Fig. 14.19.

$\mathscr{M}$	ψ_I	$\neg\psi_I$
φ_I	a_I	b_I
$\neg\varphi_I$	c_I	d_I

$4ft(\varphi_I, \psi_I, \mathscr{M})$

$\mathscr{M}$	ψ_F	$\neg\psi_F$
φ_F	a_F	b_F
$\neg\varphi_F$	c_F	d_F

$4ft(\varphi_F, \psi_F, \mathscr{M})$

Fig. 14.19 4ft-tables $4ft(\varphi_I, \psi_I, \mathscr{M})$ and $4ft(\varphi_F, \psi_F, \mathscr{M})$

The Ac4ft-quantifier $\Rightarrow_{0.2}^D$ is defined by the condition

$$\frac{a_I}{a_I+b_I} - \frac{a_F}{a_F+b_F} \geq p, \qquad (14.2)$$

see also equation 14.1 defining SD4ft-quantifier $\Rightarrow_{0.2}^D$. Please note that both definitions can be unified and related only to a couple of 4ft-tables, see also Sect. 16.6.

The analytical question

$$\mathscr{B}(Measures) \wedge \mathscr{B}(Social) \wedge \mathscr{B}(Vices, \text{Initial} \rightarrow \text{Final}) \approx \mathscr{B}(Cholesterol, \text{Extremes})$$

can be solved by a run of the Ac4ft-Miner procedure with input parameters defined according to Fig. 14.20.

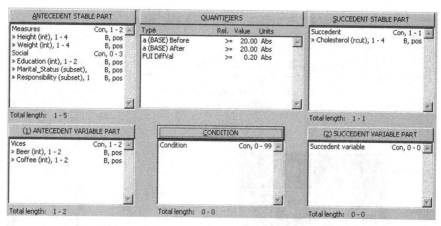

Fig. 14.20 Example of input of the Ac4ft-Miner procedure

We outline the meaning of parameters in Fig. 14.20, a more detailed description of possibilities of the Ac4ft-Miner procedure is not the goal of this book. The parameters in Fig. 14.20 means that all Ac4ft-patterns $\varphi_{St} \wedge \Phi_{Chg} \approx^* \psi_{St} \wedge \Psi_{Chg}$ satisfying conditions 1 – 5 will be generated and tested.

1. stable antecedent φ_{St} is defined by the partial antecedents **Measures** and **Social** in a similar way as in Fig. 14.11, see box <u>A</u>NTECEDENT STABLE PART
2. change of antecedent Φ_{Chg} is built from changes of coefficients of attributes *Beer* and *Coffee*, see box <u>1</u>ANTECEDENT VARIABLE PART
3. stable succedent ψ_{St} is defined as a set of relevant literals $\mathscr{B}(Cholesterol)$ given by an expression **Cholesterol(rcut), 1-4   B, pos**. This means that the basic Boolean attributes *Cholesterol*(≥ 260), *Cholesterol*(≥ 280), *Cholesterol*(≥ 300), *Cholesterol*(≥ 320), corresponding to extreme levels of cholesterol are generated, see box <u>S</u>UCCEDENT STABLE PART
4. change of succedent Ψ_{Chg} is not used
5. Ac4ft-quantifier $\Rightarrow^{D}_{p,Base_I,Base_F}$ which is defined by the condition

$$\frac{a_I}{a_I+b_I} - \frac{a_F}{a_F+b_F} \geq p \wedge a_I \geq Base_I \wedge a_F \geq Base_F$$

concerning 4ft-tables introduced in Fig. 14.19 is used. We have $p = 0.2$, $Base_I = 20$, and $Base_F = 20$, see box QUANTI<u>F</u>IERS.

Please note that it is also possible to use a condition, see box CONDITION. This leads to dealing with conditional Ac4ft-patterns and conditional association rules.

The parameters specified in Fig. 14.20 result in verification of $907 * 10^3$ Ac4ft-patterns in 4 minutes and 2 seconds, 87 true Ac4ft-patterns are found. Detail of the strongest Ac4ft-pattern (i.e. Ac4ft-pattern with the highest difference between confidences of both rules) is shown in Fig. 14.21.

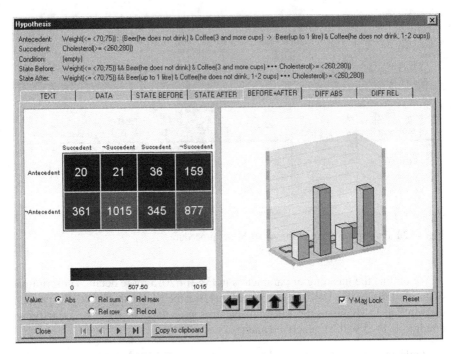

Fig. 14.21 Detail of the Ac4ft-Miner output

This Ac4ft-pattern can be written as

$$Weight(\leq 75) \wedge \Phi_{Chg} \Rightarrow^D_{0.31,20,36} Cholesterol(\geq 260),$$

where
$\Phi_{Chg} = Beer[(\text{he does not drink}) \rightarrow (\text{up to 1 litre})] \wedge$
$\qquad \wedge Coffee[(\text{3 and more cups}) \rightarrow (\text{he does not drink, 1-2 cups})].$
This Ac4ft-pattern can also be written as a couple of association rules

$$Weight(\leq 75) \wedge Beer(\text{not}) \wedge Coffee(\text{3 and more}) \Rightarrow^D_{0.49,20} Cholesterol(\geq 260)$$

and

$$Weight(\leq 75) \wedge Beer(\text{up to 1}) \wedge Coffee(\text{not, 1–2}) \Rightarrow^D_{0.18,36} Cholesterol(\geq 260)$$

where we write *Beer*(not) instead of *Beer*(he does not drink) etc. Please note that $0.49 = \frac{20}{20+21}$, $0.18 = \frac{36}{36+159}$ and $0.31 = 0.49 - 0.18$.

We can conclude that we have shown an application of the Ac4ft-Miner procedure mining for reasonable Ac4ft-patterns which can be seen as couples of association rules.

14.9 Additional GUHA Procedures in the LISp-Miner System

There are four additional GUHA procedures implemented in the LISp-Miner system [26, 81, 90]:

- KL-Miner
- CF-Miner
- SDKL-Miner
- SDCF-Miner.

All of them deal with data matrices as introduced in Chap. 2 and produce complex patterns, not association rules or couples of association rules. All of them, however, could be interesting from the point of view of additional research of observational calculi, see also Chap. 16. We shortly describe the KL-Miner procedure and outline the patterns the remaining three procedures mine for.

The KL-Miner [52, 88] procedure mines for patterns of the form

$$R \approx C/\gamma \, .$$

Here R and C are attributes, the attribute R is of type K (i.e. has K categories $1, \ldots, K$) and the attribute C is of type L (i.e. has L categories $1, \ldots, L$). Furthermore, γ is a Boolean attribute. The meaning of the expression $R \approx C/\gamma$ is that the attributes R and C are in relation given by the symbol $\approx$ when the condition given by the Boolean attribute γ is satisfied.

The symbol $\approx$ is called *KL-quantifier*. It corresponds to a condition imposed by the user on the contingency table of R and C. There are several restrictions that the user can choose to use (e.g. minimal value, sum over the table, value of the χ^2 statistic, and others). We call the expression $R \approx C/\gamma$ a *KL-pattern*. The KL-pattern $R \approx C/\gamma$ is *true* in data matrix $\mathcal{M}$ if the condition corresponding to the KL-quantifier $\approx$ is satisfied for the contingency table of R and C on the data matrix $\mathcal{M}/\gamma$.

The input of the KL-Miner procedure consists of

- the analysed data matrix $\mathcal{M}$
- a set $\mathcal{R} = \{R_1, \ldots, R_u\}$ of row attributes
- a set $\mathcal{C} = \{C_1, \ldots, C_v\}$ of column attributes
- specification of the KL-quantifier $\approx$
- specification of the set Γ of *relevant conditions* γ (i.e. derived Boolean attributes), it is specified in a same way as a set $\mathcal{A}$ of relevant antecedents of the procedure 4ft-Miner, see Sects. 14.2 and 14.4.

The KL-Miner procedure automatically generates and verifies all relevant questions

$$R \approx C/\gamma,$$

such that $R \in \mathcal{R}$, $C \in \mathcal{C}$ and $\gamma \in \Gamma$.

Pattern $R \approx C/\gamma$ is true in the data matrix $\mathcal{M}$ if a condition corresponding to the KL-quantifier $\approx$ is satisfied for the contingency table of R and C on the data matrix $\mathcal{M}/\gamma$. The data matrix $\mathcal{M}/\gamma$ consists of all rows of $\mathcal{M}$ satisfying γ.

We suppose that the contingency table of attributes R and C on the data matrix $\mathcal{M}/\gamma$ has the form of Table 14.3, where:

- $n_{k,l}$ denotes the number of rows in data matrix $\mathcal{M}/\gamma$ for which the Boolean attribute $R(k) \wedge C(l)$ is true (very informally speaking, the number of rows for which the value of R is k and the value of C is l)
- $n_{k,*} = \sum_l n_{k,l}$ denotes the number of rows in data matrix $\mathcal{M}/\gamma$ for which the Boolean attribute $R(k)$ is true
- $n_{*,l} = \sum_k n_{k,l}$ denotes the number rows in data matrix $\mathcal{M}/\gamma$ for which the Boolean attribute $C(l)$ is true
- $n = \sum_k \sum_l n_{k,l}$ denotes the number of all rows in data matrix $\mathcal{M}/\gamma$.

Table 14.3 Contingency table of R and C on $\mathcal{M}/\gamma$

$\mathcal{M}/\gamma$	1	$\cdots$	L	Σ_l
1	$n_{1,1}$	$\cdots$	$n_{1,L}$	$n_{1,*}$
$\vdots$	$\vdots$		$\vdots$	$\vdots$
K	$n_{K,1}$	$\cdots$	$n_{K,L}$	$n_{K,*}$
Σ_k	$n_{*,1}$	$\cdots$	$n_{*,L}$	n

The semantics of the KL-quantifier $\approx$ is determined by the user, who can choose to set lower / upper threshold values for several functions on the table of absolute or relative frequencies, among others to simple aggregate functions

$$\min_{k,l}\{n_{k,l}\} , \ \max_{k,l}\{n_{k,l}\} , \ \sum_{k,l} n_{k,l} , \ \frac{1}{KL} \sum_{k,l} n_{k,l} .$$

The same aggregate functions are defined also for relative frequencies $f_{k,l} = n_{k,l}/n$, $f_{k,*} = \sum_l f_{k,l} = n_{k,*}/n$ and $f_{*,l} = \sum_k f_{k,l} = n_{*,l}/n$.

In an example below we use Kendall's coefficient τ_b defined such that

$$\tau_b = \frac{2(P-Q)}{\sqrt{\left(n^2 - \sum_k n_{k,*}^2\right)\left(n^2 - \sum_l n_{*,l}^2\right)}} ,$$

where

$$P = \sum_{k,l} n_{k,l} \sum_{i>k} \sum_{j>l} n_{i,j} , \ Q = \sum_{k,l} n_{k,l} \sum_{i>k} \sum_{j<l} n_{i,j} .$$

τ_b takes values from $\langle -1, 1 \rangle$ with the following interpretation: $\tau_b > 0$ indicates positive ordinal dependence (i.e. high values of C often coincide with high values of R and low values of C often coincide with low values of R), $\tau_b < 0$ indicates negative ordinal dependence, $\tau_b = 0$ indicates ordinal independence, $|\tau_b = 1|$ indicates that C is a function of R.

Figure 14.22 contains an example of a detailed output of the KL-Miner procedure. It is the strongest KL-pattern (i.e. KL-pattern with highest value of τ_b) from the output of KL-Miner procedure with input given such that:

- the analysed data matrix is the *ENTRY* data matrix introduced in Sect. 14.3
- there is only one row attribute *Diastolic* with 7 categories, see Fig. 14.8
- there is only one column attribute *Systolic* with 9 categories, see Fig. 14.9
- the KL-quantifier is specified such that

  - the minimum number of patients satisfying a relevant condition is 100
  - the minimum value of τ_b is 0.75

- specification of the set Γ of *relevant conditions* by partial antecedents Measures, Social, and Vices in the same way as in Fig. 14.11.

This task can be understood as a task of finding a subset of patients with the strongest ordinal dependence between attributes *Diastolic* and *Systolic* among the set of patients described in the *ENTRY* data matrix. We are interested within subsets definable by the partial antecedents Measures, Social, and Vices. The value of τ_b for KL-pattern introduced in Fig. 14.22 is $\tau_b = 0.768$. For more information on the KL-Miner procedure see [52, 88].

The CF-Miner procedure mines for CF-patterns of the form

$$\approx R/\gamma.$$

Here R is an attribute of type K (i.e. has K categories $1, \dots, K$) and γ is a Boolean attribute. The meaning of the expression $R \approx C/\gamma$ is that the frequencies of attribute R are in the relation given by the symbol $\approx$ when the condition given by the Boolean attribute γ is satisfied.

The symbol $\approx$ is called *CF-quantifier* here. It corresponds to a condition imposed by the user on the frequency table of categories of R which has a form according to table 14.4, where:

- n_k denotes the number of rows in data matrix $\mathcal{M}/\gamma$ for which the Boolean attribute $R(k)$ is true (very informally speaking, the number of rows for which the value of R is k)
- $n = \sum_{k=1}^{K} n_k$ denotes the number of all rows in data matrix $\mathcal{M}/\gamma$.

Let us remember the SD4ft-patterns of the form

$$\alpha \bowtie \beta : \varphi \approx \psi / \gamma$$

introduced in Sect. 14.7. This means that the subsets of objects given by Boolean attributes α and β differ with regards to measures of interestingness of association

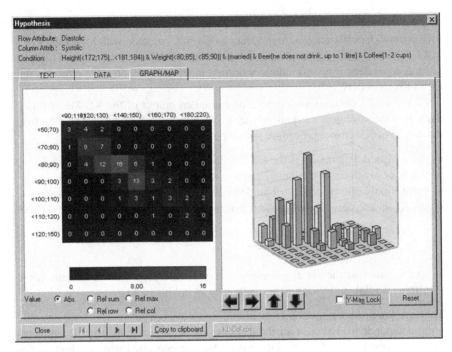

Fig. 14.22 Detail of the KL-Miner output

Table 14.4 The frequency table of categories of R on $\mathcal{M}/\gamma$

$\mathcal{M}/\gamma$	1	...	K	Σ
frequency	n_1	...	n_K	n

rule $\varphi \approx \psi$ when the condition given by Boolean attribute γ is satisfied. A measure of difference is defined by the symbol $\approx$ which is called a *SD4ft-quantifier* here.

SDKL-patterns and SDCF-patterns are defined similarly. The SDKL-pattern

$$\alpha \bowtie \beta : R \approx C/\gamma$$

means that the subsets of objects given by Boolean attributes α and β differ what concerns relations of attributes R and C when the condition given by the Boolean attribute γ is satisfied. The symbol $\approx$ is called a *SDKL-quantifier*, it corresponds to a condition concerning two contingency tables of R and C on $\mathcal{M}/(\alpha \wedge \gamma)$ and $\mathcal{M}/(\beta \wedge \gamma)$.

The SDCF-pattern

$$\alpha \bowtie \beta : \approx R/\gamma$$

means that the subsets of objects given by Boolean attributes α and β differ with regards to frequencies of attribute R when the condition given by the Boolean attribute γ is satisfied. The symbol $\approx$ is called a *SDCF-quantifier*, it corresponds to a condition concerning frequency table of categories of R in $\mathcal{M}/(\alpha \wedge \gamma)$ and $\mathcal{M}/(\beta \wedge \gamma)$.

Chapter 15
Research Projects

Experience with the GUHA procedures implemented in the LISp-Miner system [51, 52, 78, 79, 80, 81, 82, 83, 84, 85, 86, 87, 88] led to a strong belief in the importance of using domain knowledge in particular steps of GUHA applications. Also the importance of the careful presentation of results to "problem owners" was recognized together with possibilities of automatic preparation of analytical reports presenting results of analyses and disseminating them through the Semantic web. These conclusions are in accordance with the research trends in data mining formulated in [96]. This resulted in the launch of several mutually related research projects. Even though these research problems do not originate with observational calculi, their solution can profit from results concerning observational calculi. In addition, these problems can lead to the formulation of new research problems and challenges to observational calculi research. The goal of this chapter is to outline the main features of these projects.

The LMKnowledgeSource project aims to elicit and maintain useful items of domain knowledge so that they are available for additional modules and procedures of the LISp-Miner system. This concerns items of domain knowledge which are easy to understand for domain experts. The LMKnowledgeSource project is shortly introduced in Sect. 15.1.

The goal of the LAQ-Manager (Local Analytical Questions Manager) project is to assist in the formulation of reasonable analytical questions based on stored items of domain knowledge and a particular data matrix. We assume that local analytical questions are answered using GUHA procedures. The main features of the LAQ-Manager are summarized in Sect. 15.2.

The 4ft-DKFilter project aims at filtering out all association rules produced by the 4ft-Miner procedure which can be considered as consequences of a given item of domain knowledge stored by the LMKnowledgeSource project. The goal of the 4ft-DKSynthesizer project is to identify, among association rules produced by the 4ft-Miner procedure, subsets of rules which can be considered consequences of items of domain knowledge not yet stored by the LMKnowledgeSource project. The 4ft-DKFilter and 4ft-DKSynthesizer projects are introduced in Sect. 15.3.

J. Rauch: *Observational Calculi and Association Rules*, SCI 469, pp. 261–271.
DOI: 10.1007/978-3-642-11737-4_15   © Springer-Verlag Berlin Heidelberg 2013

The first goal of the SEWEBAR project is to present answers to local analytical questions. Results are to be presented in the form of analytical reports designed for domain experts and readable by WWW tools. It is also assumed that such analytical reports will be further analysed and results will be again presented in the form of analytical reports. The analytical reports presenting results of analysis of several analytical reports already presented by the SEWEBAR project are called *global analytical reports*. The SEWEBAR project is shortly introduced in Sect. 15.4.

The above mentioned projects offer the possibility to consider a system which will automatically both generate and solve reasonable analytical questions. The construction of such a system is a goal of the EverMiner project. It is introduced in Sect. 15.5.

Let us emphasize that the goal of this chapter is not to describe particular projects in a formally correct, thorough and detailed way. The goal is only to informally outline their main features and to point out their relation to research of observational calculi. Let us also emphasize that all of these projects represent relatively long-time academic research projects. Their goal is both to produce new tools for data mining and to generate new research questions and projects.

15.1   Domain Knowledge and LMKnowledgeSource project

There are various types of domain knowledge relevant to data mining, among others:

- domain knowledge concerning particular attributes, e.g.
  - basic value limits (e.g. 0° C and 100° C for temperature)
  - typical interval lengths when categorizing values (e.g. 5 or 10 years for age)
- groups of attributes which are or could be perceived by domain experts as reasonable and important, examples of such groups include groups of attributes defined in the STULONG data set [94], see also Sect. 14.3
- information on mutual influence of particular attributes, an example is the expression *Weight* ↑↑ *Systolic* saying that if weight of a patient increases then usually his/here systolic blood pressure increases too
- results of serious research projects published in scientific papers and in textbooks.

The goal of the LMKnowledgeSource project is to support elicitation and maintaining of such items of knowledge which can be used in additional research related to the LISp-Miner system. A new part of the LISp-Miner system called LMKnowledgeSource was implemented to deal with domain knowledge related to particular attributes, groups of attributes and information on mutual influence of particular attributes.

Examples of groups of attributes include groups of attributes *Measures*, *Social*, *Vices*, . . . , introduced in table 14.1. We distinguish two types of groups of attributes:

- *basic groups of attributes* BG_1, . . . , BG_N, basic groups have no common attributes and their union is the set of all attributes we are interested in. ,

- *additional groups of attributes* defined for various ad hoc analyses which can be used repeatedly, an example being the group *Problems* introduced in Table 14.1.

Several examples of items of domain knowledge concerning mutual influence of attributes are shown in Fig. 15.1. Please note that the goal of these examples is only to show several types of mutual influence, there is no relation to any actual level of knowledge.

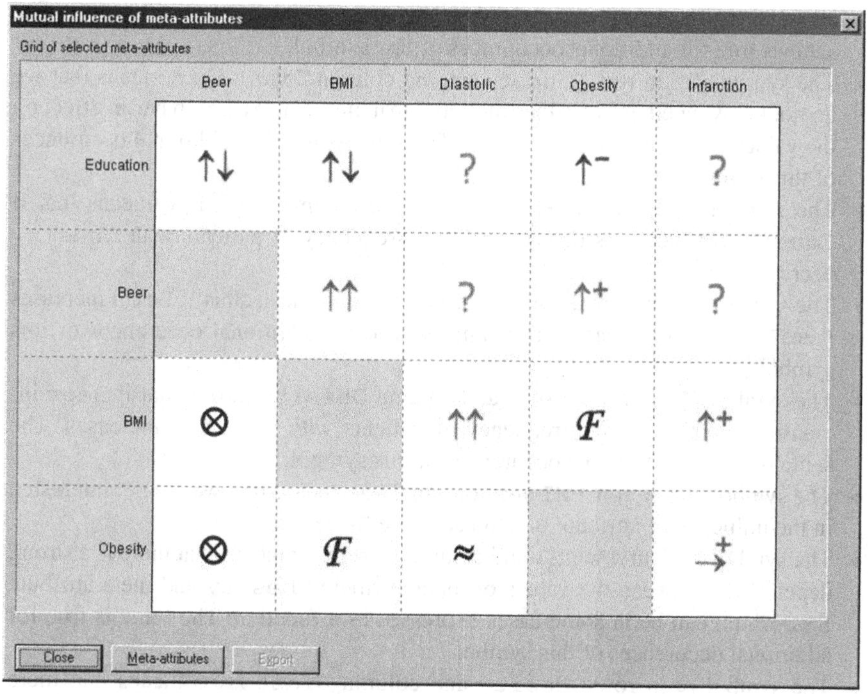

Fig. 15.1 Examples of items of domain knowledge stored in LMKnowledgeSource

The items of knowledge stored in LMKnowledgeSource refer to meta-attributes, i.e. neither to a particular data matrix nor to a particular logical calculus of associational rules. The meta-attributes shown in Fig. 15.1 concern the medical domain. The meta-attributes must be mapped to attributes belonging to a concrete data matrix or to a concrete logical calculus of associational rules. There are the following meta-attributes used in Fig. 15.1:

- `Education` is an ordinal meta-attribute, the attribute *Education* introduced in table 14.1 is one instance of the meta-attribute `Education`
- `Beer` is an ordinal meta-attribute expressing a daily beer consumption, the attribute *Beer* introduced in table 14.1 is one instance of the meta-attribute `Beer`
- `BMI` i.e. Body Mass Index is an ordinal meta-attribute

- `Diastolic` is an ordinal meta-attribute, the attribute *Diastolic* introduced in table 14.1 is one instance of the meta-attribute `Diastolic`
- `Obesity` is a Boolean meta-attribute which is considered to be true if the value of the meta-attribute `BMI` is at least 30
- `Infarction` is a Boolean meta-attribute

The following items of domain knowledge are recorded in Fig. 15.1:

- The symbol "↑↓" in row `Education` and column `Beer` means that if the meta-attribute `Education` increases then the meta-attribute `Beer` decreases. The same is true for additional occurences of this symbol.
- The symbol "?" in row `Education` and column `Diastolic` means that we do not know whether the values of meta-attribute `Education` have an effect on the values of attribute `Diastolic`. The same is true for additional occurences of this symbol.
- The symbol "↑⁻" in row `Education` and column `Obesity` means that if `Education` increases then the relative frequency of patients with `Obesity` decreases.
- The symbol "↑↑" in row `Beer` and column `BMI`   means that if `Beer` increases then `BMI`   also increases. The same is true for additional occurences of this symbol.
- The symbol "↑⁺" in row `Beer` and column `Obesity` means that if `Beer` increases then the relative frequency of patients with `Obesity` increases. The same is true for additional occurences of this symbol.
- The symbol "⊗" in row `BMI` and column `Beer` means that we are not interested in the influence of attribute `BMI` on attribute `Beer`.
- The symbol "$\mathscr{F}$" in row `BMI` and column `Obesity` means that there is a strong dependence between the values of meta-attribute `Obesity` and meta attribute `BMI` which can be in some cases expressed as a function. The same is true for additional occurences of this symbol.
- The symbol ≈ in row `Obesity` and column `Diastolic` means that there surely is some influence of `Obesity` to `Diastolic` but we do not know which influence.
- The symbol →⁺ in row `Obesity` and column `Infarction` means that the truthfulness of attribute `Obesity` increases the relative frequency of patients satisfying `Infarction`.

For more details see [85]. Let us note that similar knowledge can be also elicited and managed by the SEWEBAR project, see Sect. 15.4. It is also possible to import and export items of domain knowledge between the LISp-Miner and SEWEBAR projects. Let us again emphasize that we have only informally outlined the features of the LMKnowledgeBase project related to observational calculi. Our goal is to show that

- the stored items of domain knowledge can be used to formulate reasonable analytical questions, see Sect. 15.2

- the formulated questions can be solved using GUHA procedures which use results on observational calculi, see Sect. 15.2
- results on observational calculi, namely deduction rules can be used in an additional way to interpret results of GUHA procedures, see Sect. 15.3

Let us note that the items of knowledge stored in the LMKnowledgeBase can be enhanced in various ways. We can e.g. store which expert is the author of an item describing mutual influence of meta-attributes or which scientific paper the item was formulated in etc.

15.2 Local Analytical Questions and LAQ Manager

There are many ways of using domain knowledge managed in LMKnowledge-Source to formulate local analytical questions. Recall that local analytical questions concern a particular data matrix. We use the notion *local analytical question* to distinguish these from *global analytical questions* related to the SEWEBAR project introduced in Sect. 15.4. Several examples of local analytical questions follow.

- Each pair of groups of attributes G_A and G_B induces an analytical question *What strong relations between Boolean characteristics of groups G_A and G_B are valid in the given data matrix $\mathcal{M}$?* Symbolically we can write this as

$$\mathcal{B}(G_A) \approx \mathcal{B}(G_B)[\mathcal{M}].$$

Each such a question can be solved using the 4ft-Miner procedure.
- Each analytical question $\mathcal{B}(G_A) \approx \mathcal{B}(G_B)[\mathcal{M}]$ can be specified by choosing a concrete particular 4ft-quantifier. Usually quantifiers $\Rightarrow_{p,Base}$, $\Leftrightarrow_{p,Base}$, $\equiv_{p,Base}$, $\sim^+_{p,Base}$ are used. This leads to analytical questions

$$\mathcal{B}(G_A) \Rightarrow_{p,Base} \mathcal{B}(G_B)[\mathcal{M}], \quad \mathcal{B}(G_A) \Leftrightarrow_{p,Base} \mathcal{B}(G_B)[\mathcal{M}],$$

$$\mathcal{B}(G_A) \equiv_{p,Base} \mathcal{B}(G_B)[\mathcal{M}], \quad \mathcal{B}(G_A) \sim_{p,Base} \mathcal{B}(G_B)[\mathcal{M}].$$

- Usually it is reasonable to generate all questions $\mathcal{B}(BG_i) \approx \mathcal{B}(BG_j)[\mathcal{M}]$ where $BG_1, \ldots BG_N$ are all basic groups of attributes and $i, j = 1, \ldots N$.
- It is also possible to formulate analytical questions concerning more than two groups of attributes, see e. g. Sect. 14.4.
- Each analytical question $\mathcal{B}(G_A) \approx \mathcal{B}(G_B)[\mathcal{M}]$ can be modified using an attribute D to the question *Are there any differences among objects with various values of attribute D with regards to relations between Boolean characteristics of attributes G_A and G_B?*. Symbolically we can write

$$(D, \bowtie) : \mathcal{B}(G_A) \approx \mathcal{B}(G_B)[\mathcal{M}].$$

Such a question can be solved using the SD4ft-Miner procedure, see Sect. 14.7.

- Additional analytical questions can be formulated such that they can be solved using the procedures Ac4ft-Miner, KL-Miner, CF-Miner, SDKL-Miner, and SDCF-Miner, see Sects. 14.8 and 14.9.
- Let $\mathcal{I}_1, \ldots, \mathcal{I}_m$ be items of domain knowledge expressing mutual dependence of attributes introduced in Sects. 15.1. $A_1 \uparrow\uparrow A_2$ and $A_3 \uparrow^- A_4$ are examples of such items. The analytical question $\mathcal{B}(G_A) \approx \mathcal{B}(G_B)[\mathcal{M}]$, i.e., *"What strong relations between Boolean characteristics of groups G_A and BG_B are valid in the given data matrix $\mathcal{M}$?"*, can be modified such that we are not interested in consequences of items $\mathcal{I}_1, \ldots, \mathcal{I}_m$ of domain knowledge. Symbolically we can write the modified question as

$$\mathcal{B}(G_A) \approx \mathcal{B}(G_B)[\mathcal{M}, \text{not } Cons(\mathcal{I}_1, \ldots, \mathcal{I}_m)] \ .$$

The goal of the *LAQ Manager* is to help formulate local analytical questions and to define the initial input parameters of GUHA procedures implemented in the LISp-Miner system to solve particular local analytical questions. Below, we will use the following notation:

- Q_{4ft} denotes a local analytical question which is intended to be solved by the 4ft-Miner procedure. Such a local analytical question Q_{4ft} is called a *4ft-analytical question*. Examples of Q_{4ft} include the questions $\mathcal{B}(G_A) \approx \mathcal{B}(G_B)[\mathcal{M}]$ and $\mathcal{B}(G_A) \approx \mathcal{B}(G_B)[\mathcal{M}, \text{not } Cons(\mathcal{I}_1, \ldots, \mathcal{I}_m)]$.
- $\mathcal{P}(Q_{4ft})$ denotes input parameters of the 4ft-Miner procedure used to solve a local analytical question Q_{4ft}.
- $\mathcal{S}(\mathcal{P}(Q_{4ft}))$ denotes the output of the 4ft-Miner procedure with input parameters $\mathcal{P}(Q_{4ft})$. It is a set of association rules which is a solution of the local analytical question Q_{4ft}.
- Q_{SD4ft}, $\mathcal{P}(Q_{SD4ft})$ and $\mathcal{S}(\mathcal{P}(Q_{SD4ft}))$ denote analogous notions concerning the SD4ft-Miner procedure, similarly for additional GUHA procedures of the LISp-Miner system.

We show that there is a way to filter out all consequences of a given item of domain knowledge expressing mutual dependence of attributes and that results on observational calculi are useful in solving this task.

15.3   4ft-DKFilter and 4ft-DKSynthesizer Projects

The goal of the 4ft-DKFilter project is to filter out all consequences of given items of domain knowledge expressing mutual dependence of attributes from the output of a run of the 4ft-Miner procedure. To describe principles of this project we use a simplified situation - only two groups of attributes, one item of domain knowledge expressing mutual dependence of attributes and one of four important 4ft-quantifiers are used. Our situation can be then described in the following way:

- Q_{4ft} is a 4ft-analytical question $\mathscr{B}(G_A) \approx \mathscr{B}(G_B)[\mathscr{M}, not\ Cons(\mathscr{I})]$
- G_A and G_B are groups of attributes
- $\approx$ is one of the 4ft-quantifiers $\Rightarrow_{p,Base}, \Leftrightarrow_{p,Base}, \equiv_{p,Base}, \sim^+_{p,Base}$
- $\mathscr{I}$ is an item of domain knowledge expressing mutual dependence of attributes
- the 4ft-Miner procedure was run with input parameters $\mathscr{P}(Q_{4ft})$) and produced a set $\mathscr{S}(\mathscr{P}(Q_{4ft}))$ of association rules
- our task is to find and filter out all rules $\varphi \approx \psi \in \mathscr{S}(\mathscr{P}(Q_{4ft}))$ which can be considered consequences of the item $\mathscr{I}$ in data matrix $\mathscr{M}$.

We solve this task by defining a set $Cons_\approx(\mathscr{I}, \mathscr{M})$ of atomic consequences of the item $\mathscr{I}$ in the data matrix $\mathscr{M}$ for the 4ft-quantifier $\approx$. If the item $\mathscr{I}$ concerns attributes A and B (e.g. $\mathscr{I} = A \uparrow\uparrow B$, $\mathscr{I} = A \uparrow\downarrow B$, $\mathscr{I} = A \rightarrow^+ B$, etc.), then the set $Cons_\approx(\mathscr{I}, \mathscr{M})$ consists of simple association rules of the form $A(\tau) \approx B(\omega)$. Here $A(\tau)$ and $B(\omega)$ are basic Boolean attributes defined with assistance from a domain expert. Then the association rule $\varphi \approx \psi \in \mathscr{S}(\mathscr{P}(Q_{4ft}))$ is considered a consequence of the item $\mathscr{I}$ in data matrix $\mathscr{M}$ if there is $A(\tau) \approx B(\omega) \in Cons_\approx(\mathscr{I}, \mathscr{M})$ such that one of the following conditions is satisfied:

X) $\varphi \approx \psi$ is equal to $A(\tau) \approx B(\omega)$

Y) $\varphi \approx \psi$ logically follows from $A(\tau) \approx B(\omega)$, i.e. $\frac{A(\tau)\approx B(\omega)}{\varphi\approx\psi}$ is a sound deduction rule

Z) based on consent between the domain expert and the data mining expert, it is reasonable to consider a true rule $\varphi \approx \psi$ as a consequence of $A(\tau) \approx B(\omega)$ even if $\frac{A(\tau)\approx B(\omega)}{\varphi\approx\psi}$ is not a sound deduction rule. However, the information that the deduction rule $\frac{A(\tau)\approx B(\omega)}{\varphi\approx\psi}$ is not sound is important when dealing with results.

For more details see [73, 86], as here we only provide simple examples of important steps of this approach. Figure 15.2 shows a definition of the set $Cons_{\Rightarrow_{0.9,50}}(BMI \uparrow\uparrow Diastolic, ENTRY)$ of atomic consequences for the item $BMI \uparrow\uparrow Diastolic$ which concerns attributes BMI and $Diastolic$ of the data matrix $ENTRY$ introduced in Sect. 14.3 and for 4ft-quantifier $\Rightarrow_{0.9,50}$.

The set $Cons_{\Rightarrow_{0.9,50}}(BMI \uparrow\uparrow Diastolic, ENTRY)$ is defined in Fig. 15.2 such that it consists of all the following rules:

- $BMI(\tau_{small}) \Rightarrow_{p,Base} Diastolic(\omega_{small})$
- $BMI(\tau_{medium}) \Rightarrow_{p,Base} Diastolic(\omega_{medium})$
- $BMI(\tau_{high}) \Rightarrow_{p,Base} Diastolic(\omega_{high})$

where $p \geq 0.9$, $Base \geq 50$ and the following conditions are satisfied

- $\tau_{small} \subseteq \{\langle 16;21\rangle, \langle 21;22\rangle, \ldots, \langle 24;25\rangle\}$ and
  $\omega_{small} \subseteq \{\langle 50;70\rangle, \langle 70;80\rangle, \langle 80;90\rangle\}$
- $\tau_{medium} \subseteq \{\langle 23;24\rangle, \ldots, \langle 28;29\rangle\}$ and
  $\omega_{medium} \subseteq \{\langle 80;90\rangle, \ldots, \langle 100;110\rangle\}$
- $\tau_{high} \subseteq \{\langle 27;28\rangle, \ldots, \langle 30;31\rangle, > 32\}$ and
  $\omega_{high} \subseteq \{\langle 90;100\rangle, \ldots, \langle 110;120\rangle, \langle 120;150\rangle\}$,

see also Fig. 14.8.

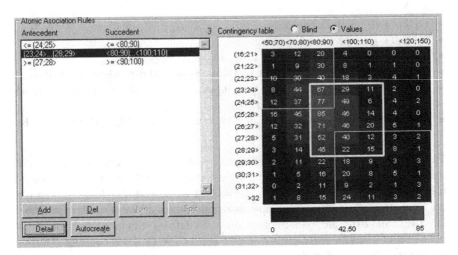

Fig. 15.2 Definition of the set $Cons_{\Rightarrow 0.9,50}(BMI \uparrow\uparrow Diastolic, ENTRY)$ of atomic consequences

Let us note that the true rule

$$BMI(\langle 16;21\rangle, \langle 21;22\rangle) \Rightarrow_{0.93,83} Diastolic(\langle 70;80\rangle, \langle 80;90\rangle, \langle 90;100\rangle)$$

i.e. $BMI\langle 16;22\rangle \Rightarrow_{0.93,83} Diastolic\langle 70;100\rangle$ is a consequence of $BMI \uparrow\uparrow Diastolic$ according to condition Y) because

$$\frac{BMI(\langle 16;21\rangle, \langle 21;22\rangle) \Rightarrow_{0.93,83} Diastolic(\langle 70;80\rangle, \langle 80;90\rangle)}{BMI(\langle 16;21\rangle, \langle 21;22\rangle) \Rightarrow_{0.93,83} Diastolic(\langle 70;80\rangle, \langle 80;90\rangle, \langle 90;100\rangle)}$$

is a sound deduction rule and

$$BMI(\langle 16;21\rangle, \langle 21;22\rangle) \Rightarrow_{0.93,83} Diastolic(\langle 70;80\rangle, \langle 80;90\rangle) \in$$

$$\in Cons_{\Rightarrow 0.9,50}(BMI \uparrow\uparrow Diastolic, ENTRY)$$

due to $0.93 \geq 0.9$, $83 \geq 50$, $\{\langle 16;21\rangle, \langle 21;22\rangle\} \subseteq \{\langle 16;21\rangle, \langle 21;22\rangle, \ldots, \langle 24;25\rangle\}$ and $\{\langle 70;80\rangle, \langle 80;90\rangle\} \subseteq \{\langle 50;70\rangle, \langle 70;80\rangle, \langle 80;90\rangle\}$.

Let us also note that it could be reasonable to consider each true rule

$$BMI(\tau) \wedge Education(\alpha) \Rightarrow_{p,Base} Diastolic(\omega) \,,$$

where $BMI(\tau) \Rightarrow_{p,Base} Diastolic(\omega)$ is a consequence of $(BMI \uparrow\uparrow Diastolic, ENTRY)$, also to be a consequence of $BMI \uparrow\uparrow Diastolic$. Then, according to condition Z) we can also consider the true rule

$$BMI\langle 21;24\rangle \wedge Educations(\text{university}) \Rightarrow_{0.93,89} Diastolic\langle 70;100\rangle$$

a consequence of $(BMI \uparrow\uparrow Diastolic, ENTRY)$.

The project 4ft-DKSynthesizer tries to find a sufficiently large subset of rules $\varphi \approx \psi \in \mathscr{S}(\mathscr{P}(Q_{4ft}))$ which can be considered consequences of an item $\mathscr{I}$ of domain knowledge expressing a yet unknown mutual dependence of attributes. Both projects are still under development and their more detailed description is not the goal of this book. However, it is clear that deduction rules are important for these projects.

15.4 SEWEBAR Project

An application of 4ft-Miner is a complex process involving formulation of an analytical question Q_{4ft}, converting the question Q_{4ft} into input parameters $\mathscr{P}(Q_{4ft})$ and interpretation of a resulting set $\mathscr{S}(\mathscr{P}(Q_{4ft}))$ of association rules. Formalized domain knowledge can also be used. It is very important to present the results in a suitable form.

One way of meeting this requirement is to arrange results of data mining into an analytical report structured both according to an analysed problem and to the needs of users. An attempt to produce analytical reports automatically is described in [54], these possibilities are also mentioned in [65]. Such analytical reports are natural candidates for the Semantic Web. This has led to the launch of the SEWEBAR project [85]. Its principle is outlined in Fig. 15.3.

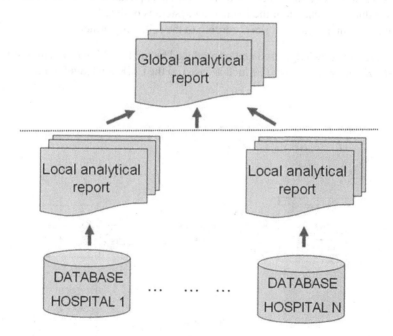

Fig. 15.3 Principle of the SEWEBAR project

We assume there are hospitals storing data concerning patients in their databases. Local analytical reports give answers to local analytical questions concerning patients, their illnesses and treatments in particular hospitals. There are also global analytical reports. Each global analytical report summarizes two or more local or additional global analytical reports.

It is supposed that the contents of analytical reports are indexed, among other, by logical formulas corresponding to patterns resulting from data mining instead of usual key words. The possibility of indexing the contents of analytical reports by logical formulas is outlined in [65], see also [50, 51]. Observational calculi are natural tools for the SEWEBAR project. There is intensive research on the application of Semantic web technologies in the SEWEBAR project [2, 43, 44, 45, 46].

15.5 EverMiner Project

The particular projects described in Sects. 15.1 – 15.3 open the possibility to start building the EverMiner system, the goal of which is the automatic generation and solution of reasonable analytical questions. The idea of the EverMiner project is inspired also by the GUHA80 project [19, 22], which was never realized. The architecture, particular software and theoretical components and the principles of the mining process management used in EverMiner differ from the one used in the GUHA80. However both projects are based on the application of GUHA data mining procedures. A schema of the EverMiner system is outlined in Fig. 15.4.

The EverMiner system works according to the following principles:

- All relevant knowledge used and/or produced by the EverMiner system is stored in the *Knowledge repository* which is based on the LMKnowledgeSource project.

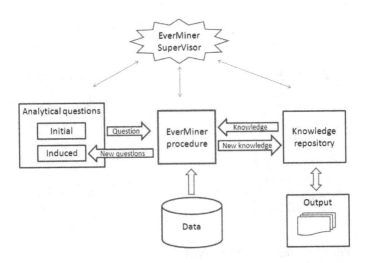

Fig. 15.4 Architecture of the EverMiner system

Before beginning, the initial knowledge is stored in the Knowledge repository in the form of basic groups of attributes.

- At the beginning a set of initial analytical questions $\mathscr{B}(BG_i) \approx \mathscr{B}(BG_j)[\mathscr{M}]$ is generated. Here BG_i and BG_j where $i, j = 1, \ldots N$ are basic groups of attributes, see Sect. 15.2.
- Each analytical question is solved by an EverMiner procedure. The core of each EverMiner procedure is one of the GUHA procedures implemented in the LISp-Miner system. The GUHA procedure is run iteratively until the solved analytical question is answered in the best possible way. Results of the 4ft-DKFilter and 4ft-DKSynthesizer projects introduced in Sect. 15.3 are used.
- There are various ways of formulating additional analytical questions using various items of knowledge stored in the Knowledge repository. Some of them are outlined in Sect. 15.2.
- Both the initial and the newly formulated analytical questions are stored in the list of analytical questions.
- The whole process is managed by the EverMiner Supervisor. There is also an Output module that produces output reports on the discovered knowledge according to the user's requests. The output reports can be presented using the SEWEBAR project.

For more information see [77, 91].

Let us emphasize that both the SEWEBAR project and the EverMiner project are rather long-time academic research projects, the focus of which is also to involve students into research. Their detailed description is not the goal of this book. We only present the projects here to demonstrate various possibilities of applications of observational calculi.

Chapter 16
Additional Results and Open Problems

Various results on associational rules and 4ft-quantifiers are presented in parts I –
III. However, there are additional important and interesting results on observational
calculi tackling problems not closely related to association rules. In addition, there
exist only partly solved problems as well as only informally formulated problems
related to observational calculi and data mining. There are also results which can
be seen as initial steps opening new research issues. Some of them are inspired by
the research projects introduced in Chap. 15 and some of them have already been
formulated a long time ago. Some problems seem to be easy and some can be very
hard.

The goal of this chapter is to give a short overview of results not presented in
previous chapters and to provide an inventory of research challenges and partly
solved problems related to data mining and observational calculi.

Many 4ft-quantifiers are introduced in Sects. 4.1 and 4.3, and 4ft-quantifiers introduced in Sect. 4.3 are defined on the basis of measures of interestingness of association rules. There are additional measures of interestingness of association rules
which can be used to define new 4ft-quantifiers, see Sect. 16.1.

Various classes of 4ft-quantifiers are studied in part II. Additional possibilities
to define new reasonable classes of association rules are introduced in Sect. 16.2.
New classes of association rules bring new challenges concerning deduction rules,
see Sect. 16.3. However, there are also some additional results on deduction rules
concerning association rules studied in part II. These results are also mentioned in
Sect. 16.3. An important feature of deduction rules is their transparency. Results
concerning this topic are introduced in Sect. 16.4.

Practically important conditional association rules are introduced in Sect. 14.2.
Important theoretical features of conditional association rules and related challenges
are mentioned in Sect. 16.5. Additional practically important patterns include the
SD4ft-rules and Ac4ft-rules introduced in Sects. 14.7 and 14.8. These patterns can
be understood as couples of association rules and thus it seems reasonable to develop observational calculi of couples of association rules. Some related results are
introduced in Sect. 16.6.

J. Rauch: *Observational Calculi and Association Rules*, SCI 469, pp. 273–282.
DOI: 10.1007/978-3-642-11737-4_16 © Springer-Verlag Berlin Heidelberg 2013

Logic of discovery is developed in [18]. Research projects introduced in Chap. 15 have led to an attempt to develop a formal framework for data mining of association rules called FOFRADAR which can serve as a theoretical background covering several stages described in the CRISP-DM methodology. Some introductory remarks to this challenge are presented in Sect. 16.7. The intention of applying the GUHA method to data in databases has led to the study of many-sorted observational predicate calculi. The principles of their definition and related challenges are introduced in Sect. 16.8.

The results and open theoretical problems mentioned in Sects. 16.1 – 16.8 are closely related to results and projects presented in previous chapters. Several additional related research topics are shortly introduced in Sect. 16.9.

16.1 Additional 4ft-quantifiers

There are 29 measures of interestingness of association rules in table 4.5. These are used in Sect. 4.3 to define 4ft-quantifiers. Here we present five additional important measures of interestingness of association rules [9] which can be used to define additional 4ft-quantifiers in a similar way. The measures are defined using probabilities, see also Sect. 4.2 and 4ft-table $4ft(A, B, \mathcal{M})$ in Fig. 4.1:

- *Collective strength* defined by the formula

$$\frac{P(AB) + P(\neg B|\neg A)}{P(A)P(B) + P(\neg A)P(\neg B)} * \frac{1 - P(A)P(B) - P(\neg A)P(\neg B)}{1 - P(AB) - P(\neg B|\neg A)}$$

- *Gini index* defined by the formula

$$P(A)[P(B|A)^2 + P(\neg B|A)^2] + P(\neg A)[P(B|\neg A)^2 + P(\neg B|\neg A)^2] - P(B)^2 - P(\neg B)^2$$

- *J-measure* defined by the formula

$$P(AB)\log(\frac{P(B|A)}{P(B)}) + P(A\neg B)\log(\frac{P(\neg B|A)}{P(\neg B)})$$

- *two-way support variation* defined by the formula

$$P(AB)\log_2\frac{P(AB)}{P(A)P(B)} + P(A\neg B)\log_2\frac{P(A\neg B)}{P(A)P(\neg B)} +$$

$$+P(\neg AB)\log_2\frac{P(\neg AB)}{P(\neg A)P(B)} + P(\neg A\neg B)\log_2\frac{P(\neg A\neg B)}{P(\neg A)P(\neg B)}$$

- *ø-Coefficient (linear correlation coefficient)* defined by the formula

$$\frac{P(AB) - P(A)P(B)}{\sqrt{P(A)P(B)P(\neg A)P(\neg B)}}.$$

The measures can of course also be written using frequencies from 4ft-table $4ft(\varphi, \psi, \mathcal{M})$ in Fig. 4.1.

It is not the goal of this book to give a complete overview of measures of interestingness of association rules which can be used to define 4ft-quantifiers. Let us only note that we can find additional suitable measures of interestingness of associational rules in [9, 29, 30, 55, 93].

16.2 Additional Classes of 4ft-quantifiers

Classes of 4ft-quantifiers are defined using TPC's - truth preservation conditions, see Sect. 6.1. Each TPC is defined such that it preserves truth of an important 4ft-quantifier. For example, the truth preservation conditions $TPC_\Rightarrow$ for implicational quantifiers is defined to preserve truth for the 4ft-quantifier $\Rightarrow_p$ of p-implication. The simple form of $TPC_\Rightarrow$ is defined as $a' \geq a \wedge b' \leq b$, see definition 6.2.

It is possible to define additional classes of 4ft-quantifiers this way. Two examples of new classes defined this way are given in [74]. The first one concerns 4ft-quantifier $\Rightarrow_p^R$ of recall, see rows 5 in tables 4.5 and 4.6. 4ft-quantifier $\Rightarrow_p^R$ is defined by the condition $\frac{a}{a+c} \geq p$ and leads to the definition of the class of *recall-like 4ft-quantifiers* by the truth preservation condition TPC_{recc} defined as $a' \geq a \wedge c' \leq c$.

The second example concerns 4ft-quantifier $\Rightarrow_p^S$ of specificity, see rows 6 in tables 4.5 and 4.6. 4ft-quantifier $\Rightarrow_p^S$ is defined by the condition $\frac{d}{b+d} \geq p$ and leads to the definition of the class of *specificity-like 4ft-quantifiers* by the truth preservation condition TPC_{spec} defined as $d' \geq d \wedge b' \leq b$.

This approach can be applied to additional 4ft-quantifiers, among others to 4ft-quantifiers introduced in tables 4.2 and 4.6. However, only those 4ft-quantifiers which do not belong to known classes of 4ft-quantifiers can be used. This means that 4ft-quantifiers with no information in columns *Class* in tables 4.4 and 4.6 with respect to known classes of 4ft-quantifiers must be studied.

Let us also mention the class of 4ft-quantifiers with the F^+-property. The class of 4ft-quantifiers with the F-property is defined and studied in Chap. 10. In Chap. 11 there are results on deduction rules of the form $\frac{\varphi \approx \psi}{\varphi' \approx \psi'}$ where $\varphi \approx \psi$ and $\varphi' \approx \psi'$ are association rules. The results concern implicational quantifiers, Σ-double implicational quantifiers and Σ-equivalence quantifiers, no results on deduction rules for 4ft-quantifiers with the F-property are presented. It is possible to get similar results for a class of 4ft-quantifiers with the F^+-property which is a subclass of 4ft-quantifiers with the F-property, see Sect. 16.3.

16.3 Additional Deduction Rules

Chap. 11 is devoted to deduction rules. There are various additional results on deduction rules not presented in Chap. 11 as well as several related research challenges. Some of the additional results are related to deduction rules of the form $\frac{\varphi \approx \psi}{\varphi' \approx \psi'}$ where $\varphi \approx \psi$ and $\varphi' \approx \psi'$ are association rules. There are known criteria of soundness of such deduction rules for strong double implicational quantifiers, strong equivalence quantifiers and pure equivalence quantifiers. The corresponding theorems are formulated and proved in [67]. However, the criteria are complex and the proofs of the theorems are cumbersome. In addition, these theorems seem to be less important.

The criterion of soundness of a deduction rule $\frac{\varphi \approx \psi}{\varphi' \approx \psi'}$ for the class of 4ft-quantifiers with the F^+-property mentioned in Sect. 16.2 seems to be more important. The corresponding theorem is formulated and proved in [63] for an observational predicate calculus. This version of the theorem is also published without a proof in [74]. Both the theorem and the proof are again a bit cumbersome.

A criterion of soundness of deduction rules of the form $\frac{\varphi \approx \psi}{\varphi' \approx \psi'}$ for new classes of 4ft-quantifiers introduced in Sect. 16.2 is still a challenge. However, in some cases dealing with new classes will be analogous to dealing with already known classes, see also [67, 74].

Please note that there are also results on various additional deduction rules. An example is the fact that deduction rule $\frac{\varphi \Rightarrow_{p,Base} \psi, \ \psi \Rightarrow_{p,Base} \varphi}{\varphi' \Leftrightarrow_{p,Base} \psi'}$ is not sound. These results are mostly of only of theoretical interest. For more details see [67].

16.4 Transparent Deduction Rules

Deduction rules of the form $\frac{\varphi \approx \psi}{\varphi' \approx \psi'}$ are used in the output of the ASSOC procedure when dealing with prime association rules, see Sect. 1.2.2. In this context it is important to have a transparent deduction rule. However, it is not too clear what it means that the deduction rule $\frac{\varphi \approx \psi}{\varphi' \approx \psi'}$ is transparent.

An approach to define formally transparent deduction rules is made in [6]. Two definitions of transparency are given. The core of each of them is a definition of basic transparent transformations. The first definition is based on *atomic* basic transparent transformations. A transformation is considered atomic if it cannot be reached by other transformations. A transparent transformation consists of a chain of basic transparent transformations. This chain is limited by its length. Examples of atomic basic transparent transformations are *addition of a category to a basic Boolean attribute, extraction of a category from a basic Boolean attribute, addition of a basic Boolean attribute to an antecedent*, etc.

The second definition is based on basic transparent transformations which are not atomic. The emphasis is placed on the structure of φ and ψ in this case. A normal form (CNF or DNF) is expected. For more information see [6].

16.5 Conditional Association Rules

The 4ft-Miner procedure mines also for conditional association rules, see Sect. 14.2. A conditional association rule is an expression $\varphi \approx \psi/\chi$ where φ, ψ, and χ are Boolean attributes. It is verified in a given data matrix $\mathcal{M}$ such that we consider the rule $\varphi \approx \psi/\chi$ true in $\mathcal{M}$ if the rule $\varphi \approx \psi$ is true in a data matrix $\mathcal{M}/\chi$. The data matrix $\mathcal{M}/\chi$ consists of all rows of $\mathcal{M}/\chi$ satisfying χ.

It is a challenge to develop and study a logical calculus whose formulas are both association rules and conditional association rules. Such a calculus will be an enhancement of the logical calculus of association rules developed in Chap. 3. However, one first needs to be aware of at least the two following problems.

The first problem is related to the possibility that there is no row of a given data matrix $\mathcal{M}$ satisfying χ. This could be solved by enhancing definition 3.8 of associated function $F_\approx$ of the 4ft-quantifier $\approx$ so that it is defined for all quadruples $\langle a,b,c,d \rangle$ of non-negative integer numbers, not only for quadruples satisfying $a+b+c+d > 0$.

The second problem is that there is no relation between the value $Val(\varphi \approx \psi, \mathcal{M})$ of association rule $\varphi \approx \psi$ and the value $Val(\varphi \approx \psi/\chi, \mathcal{M})$ of conditional association rule $\varphi \approx \psi/\chi$ in data matrix $\mathcal{M}$, see also [79]. We outline an example. Let us have data matrices $\mathcal{M}_{01}$ and $\mathcal{M}_{10}$.

The values of φ, ψ and χ in data matrix $\mathcal{M}_{01}$ are given in Fig. 16.1 together with 4ft-tables $4ft(\varphi, \psi, \mathcal{M}_{01})$ of φ and ψ in data matrix $\mathcal{M}_{01}$ and 4ft-table $4ft(\varphi, \psi, \mathcal{M}_{01}/\chi)$ of φ and ψ in data matrix $\mathcal{M}_{01}/\chi$.

rows	φ	ψ	χ
$o_1 - o_{100}$	1	1	1
o_{101}	1	0	1
$o_{102} - o_{200}$	1	0	0
o_{201}	0	1	1
$o_{202} - o_{301}$	0	0	1

$\mathcal{M}_{01}$	ψ	$\neg\psi$
φ	100	100
$\neg\varphi$	1	100

$\mathcal{M}_{01}/\chi$	ψ	$\neg\psi$
φ	100	1
$\neg\varphi$	1	100

Data matrix $\mathcal{M}_{01}$ $\qquad$ $4ft(\varphi, \psi, \mathcal{M}_{01})$ $\qquad$ $4ft(\varphi, \psi, \mathcal{M}_{01}/\chi)$

Fig. 16.1 Values of φ, ψ and χ in $\mathcal{M}_{01}$ and 4ft-tables $4ft(\varphi, \psi, \mathcal{M}_{01})$ and $4ft(\varphi, \psi, \mathcal{M}_{01}/\chi)$

The associated function $F_{\Leftrightarrow p,Base}$ is defined such that $F_{\Leftrightarrow p,Base}(a,b,c,d) = 1$ if and only if $\frac{a}{a+b+c} \geq p \wedge a \geq Base$, see theorem 8.5. We can conclude

- $Val(\varphi \Leftrightarrow_{0.9,100} \psi, \mathcal{M}_{01}) = 0$ because of $\frac{100}{100+100+1} < 0.9$
- $Val(\varphi \Leftrightarrow_{0.9,100} \psi/\chi, \mathcal{M}_{01}) = 1$ because of $\frac{100}{100+1+1} \geq 0.9 \wedge 100 \geq 100$.

The values of φ, ψ and χ in data matrix $\mathcal{M}_{10}$ are given in Fig. 16.2 together with 4ft-tables $4ft(\varphi, \psi, \mathcal{M}_{10})$ of φ and ψ in data matrix $\mathcal{M}_{10}$ and 4ft-table $4ft(\varphi, \psi, \mathcal{M}_{10}/\chi)$ of φ and ψ in data matrix $\mathcal{M}_{10}/\chi$.

We can conclude

- $Val(\varphi \Leftrightarrow_{0.9,100} \psi, \mathcal{M}_{10}) = 1$ because of $\frac{100}{100+1+1} \geq 0.9 \wedge 100 \geq 100$
- $Val(\varphi \Leftrightarrow_{0.9,100} \psi/\chi, \mathcal{M}_{10}) = 0$ because of $\frac{1}{1+1+1} < 0.9$.

Similar couples of data matrices can be found for additional 4ft-quantifiers. Implicational 4ft-quantifiers deserve special attention in this regard.

rows	φ	ψ	χ
$o_1 - o_{99}$	1	1	0
o_{100}	1	1	1
o_{101}	1	0	1
o_{102}	0	1	1
$o_{103} - o_{202}$	0	0	1

$\mathscr{M}_{10}$	ψ	$\neg\psi$
φ	100	1
$\neg\varphi$	1	100

$\mathscr{M}_{10}/\chi$	ψ	$\neg\psi$
φ	1	1
$\neg\varphi$	1	100

Data matrix $\mathscr{M}_{10}$ $4ft(\varphi,\psi,\mathscr{M}_{10})$ $4ft(\varphi,\psi,\mathscr{M}_{10}/\chi)$

Fig. 16.2 Values of φ, ψ and χ in $\mathscr{M}_{10}$ and 4ft-tables $4ft(\varphi,\psi,\mathscr{M}_{10})$ and $4ft(\varphi,\psi,\mathscr{M}_{10}/\chi)$

16.6   Observational Calculi of Couples of Association Rules

The procedures SD4ft-Miner and Ac4ft-Miner introduced in Sects. 14.7 and 14.8 mine for useful patterns which can be seen as couples of association rules. The SD4ft-pattern $\alpha \bowtie \beta : \varphi \approx \psi/\gamma$ means that the subsets of patients given by Boolean attributes α and β differ with regards to measures of interestingness of association rule $\varphi \approx \psi$ when the condition given by Boolean attribute γ is satisfied. It can be seen as a couple of conditional association rules $\varphi \approx_\alpha \psi / (\gamma \wedge \alpha)$ and $\varphi \approx_\beta \psi / (\gamma \wedge \beta)$.

The Ac4ft-pattern $\varphi_{St} \wedge \Phi_{Chg} \approx^* \psi_{St} \wedge \Psi_{Chg}$ describes what happens when we change the values of attributes occurring in Φ_{Chg}. The effect of the change is described by two association rules $\varphi_I \approx_I \psi_I$ and $\varphi_F \approx_F \psi_F$. The first rule characterizes the initial state and the second rule $\mathscr{R}_F$ describes the final state induced by the change. Suitable conditions can also be used.

Thus it is reasonable to try to modify logical calculi of association rules defined and studied in this book so that their formulas will correspond to these couples of association rules. A lot of work in this direction is done in [47]. Several classes of SD4ft-patterns are defined and studied. Logical calculus of SD4ft-patterns is defined and deduction rules are studied together with possibilities of dealing with missing information. Various important results are achieved.

However, no analogous work has been done for Ac4ft-patterns. It is a challenge to develop a logical calculus which will give a unified view of both SD4ft-patterns and Ac4ft-patterns.

16.7   FOFRADAR – A Formal Framework for Data Mining

The research projects LMKnowledgeSource, LAQ-Manager, 4ft-DKFilter, 4ft-DKSynthesizer, SEWEBAR and EverMiner introduced in Chap. 15 deal with formal aspects of domain knowledge, analytical questions, interpretation of results of GUHA procedures and presentation of results. Experience with these projects and a logic of discovery developed in [18] has led to the launch of a theoretical research project with a goal of developing a theoretical framework covering more or less the whole process of data mining of association rules. This frame is denoted as FOFRADAR i.e. FOrmal FRAme for Data mining of Association Rules. First considerations related to this project are located in [75, 76].

The main features of the FOFRADAR project are outlined below in this section. The main parts of FOFRADAR are:

- enhanced logical calculus of association rules
- language and procedures for dealing with items of domain knowledge
- language for expressing items of knowledge related to analysed data
- language and procedures for dealing with analytical questions
- tools for interpretation of output of GUHA procedures
- tools for preparing analytical reports.

All parts are assumed to be developed in close relation to GUHA procedures 4ft-Miner, SD4ft-Miner and Ac4ft-Miner and the research projects introduced in Chaps. 14 and 15. However, the goal is to develop a theoretical framework independent on any particular software.

Enhanced logical calculus of association rules is assumed to be based on a logical calculus $\mathscr{LC}_{\mathscr{T}}$ of association rules of type $\mathscr{T} = \langle t_1, \ldots, t_K \rangle$, see definition 3.9. We assume language $\mathscr{L}_{\mathscr{T}}$ of $\mathscr{LC}_{\mathscr{T}}$ has attributes $A_1, \ldots A_K$. Logical calculus $\mathscr{LC}_{\mathscr{T}}$ is enhanced by conditional association rules, see Sect. 16.5, and by means of observational calculus of couples of association rules, see Sect. 16.6.

Language of items of domain knowledge allows the expression of both types of items of knowledge studied in the LMKnowledgeSource project – groups of attributes and items expressing mutual influence among particular attributes, see Sect. 15.1.

- Groups of attributes are defined as sets of attributes of language $\mathscr{L}_{\mathscr{T}}$ of calculus $\mathscr{LC}_{\mathscr{T}}$. An example of a definition of a group of attributes is $G = \{A_{i_1}, \ldots, A_{i_k}\}$ where $A_{i_j} \in \{A_1, \ldots A_K\}$ for $j = 1, \ldots, k$.
- Mutual influence of attributes is represented by expressions such as *Beer* $\uparrow\uparrow$ *BMI*, see Sect. 15.1. This means that language of FOFRADAR includes expressions $A_i \uparrow\uparrow A_j$, $A_i \uparrow\downarrow A_j$, $A_i \uparrow^+ A_j(\alpha_j)$, $A_i(\alpha_i) \rightarrow^+ A_j(\alpha_j)$ etc.

Procedures producing these items of knowledge will be formalized in a suitable way to include both initial definitions of items of knowledge by the user and procedures implemented in the 4ft-DKSynthesizer project, see Sect. 15.3.

Language for expressing items of knowledge related to analysed data includes tools for expressing the type of each attribute (nominal or ordinal), frequencies of particular categories of each attribute and additional items of knowledge which can be used in analysis.

Language for dealing with analytical questions makes it possible to express analytical questions, e.g. 4ft-analytical questions $\mathscr{B}(G_A) \Rightarrow_{p,Base} \mathscr{B}(G_B)[\mathscr{M}]$ and $\mathscr{B}(G_A) \approx \mathscr{B}(G_B)[\mathscr{M}, not\ Cons(\mathscr{I}_1, \ldots, \mathscr{I}_m)]$ and SD4ft-analytical question $(D, \bowtie): \mathscr{B}(G_A) \approx \mathscr{B}(G_B)[\mathscr{M}]$ introduced in Sect. 15.2. There are procedures for generating reasonable analytical questions and procedures for converting analytical questions into definitions of sets of relevant patterns. The definition of a set of relevant patterns can be understood as the set of input parameters of an appropriate

GUHA procedure. For example, the LAQ Manager introduced in Sect. 15.2 is assumed to assign, to a given local analytical question Q_{4ft}, input parameters $\mathscr{P}(Q_{4ft})$ of the 4ft-Miner procedure.

Tools for interpretation of output of GUHA procedures can be understood as the theoretical background for 4ft-DKFilter and 4ft-DKSynthesizer projects introduced in Sect. 15.3.

Tools for preparing analytical reports are assumed to involve suitable basic partial expressions which can explain the whole chain *initial knowledge – analytical question – input parameters – interpretation – conclusions* to an end use of results. We have to assume iterations in the data mining process, and these iterations also need to be explained. A strong association rule which cannot be understood as a consequence of known items of domain knowledge is an example of a basic partial expression. Another example would be assertions such as *all strong association rules true in given data matrix can be understood as a consequence of a known item of domain knowledge*. The conversion of association rules into natural language can also be considered [92].

16.8   Many-Sorted Observational Calculi

The study of many-sorted observational calculi (MSOC for short) began by the realization of the potential of applying the GUHA method on data stored in databases, see e.g. [62, 64]. The idea of applying the GUHA method to data stored in databases has led to the definition and study of *Codasyl observational predicate calculi*, formulas of which concern data stored in databases with the CODASYL database model [7]. Interesting results on Codasyl observational predicate calculi were achieved in [63].

We outline the way of defining association rules in Codasyl observational predicate calculi, see also [70]. The principle is shown in Fig. 16.3. There are two data matrices $\mathscr{A}$ and $\mathscr{B}$ that are connected by a function $\mathscr{F}$. We can assume that these data matrices correspond to relational tables in a database and that there is a 1:N relation between these tables corresponding to the function $\mathscr{F}$. This relation is implemented by suitable database keys. Data matrices $\mathscr{A}$ and $\mathscr{B}$ can also be understood

	data matrix $\mathscr{A}$					data matrix $\mathscr{B}$		
row	attributes	multi-relational attributes				row of $\mathscr{B}$	attributes $B_1 \dots B_L$	
of $\mathscr{A}$	$A_1 \dots A_K$	$[B_1 \wedge B_2 \Rightarrow^+_{0.5,30} B_3]/\mathscr{F}$	$[B_5 \Rightarrow_{0.9,20} \neg B_7]/\mathscr{F}$		of $\mathscr{A}$			
a_1	$1 \dots 0$	1	0		a_1	b_1	$1 \dots 0$	
a_2	$1 \dots 1$	0	1		$\vdots$	$\vdots$	$\vdots \ddots \vdots$	
$\vdots$	$\vdots \ddots \vdots$	$\vdots$	$\vdots$		a_1	b_{200}	$0 \dots 1$	
a_n	$0 \dots 1$	1	1		a_2	b_{201}	$1 \dots 1$	
					$\vdots$	$\vdots$	$\vdots \ddots \vdots$	
					a_n	b_m	$0 \dots 0$	

Fig. 16.3 Multi-relational attributes

as data structures interpreting two different sorts of a many-sorted observational calculus.

A typical example is the situation where $\mathscr{A}$ corresponds to patients and $\mathscr{B}$ corresponds to their examinations. Another example is a situation where $\mathscr{A}$ corresponds to clients of a bank and $\mathscr{B}$ corresponds to transactions on their accounts. The relation between data matrices can be understood as a function assigning a row of $\mathscr{A}$ to each row of $\mathscr{B}$. It is e.g. $\mathscr{F}(b_i) = a_1$ for $i = 1,\dots,200$ in Fig. 16.3.

If we search for interesting patterns concerning clients of a bank, we have to consider the characteristics of behavior of their accounts. If we search for interesting patterns concerning patients we have to consider relations among results of their examinations. This means that we have to consider new attributes concerning data matrix $\mathscr{A}$ which are derived from single table patterns concerning sets of rows of data matrix $\mathscr{B}$ related to particular rows of $\mathscr{A}$ through the function $\mathscr{F}$. We can call these attributes *multi–relational attributes*. Please note that there are many multi–relational attributes which can be computed using the SQL language in various database systems (e.g. using SUM or AVERAGE).

However, we are interested in deeper characteristics than those provided by SQL. There are two examples of such multi–relational attributes in Fig. 16.3. The first example is

$$[B_1 \wedge B_2 \Rightarrow^+_{0.5,30} B_3]/\mathscr{F} .$$

This is a Boolean attribute defined on $\mathscr{A}$. It is true for the row a_1 if the association rule $B_1 \wedge B_2 \Rightarrow^+_{0.5,30} B_3$ is true in the data matrix $\mathscr{B}/\mathscr{F}[a_1]$, see Fig. 16.4. Data matrix $\mathscr{B}/\mathscr{F}[a_1]$ is a sub-matrix of data matrix $\mathscr{B}$ consisting of all rows b of $\mathscr{B}$ such that $\mathscr{F}(b) = a_1$ (i.e. of rows $b_1,\dots,b_{200}$).

If we consider rows of $\mathscr{A}$ as clients of a bank, then we can say that the multi–relational attribute $[B_1 \wedge B_2 \Rightarrow^+_{0.5,30} B_3]/\mathscr{F}$ is true for the client a_1 if the relative frequency of his transactions satisfying B_3 among the transactions satisfying $B_1 \wedge B_2$ is at least 50 per-cent higher than the relative frequency of transactions satisfying B_3 among all his transactions and that there are at least 30 of his transactions satisfying both $B_1 \wedge B_2$ and B_3. The value of $[B_1 \wedge B_2 \Rightarrow^+_{0.5,30} B_3]/\mathscr{F}$ for other rows of $\mathscr{A}$ is defined analogously. The second example of a multi–relational attribute given in Fig. 16.3 is the attribute $[B_5 \Rightarrow_{0.9,20} \neg B_7]/\mathscr{F}$ which is defined analogously. Please note the presented approach differs from the approach introduced in [8].

A "many-sorted" GUHA procedure ASSOC was suggested in [63]. This "many-sorted" GUHA procedure was later implemented and experimentally applied

row of $\mathscr{B}$	of $\mathscr{A}$	attributes $B_1\ B_2\ \dots\ B_L$
a_1	b_1	1 0 … 1
$\vdots$	$\vdots$	$\vdots$ $\vdots$ $\ddots$ $\vdots$
a_1	b_{200}	0 1 … 0

Fig. 16.4 Data matrix $\mathscr{B}/\mathscr{F}[a_1]$

[42, 49, 90]. The main experience is that the application of such procedures requires even more intensive use of domain knowledge than the application of the classical "one-sorted" GUHA procedure.

The problem is that the observational calculi defined and studied in [63] are results of modifications of many-sorted classical predicate calculi such that the generalized quantifiers are added and only final data structures are allowed as models. The same method is used in [18] to define observational predicate calculi by modifications of classical (one-sorted) predicate calculi. An example of a resulting calculus is located in Sect. 13.1.

It is a challenge is to modify the Codasyl observational predicate calculi defined in [63] so that the resulting calculus will be a "many-sorted" version of the calculus of association rules, as defined in Chap. 3.

16.9   Additional Related Topics and Challenges

The results and open problems mentioned in Sects. 16.1 – 16.8 are related to results and projects presented in Chaps. 1 – 15. There are additional interesting topics related to observational calculi and the GUHA method. We mention two of them – decidability of observational calculi and fuzzy approach to GUHA method.

Decidability of observational calculi is studied in [18]. The decidability of many-sorted observational calculi is studied in [63]. A short overview of results on the decidability of many-sorted observational calculi is presented [70].

Fuzzy approach to GUHA method is studied in several works. Fuzzy hypothesis testing should be pointed out [31, 32, 33, 34, 35, 36, 37]. Another approach is published in [40]. An extensive overview of fuzzy association rules is presented in [57] where an implementation of a fuzzy version of the ASSOC procedure is also described.

Additional topics related to observational calculi and the GUHA method can be found in [18] and in a survey paper [26]. It is of course also possible to formulate additional research topics and challenges related to Chaps. 1 – 15 which are not listed in Sects. 16.1 – 16.8. For examples, see notes 9.1, 13.3, 13.4, 13.5, 13.6, and 13.7.

It is also a challenge to define and study logical calculi for optimistic completion and deletion of missing information, see Sect. 12.7. It is also a great challenge to develop observational calculi related to additional GUHA procedures which are only shortly introduced in Sect. 14.9.

References

1. Agrawal, R., Imielinski, T., Swami, A.: Mining Associations between Sets of Items in Large Databases. In: Buneman, P., Jajodia, S. (eds.) Proceedings of the 1993 ACM SIGMOD International Conference on Management of Data, pp. 207–216. ACM Press, Fort Collins (1993)
2. Balhar, J., Kliegr, T., Šťastný, D., Vojíř, S.: Elicitation of Background Knowledge for Data Mining. In: Smrž, P. (ed.) Znalosti 2010, pp. 167–170. Oeconomica, Praha (2010)
3. Berka, P.: ETree Miner: A New GUHA Procedure for Building Exploration Trees. In: Kryszkiewicz, M., Rybinski, H., Skowron, A., Raś, Z.W. (eds.) ISMIS 2011. LNCS, vol. 6804, pp. 96–101. Springer, Heidelberg (2011)
4. Berka, P., Ivánek, J.: Automated Knowledge Acquisition for PROSPECTOR-like Expert Systems. In: Bergadano, F., De Raedt, L. (eds.) ECML 1994. LNCS, vol. 784, pp. 337–342. Springer, Heidelberg (1994)
5. Burian, J.: Data mining and AA (Above Average) quantifier. In: Svátek, V. (ed.) Znalosti 2003, pp. 297–302. VŠB TU Ostrava, Ostrava (2003) (in Czech)
6. Chrz, M.: Transparent deduction rules for the GUHA procedures. Master thesis, Faculty of Mathematics and Physics, Charles University, Prague (2007)
7. Date, C.J.: An Introduction to Database Systems. Addison–Wesley Publishing Company, Boston (1976)
8. Dehaspe, L., Toivonen, H.: Discovery of Relational Association Rules. In: Džeroski, S., Lavrač, N. (eds.) Relational Data Mining, pp. 189–208. Springer, Heidelberg (2001)
9. Geng, L., Hamilton, H.J.: Interestingness Measures for Data Mining: A survey. ACM Comput. Surv. 38, 1–32 (2006)
10. Hájek, P. (guest ed.): International Journal of Man-Machine Studies, special issue on GUHA 10 (1978)
11. Hájek, P. (guest ed.): International Journal of Man-Machine Studies, second special issue on GUHA 15 (1981)
12. Hájek, P.: Logics for Data Mining (GUHA rediviva). In: Workshop of Japanese Society for Artificial Intelligence, Tokyo, JSAI, pp. 27–34 (1998); Reprinted: Neural Network World 10, 301–311 (2000)
13. Hájek, P.: The GUHA Method and Mining Association Rules. In: Kuncheva, L. (ed.) Computational Intelligence: Methods and Applications, pp. 533–539. ICSC Academic Press, Canada (2001)

14. Hájek, P.: Relations in GUHA Style Data Mining. In: de Swart, H. (ed.) Proceedings RelMiCS'6 and the First International Workshop of COST Action 274 Theory and Application of Relational Structures as Knowledge Instruments TARSKI, pp. 91–96. Katholieke Universiteit Brabant, Oisterwijk (2001)

15. Hájek, P.: On generalized quantifiers, finite sets and data mining. In: Klopotek, M.A., Wierzchon, S.T., Trojanowski, K. (eds.) Intelligent Information Processing and Web Mining, Proceedings of IIPWM 2003, pp. 489–496. Springer, Heidelberg (2003)

16. Hájek, P.: Relations in GUHA Style Data Mining II. In: Berghammer, R., Möller, B., Struth, G. (eds.) RelMiCS 2003. LNCS, vol. 3051, pp. 163–170. Springer, Heidelberg (2004)

17. Hájek, P.: The GUHA in the Last Century and Today. In: Snášel, V. (ed.) ZNALOSTI 2004, pp. 10–20. VŠB TU, Ostrava (2004) (in Czech)

18. Hájek, P., Havránek, T.: Mechanising Hypothesis Formation - Mathematical Foundations for a General Theory. Springer, Heidelberg (1978), http://www.cs.cas.cz/hajek/guhabook/ (cited August 15, 2011)

19. Hájek, P., Havránek, T.: GUHA 80: An Application of Artificial Intelligence to Data Analysis. Computers and Artificial Intelligence 1, 107–134 (1982)

20. Hájek, P., Holeňa, M.: Formal Logics of Discovery and Hypothesis Formation by Machine. In: Arikawa, S., Motoda, H. (eds.) DS 1998. LNCS (LNAI), vol. 1532, pp. 291–302. Springer, Heidelberg (1998)

21. Hájek, P., Holeňa, M.: Formal Logics of Discovery and Hypothesis Formation by Machine. Theoretical Computer Science 292, 345–357 (2003)

22. Hájek, P., Ivánek, J.: Artificial Intelligence and Data Analysis. In: Caussinus, H., Ettinger, P., Tomassone, R. (eds.) Proceedings COMPSTAT 1982, pp. 54–60. Physica Verlag, Wien (1982)

23. Hájek, P., Havel, I., Chytil, M.: The GUHA method of automatic hypotheses determination. Computing 1, 293–308 (1966)

24. Hájek, P., Havránek, T., Chytil, M.: Metoda GUHA. Academia, Praha (1983) (in Czech)

25. Hájek, P., Sochorová, A., Zvárová, J.: GUHA for personal computers. Computational Statistics & Data Analysis 19, 149–153 (1995)

26. Hájek, P., Holeňa, M., Rauch, J.: The GUHA method and its meaning for data mining. J. Comput. Syst. Sci. 76, 34–48 (2010)

27. Havránek, T.: The statistical interpretation and modification of GUHA method. Kybernetika 7, 13–21 (1971)

28. Havránek, T.: The present state of the GUHA software. Int. J. Man Mach. Stud. 15, 253–264 (1981)

29. Hébert, C., Crémilleux, B.: A Unified View of Objective Interestingness Measures. In: Perner, P. (ed.) MLDM 2007. LNCS (LNAI), vol. 4571, pp. 533–547. Springer, Heidelberg (2007)

30. Hilderman, R., Hamilton, H.: Knowledge Discovery and Measures of Interest. Kluwer, Boston (2001)

31. Holeňa, M.: Fuzzy hypotheses testing and Guha implicational quantifiers. Bull. Stud. Exchanges Fuzz. Appl. 63, 1015 (1995)

32. Holeňa, M.: Exploratory data processing using a fuzzy generalization of the Guha approach. In: Baldwin, J.F. (ed.) Fuzzy Logic, pp. 213–229. John Wiley and Sons, New York (1996)

33. Holeňa, M.: A method for approximate reasoning in exploratory data analysis. In: Trapl, R. (ed.) Proceedings of the 13th European Meeting on Cybernetics and Systems Research, vol. 1, pp. 329–334. ASCS, Vienna (1996)

34. Holeňa, M.: Fuzzy hypotheses for Guha implications. Fuzzy Sets and Systems 98, 101–125 (1998)
35. Holeňa, M.: A fuzzy logic framework for testing vague hypotheses with empirical data. In: Proceedings of the Fourth International ICSC Symposium on Soft Computing and Intelligent Systems for Industry, pp. 401–407. Academic Press, Sliedrecht (2001)
36. Holeňa, M.: A fuzzy logic generalization of a data mining approach. Neural Netw. World 11, 595–610 (2001)
37. Holeňa, M.: Fuzzy hypotheses testing in the framework of fuzzy logic. Fuzzy Sets and Systems 145, 229–252 (2004)
38. Ivánek, J.: Using Fuzzy Logic Operators for Construction of Data Mining quantifiers. Neural Netw. World 14, 403–410 (2004)
39. Ivánek, J.: Construction of implicational quantifiers from fuzzy implications. Fuzzy Sets and Systems 151, 381–391 (2005)
40. Ivánek, J.: Combining implicational quantifiers for equivalence ones by fuzzy connectives. Int. J. Intell. Syst. 21, 325–334 (2006)
41. Ivánek, J.: On the Correspondence between Classes of Implicational and Equivalence Quantifiers. In: Żytkow, J.M., Rauch, J. (eds.) PKDD 1999. LNCS (LNAI), vol. 1704, pp. 116–124. Springer, Heidelberg (1999)
42. Karban, T.: Relational Data Mining and GUHA. In: Richta, K., Snášel, V., Pokorný, J. (eds.) Proceedings of the Dateso 2005 Workshop, pp. 103–112, Faculty of Electrical Engineering, Czech Technical University in Prague, Prague (2005), http://www.cs.vsb.cz/dateso/2005/ (cited August 15, 2011)
43. Kliegr, T., Rauch, J.: An XML Format for Association Rule Models Based on the GUHA Method. In: Dean, M., Hall, J., Rotolo, A., Tabet, S. (eds.) RuleML 2010. LNCS, vol. 6403, pp. 273–288. Springer, Heidelberg (2010)
44. Kliegr, T., Svátek, V., Ralbovský, M., Šimůnek, M.: SEWEBAR-CMS: semantic analytical report authoring for data mining results. J. Intell. Inf. Syst. (2010)
45. Kliegr, T., Chudán, D., Hazucha, A., Rauch, J.: SEWEBAR-CMS: A System for Postprocessing Association Rule Models. In: Palmirani, M., Shafiq, M.O., Francesconi, E., Vitali, F. (eds.) RuleML-2010 Challenge. CEUR (2010), http://sunsite.informatik.rwth-aachen.de/Publications/CEUR-WS/Vol-649/paper9.pdf (cited August 15, 2011)
46. Kliegr, T., Svátek, V., Šimůnek, M., Šťastný, D., Hazucha, A.: XML Schema and Topic Map Ontology for Formalization of Background Knowledge in Data Mining. In: d'Amato, C., Fanizzi, N., Grobelnik, M., Lawrynowicz, A., Svátek, V. (eds.) Inductive Reasoning and Machine Learning for the Semantic Web. CEUR (2010), http://sunsite.informatik.rwth-aachen.de/Publications/CEUR-WS/Vol-611/paper8.pdf (cited August 15, 2011)
47. Kodym, J.: Classes of SD4ft-patterns. Master thesis, Faculty of Mathematics and Physics, Charles University, Prague (2007)(in Czech)
48. Kupka, D.: User support of the 4ft-Miner procedure for knowledge discovery in databases. Master thesis, Faculty of Mathematics and Physics, Charles University, Prague (2007)(in Czech)
49. Kuzmin, A.: Relational GUHA procedures. Master thesis, Faculty of Mathematics and Physics, Charles University, Prague (2007) (in Czech)
50. Lín, V., Rauch, J., Svátek, V.: Content–based Retrieval of Analytical Reports. In: Schroeder, M., Wagner, G. (eds.) Rule Markup Languages for Business Rules on the Semantic Web 2002. CEUR (2002), http://sunsite.informatik.rwth-aachen.de/Publications/CEUR-WS/Vol-60/lin.pdf (cited August 15, 2011)

51. Lín, V., Rauch, J., Svátek, V.: Mining and Querying in Association Rule Discovery. In: Klemettinen, M., Meo, R. (eds.) Proceedings of the First International Workshop on Inductive Databases, pp. 97–98. University of Helsinki, Helsinki (2002)

52. Lín V., Dolejší, P., Rauch J.,Šimůnek, M.: The KL-Miner Procedure for Datamining. Neural Netw. World 14, 411–420 (2004)

53. Louie, E., Lin, T.Y.: Finding Association Rules Using Fast Bit Computation: Machine-Oriented Modeling. In: Raś, Z.W., Ohsuga, S. (eds.) ISMIS 2000. LNCS (LNAI), vol. 1932, pp. 486–494. Springer, Heidelberg (2000)

54. Matheus, C., Piatetsky-Shapiro, G., Mc-Neill, D.: Selecting and Reporting What is Interesting: The KEFIR Application to Healthcare Data. In: Fayyad, U.M., Piatetsky-Shapiro, G., Smyth, P., Uthurusamy, R. (eds.) Advances in Knowledge Discovery and Data Mining, pp. 495–515. AAAI Press/The MIT Press (1996)

55. Piatetski-Shapiro, G.: Discovery, Analysis, and Presentation of Strong Rules. In: Piatetski-Shapiro, G., Frawley, W.J. (eds.) Knowledge Discovery in Databases, pp. 229–248. AAI/MIT Press (1991)

56. Ralbovský, M., Kuchař, T.: Using Disjunctions in Association Mining. In: Perner, P. (ed.) ICDM 2007. LNCS (LNAI), vol. 4597, pp. 339–351. Springer, Heidelberg (2007)

57. Ralbovský, M.: Fuzzy GUHA. Dissertation, Faculty of Informatics and Statistics, The University of Economics, Prague (2009)

58. Raś, Z.W., Wieczorkowska, A.: Action-Rules: How to Increase Profit of a Company. In: Zighed, D.A., Komorowski, J., Żytkow, J. (eds.) PKDD 2000. LNCS (LNAI), vol. 1910, pp. 587–592. Springer, Heidelberg (2000)

59. Rauch, J.: Application of the three valued logic for the GUHA method. Master thesis, Faculty of Mathematics and Physics, Charles University, Prague (1971) (in Czech)

60. Rauch, J.: Ein Beitrag zu der GUHA method in der dreivertigen logic. Kybernetika 11, 101–113 (1975)

61. Rauch, J.: Some Remarks on Computer Realizations of GUHA Procedures. Int. J. Man Mach. Stud. 10, 23–28 (1978)

62. Rauch, J.: Query languages and mechanizing hypothesis formation. In: Proceedings SOFSEM 1979, pp. 388–389. VVS, Bratislava (1979) (in Czech)

63. Rauch, J.: Logical Foundations of Hypothesis Formation from Databases. Dissertation, Mathematical Institute of the Czechoslovak Academy of Sciences, Prague (1986) (in Czech)

64. Rauch, J.: Logical Problems of Statistical Data Analysis in Data Bases. In: Proc. Eleventh International Seminar on Data Base Management Systems, Seregélyes, Hungary, pp. 53–63 (1988)

65. Rauch, J.: Logical Calculi for Knowledge Discovery in Databases. In: Komorowski, J., Żytkow, J.M. (eds.) PKDD 1997. LNCS, vol. 1263, pp. 47–57. Springer, Heidelberg (1997)

66. Rauch, J.: Classes of Four-Fold Table Quantifiers. In: Żytkow, J.M., Quafafou, M. (eds.) PKDD 1998. LNCS, vol. 1510, pp. 203–211. Springer, Heidelberg (1998)

67. Rauch, J.: Contribution to Logical Foundations of KDD. Assoc. Prof. Thesis, Faculty of Informatics and Statistics, The University of Economics, Prague (1998) (in Czech)

68. Rauch, J.: Logic of Association Rules. Appl. Intell. 22, 9–28 (2005)

69. Rauch, J.: Definability of Association Rules in Predicate Calculus. In: Lin, T.Y., Ohsuga, S., Liau, C.J., Hu, X. (eds.) Foundations and Novel Approaches in Data Mining, pp. 23–40. Springer, Heidelberg (2006)

70. Rauch, J.: Many Sorted Observational Calculi for Multi-Relational Data Mining. In: Workshops Proceedings of the 6th IEEE International Conference on Data Mining, pp. 417–422. IEEE Computer Society (2006)

71. Rauch, J.: Classes of Association Rules: An Overview. In: Lin, T.Y., Xie, Y., Wasilewska, A., Liau, C.J. (eds.) Data Mining: Foundations and Practice, pp. 315–337. Springer, Heidelberg (2008)

72. Rauch, J.: Definability of Association Rules and Tables of Critical Frequencies. In: Lin, T.Y., Xie, Y., Wasilewska, A., Liau, C.J. (eds.) Datamining: Foundations and Practice, pp. 299–314. Springer, Heidelberg (2008)

73. Rauch, J.: Considerations on Logical Calculi for Dealing with Knowledge in Data Mining. In: Ras, Z.W., Dardzinska, A. (eds.) Advances in Data Management. SCI, vol. 223, pp. 177–199. Springer, Heidelberg (2009)

74. Rauch, J.: Logical Aspects of the Measures of Interestingness of Association Rules. In: Koronacki, J., Raś, Z.W., Wierzchoń, S.T., Kacprzyk, J. (eds.) Advances in Machine Learning II. SCI, vol. 263, pp. 175–203. Springer, Heidelberg (2010)

75. Rauch, J.: Modifying Logic of Discovery for Dealing with Domain Knowledge in Data Mining. In: Kryszkiewicz, M., Obiedkov, S. (eds.) Concept Lattices and their Applications, pp. 174–186. University of Sevilla, Sevilla (2010)

76. Rauch, J.: Logic of Discovery, Data Mining and Semantic Web - Position Paper. In: Fred, A.L.N., Filipe, J. (eds.) KDIR 2010 - Proceedings of the International Conference on Knowledge Discovery and Information Retrieval, pp. 342–351. SciTePress (2010)

77. Rauch, J.: EverMiner - Consideration on Knowledge Driven Permanent Data Mining Process. To appear in Int. J. Data Mining, Modeling and Management

78. Rauch, J., Berka, P.: Mining in Financial Data - a Case Study. Neural Netw. World 7, 427–437 (1997)

79. Rauch, J., Šimůnek, M.: Mining for 4ft Association Rules. In: Arikawa, S., Morishita, S. (eds.) DS 2000. LNCS (LNAI), vol. 1967, pp. 268–272. Springer, Heidelberg (2000)

80. Rauch, J., Šimůnek, M.: An Alternative Approach to Mining Association Rules. In: Lin, T.Y., Ohsuga, S., Liau, C.J., Tsumoto, S. (eds.) Foundations of Data Mining and Knowledge Discovery. SCI, vol. 6, pp. 211–231. Springer, Heidelberg (2005)

81. Rauch, J. Šimůnek, M.: GUHA Method and Granular Computing. In: Hu, X., Liu, Q., Skowron, A., Lin, T.Y., Yager, R.R., Zhang, B. (eds.) IEEE International Conference on Granular Computing, pp. 630–635. IEEE Computer Society (2005)

82. Rauch, J., Šimůnek, M.: Semantic Web Presentation of Analytical Reports from Data Mining – Preliminary Considerations. In: 2007 IEEE/WIC/ACM International Conference on Web Intelligence, pp. 3–7. IEEE Computer Society (2007)

83. Rauch, J., Šimůnek, M.: LAREDAM – Considerations on System of Local Analytical Reports from Data Mining. In: An, A., Matwin, S., Raś, Z.W., Ślęzak, D. (eds.) ISMIS 2008. LNCS (LNAI), vol. 4994, pp. 143–149. Springer, Heidelberg (2008)

84. Rauch, J., Šimůnek, M.: Action Rules and the GUHA Method: Preliminary Considerations and Results. In: Rauch, J., Raś, Z.W., Berka, P., Elomaa, T. (eds.) ISMIS 2009. LNCS, vol. 5722, pp. 76–87. Springer, Heidelberg (2009)

85. Rauch, J., Šimůnek, M.: Dealing with Background Knowledge in the SEWEBAR Project. In: Berendt, B., Mladenič, D., de Gemmis, M., Semeraro, G., Spiliopoulou, M., Stumme, G., Svátek, V., Železný, F. (eds.) Knowledge Discovery Enhanced with Semantic and Social Information. SCI, vol. 220, pp. 89–106. Springer, Heidelberg (2009)

86. Rauch, J., Šimůnek, M.: Applying Domain Knowledge in Association Rules Mining Process – First Experience. In: Kryszkiewicz, M., Rybinski, H., Skowron, A., Raś, Z.W. (eds.) ISMIS 2011. LNCS, vol. 6804, pp. 113–122. Springer, Heidelberg (2011)

87. Rauch, J., Tomečková, M.: System of Analytical Questions and Reports on Mining in Health Data – a Case Study. In: Roth, J., Gutiérrez, J., Abraham, A.P. (eds.) Proceedings of the IADIS European Conference on Data Mining, pp. 176–181. IADIS Press (2007)

88. Rauch, J., Šimůnek, M., Lín, V.: Mining for Patterns Based on Contingency Tables by KL-Miner - First Experience. In: Lin, T.Y., Ohsuga, S., Liau, C.J., Hu, X. (eds.) Foundations and Novel Approaches in Data Mining, pp. 155–167. Springer, Heidelberg (2005)

89. Šimůnek, M.: Academic KDD Project LISp-Miner. In: Abraham, A., Franke, K., Koppen, K. (eds.) Advances in Soft Computing - Intelligent Systems Desing and Applications, pp. 263–272. Springer, Heidelberg (2003)

90. Šimůnek, M.: LISp-Miner, academic system for knowledge discovery in databases, history and description of use. Oeconomica, Prague (2010) (in Czech)

91. Šimůnek, M., Rauch, J.: EverMiner - Towards Fully Automated KDD Process. In: Funatsu, K., Hasegava, K. (eds.) New Fundamental Technologies in Data Mining, pp. 221–240. InTech, Rijeka (2011)

92. Strossa, P., Černý, Z., Rauch, J.: Reporting Data Mining Results in a Natural Language. In: Lin, T.Y., Ohsuga, S., Liau, C.J., Tsumoto, S. (eds.) Foundations of Data Mining and Knowledge Discovery. SCI, vol. 6, pp. 347–361. Springer, Heidelberg (2005)

93. Tan, P.N., Kumar, V., Srivastava, J.: Selecting the Right Objective Measure for Association Analysis. Inf. Syst. 29, 293–313 (2004)

94. Tomečková, M., Rauch, J., Berka, P.: STULONG - Data from a Longitudial Study of Atherosclerosis Risk Factors. In: Berka, P. (ed.) Discovery Challenge Workshop Notes. ECML/PKDD - 2002, University of Helsinki, Helsinki (2002)

95. Wu, X., et al.: Top 10 Algorithms in Data Mining. Knowl. Inf. Syst. 14, 1–37 (2008)

96. Yang, Q., Wu, X.: 10 Challenging Problems in Data Mining Research. Int. J. Inf. Technol. Decis. Mak. 5, 597–604 (2006)

97. Zembowicz, R., Zytkow, J.: From Contingency Tables to Various Forms of Knowledge in Databases. In: Fayyad, U.M., Piatetsky-Shapiro, G., Smyth, P., Uthurusamy, R. (eds.) Advances in Knowledge Discovery and Data Mining, pp. 329–349. AAAI Press/The MIT Press (1996)

Glossary

4ft-quantifier $\approx$ is a basic symbol of a language of association rules. It is used to express a relation of two Boolean attributes.

4ft-table $4ft(\varphi,\psi, \mathcal{M})$ of Boolean attributes φ and ψ in data matrix $\mathcal{M} = \langle M, f_1, \ldots, f_K \rangle$ is a quadruple $\langle a, b, c, d \rangle$ of non-negative integers. Here a is the number of rows of $\mathcal{M}$ satisfying both φ and ψ, b is the number of rows of $\mathcal{M}$ satisfying φ and not satisfying ψ, c is the number of rows of $\mathcal{M}$ satisfying ψ and not satisfying φ, and d is the number of rows of $\mathcal{M}$ satisfying neither φ nor ψ.

Associated function $F_{\approx}$ of 4ft-quantifier $\approx$ is a $\{0; 1\}$ - valued function defined for all quadruples $\langle a, b, c, d \rangle$ of non-negative integers satisfying $a + b + c + d > 0$.

Association rule is an expression $\varphi \approx \psi$ where φ and ψ are Boolean attributes and $\approx$ is a 4ft-quantifier. Each association rule corresponds to a general relation of two Boolean attributes derived from columns of data matrices.

Basic attribute A_i is a symbol of language $\mathcal{L}_{\mathcal{T}}$ of association rules of type $\mathcal{T}$. It is used to talk about columns of data matrices. Basic attribute A_i corresponds to a column f_i of a data matrix $\mathcal{M} = \langle M, f_1, \ldots, f_K \rangle$ for $i = 1, \ldots, K$.

Basic Boolean attribute is an expression $A(\alpha)$ where A is a basic attribute and α is a subset of categories of A.

Basic symbols of language $\mathcal{L}_{\mathcal{T}}$ of association rules of type $\mathcal{T} = \langle t_1, \ldots, t_K \rangle$ are basic attributes $A_1, \ldots, A_K$, categories $1, \ldots, \max\{t_1, \ldots, t_K\}$, propositional connectives $\wedge$, $\vee$, and $\neg$ and 4ft-quantifiers $\approx_1, \ldots \approx_Q$.

Boolean attribute is derived from basic Boolean attributes using propositional connectives $\wedge$, $\vee$ and $\neg$ in the usual way.

Category is a possible value of a basic attribute. Basic attribute A_i of language $\mathcal{L}_{\mathcal{T}}$ of association rules of type $\mathcal{T} = \langle t_1, \ldots, t_K \rangle$ has categories $1, \ldots, t_i$.

Coefficient of a basic Boolean attribute $A(\alpha)$ is the set α.

Data matrix $\mathscr{M} = \langle M, f_1, \ldots, f_K \rangle$ **of type** $\mathscr{T} = \langle t_1, \ldots, t_K \rangle$ consists of a non-empty set M of rows of $\mathscr{M}$ and unary functions $f_1, \ldots, f_K$ i.e. columns of $\mathscr{M}$. The function f_i maps the set M to the set $1, \ldots, t_i$ of integers for $i = 1, \ldots, K$.

Language $\mathscr{L}_{\mathscr{T}}$ **of association rules of type** $\mathscr{T} = \langle t_1, \ldots, t_K \rangle$ is used to define association rules as sentences of a semantic system. It consists of basic symbols, Boolean attributes and association rules.

Logical calculus $\mathscr{LC}_{\mathscr{T}}$ **of association rules of type** $\mathscr{T}$ is defined such that association rules correspond to sentences of a semantic system, data matrices correspond to models of this semantic system and abstract values are 1 i.e. *true* and 0 i.e. *false*.

Semantic system $\langle \text{Sent}, \text{M}, V, Val \rangle$ consists of a non-empty set Sent of sentences, a non-empty set M of models, a non-empty set V of abstract values, and an evaluating function $Val: \text{Sent} \times \text{M} \to V$. If $\Omega \in \text{Sent}$ and $\mathscr{M} \in \text{M}$ then $Val(\Omega, \mathscr{M})$ is the value of Ω in $\mathscr{M}$.

Type of a logical calculus of association rules is a K-tuple $\mathscr{T} = \langle t_1, \ldots, t_K \rangle$ where $K \geq 2$ is an integer number and $t_i \geq 2$ are integer numbers for $i = 1, \ldots, K$. Type $\mathscr{T}$ determines number of categories of particular basic attributes of a language of association rules.

Value $Val(\varphi \approx \psi, \mathscr{M})$ **of association rule** $\varphi \approx \psi$ in data matrix $\mathscr{M} = \langle M, f_1, \ldots, f_K \rangle$ is defined as $Val(\varphi \approx \psi, \mathscr{M}) = F_{\approx}(a, b, c, d)$ where $F_{\approx}$ is an associated function of 4ft-quantifier $\approx$ and $\langle a, b, c, d \rangle$ is a 4ft-table of φ and ψ in $\mathscr{M}$. Association rule $\varphi \approx \psi$ is true in data matrix $\mathscr{M}$ if $Val(\varphi \approx \psi, \mathscr{M}) = 1$, otherwise association rule $\varphi \approx \psi$ is false in data matrix $\mathscr{M}$.

Index